Lectures on Current Algebra
and Its Applications

Lectures on Current Algebra and Its Applications

Sam B. Treiman

Roman Jackiw

David J. Gross

Princeton Series in Physics

Princeton University Press

Princeton, New Jersey 1972

LC Card: 70-181519
ISBN: 0-691-08118-2 (hard cover edition)
ISBN: 0-691-08107-7 (paperback edition)

Printed in the United States of America
by Princeton University Press

FOREWORD

The three sections of this book comprise three lectures which we delivered at the Brookhaven Summer School in Theoretical Physics during the summer of 1970. "Current Algebra and PCAC" summarizes the "classical" results of the theory. The general methodology is introduced and then applied to derivations of various sum rules and low-energy theorems. In "Field Theoretic Investigations in Current Algebra" calculations are presented which have exhibited unexpected dynamical dependence of the predictions of current algebra. In models many predictions fail due to anomalies which are present as a consequence of the divergences of local quantum field theory, at least in perturbation theory. Finally "High-Energy Behavior of Weak and Electromagnetic Processes" concerns itself with the recent uses of current algebra for high-energy processes, especially the "deep inelastic" ones. Much of this development is motivated by the MIT-SLAC experiments on electron-nucleon inelastic scattering and represents the new direction for the theory.

For permission to reprint several figures that appear in the third section, we thank R. Diebold (Figure 11); F. Gilman (Figure 12); C. E. Dick and J. W. Mock (Figure 15, which originally appeared in Physics Review 171, 75 [1968]); E. Bloom et al (Figures 16 - 21, which originally appeared in Physics Review Letters 23, 935 [1969]); and I. Bugadov et al. (Figure 22, which originally appeared in Physics Letters 30B, 364 [1969]).

We wish to thank Dr. Ronald F. Peierls for giving us the opportunity of presenting these lectures at the Summer School. Mrs. Patricia Towey, and her secretarial staff, carried out the initial production of the manuscript with admirable efficiency for which we are most grateful.

S. Treiman

R. Jackiw

D. J. Gross

October, 1971

CONTENTS

CURRENT ALGEBRA AND PCAC by Sam B. Treiman

FIELD THEORETIC INVESTIGATIONS IN CURRENT ALGEBRA by Roman Jackiw

THE HIGH ENERGY BEHAVIOR OF WEAK AND ELECTROMAGNETIC PROCESSES by David J. Gross

Lectures on Current Algebra
and Its Applications

CURRENT ALGEBRA AND PCAC

S. B. Treiman

I. INTRODUCTION

The basic ideas for the subject of current algebra were introduced by Gell-Mann[1] as long ago as 1961. But the development proceeded slowly for several years, until 1965, when Fubini and Furlan[2] suggested the appropriate techniques for practical applications and Adler[3] and Weisberger[4] derived their remarkable formula relating β decay parameters to pion-nucleon scattering quantities. This inaugurated the golden age and the literature soon reflected what always happens when a good idea is perceived. In 1967 Renner[5] counted about 500 papers, and the number may well have doubled by now. Of course the number of really distinct advances in understanding is somewhat smaller than may be suggested by the counting of publications. Indeed, the major theoretical outlines have not changed much since 1967, by which time most of the "classical" applications had been worked out. During the last few years numerous variations on the earlier themes have inevitably appeared. But there is also developing an increasingly sophisticated concern about the validity of some of the formal manipulations that people have indulged in to get results. These interesting matters will be discussed by Professor Jackiw in his lectures at the Summer School.

On the experimental side the situation is still unsettled for a number of processes that may well be decisive for the subjects under discussion here. This is the case with $K_{\ell 3}$ decay, where there are lots of data but not all of it consistent; and with $K_{\ell 4}$ decay, where the data is still

The lecture notes were prepared in cooperation with Mrs. Glennys Farrar.

limited. As for the interesting and perhaps crucial applications to $\pi - \pi$ scattering and to high energy neutrino reactions, relevant experimental information is altogether lacking at present. The neutrino reactions in particular have been attracting a great deal of attention in recent times, and not only from the point of view of what current algebra has to say about them. This subject will be reviewed in the lectures to be given by Professor Gross.

The primary ingredients of current algebra are a set of equal-time commutation relations conjectured by Gell-Mann for the currents that arise in the electromagnetic and weak interactions of hadrons. As will be described and qualified more fully below, the Gell-Mann scheme may be taken to refer to commutators involving the time components of the currents. Various conjectures for the space-space commutators have also been suggested subsequently. It is especially in connection with the latter that possibly dangerous manipulations come into play in the applications. The dangers and the applications will however be left to the other lecturers. In any event, commutation relations imply sum rules on the matrix elements of the operators which are involved. But the matrix elements which arise are usually not physically accessible in a practical way. It therefore takes some inventiveness to extract physically useful results; and inevitably this requires approximations or extra assumptions. Tastes can differ here! But one might well wonder what is being tested when, say, an infinite sum is truncated arbitrarily at one or two terms chosen more or less for pure convenience. On the other hand one is likely to be more charitable to the rather discrete demand that a certain dispersion integral shall merely converge, where this doesn't contradict known facts or principles. Even at the very best, however, a high order of rigor is not to be expected. Still, some topics are cleaner than others; and, even though this encroaches on Professor Gross' subject, I shall want later on to discuss the Adler sum rule for neutrino reactions, as exemplifying an especially model insensitive test of the Gell-Mann conjectures.

It will be noticed in any listing of current algebra applications that processes involving pions make a disproportionate appearance. This comes about because the commutation relation hypotheses are nicely matched to another and independent set of ideas about the weak interaction currents. These form the so-called PCAC notion of pion pole dominance for the divergence of the strangeness conserving axial vector current.[6] This too has its independent tests, i.e., independent of the ideas of current algebra, and I will want to take these up. But it is in combination that the two sets of ideas, current algebra and PCAC, display their most impressive predictive powers. A contrast must however be noted. The Gell-Mann conjectures seem to be clearly and consistently posed, so that for them the question is whether they happen to be true for the real world. The PCAC notion, on the other hand, has so far not been sharply stated; so for it the question is not only whether it's true, but—*what is it?*

These lectures will be focused mainly on concrete applications of current algebra and PCAC, with examples taken from the "classical" portion of these subjects. The question, what is PCAC, will no doubt be unduly belabored. But in general, the fare will be standard. It is addressed to people who have not yet had occasion to make themselves experts, and it is intended to serve also as introduction for the more up-to-date material which will appear in the other lecture series here. Luckily, an outstanding published review is provided in the book by Adler and Dashen: *Current Algebras* (W. A. Benjamin Publ., New York, 1968). For more general matters concerning the weak interactions, reference may be made to the book by Marshak, Riazuddin, and Ryan: *Theory of Weak Interactions In Particle Physics* (Wiley-Interscience, New York, 1969). For SU_3 matters, see Gell-Mann and Ne'eman *The Eightfold Way* (W. A. Benjamin Publ., New York, 1964).

Owing to the existence of these excellent works, the present notes will be sparing in references.

II. THE PHYSICAL CURRENTS

The Adler-Weisberger formula makes an improbable connection between strong interaction quantities and a weak interaction parameter. By generalizing the formula to cover pion scattering on various hadron targets, one can eliminate the weak interaction parameter and obtain connections between purely strong interaction quantities. This remarkable circumstance, and others like it, arises from the PCAC hypothesis concerning a certain physical weak interaction current. In retrospect it would be possible to formulate matters in such a way that no reference is made to the weak interactions: the current in question could be introduced as a purely mathematical object. However, that's not how things happened; and anyway the physical currents that constitute our subject are interesting in their own right. So we begin with a brief review of the way in which these currents arise in the present day description of the weak and electromagnetic interactions of hadrons. The Heisenberg picture is employed throughout for quantum mechanical states and operators. Our metric corresponds to

$$a.b = \vec{a}.\vec{b} - a_0 b_0.$$

(i) Electromagnetic Hadron Currents: –

Of all the currents to be dealt with, the electromagnetic is no doubt the most familiar. The coupling of hadrons to the electromagnetic field aperator a_λ is described, to lowest order in the unit of electric charge e, by an interaction Hamiltonian density

(2.1) $$\mathcal{H}^{em} = e j_\lambda^{em} a_\lambda ,$$

where j_λ^{em} is the hadron electromagnetic current. It is a vector operator formed out of hadron fields in a way whose details must await a decision about fundamental matters concerning the nature of hadrons. It is in line with the trend of contemporary hadron physics to put off such a decision and instead concentrate on symmetry and other forms of characterization which are supposed to transcend dynamical details. Thus, reflecting conservation of electric charge, we assert that the charge operator

$$Q^{em} = \int d^3x j_0 (\vec{x}, x_0)$$

is independent of time x_0 and that the current j_λ^{em} is conserved:

$$\partial j_\lambda^{em} / \partial x_\lambda = 0.$$

It is customarily assumed that j_λ^{em} is odd under the charge conjugation symmetry transformation defined by the strong interactions. In connection with the discovery of CP violation it has however been suggested that the electromagnetic current may also have a piece which is *even* under charge conjugation; but for present purposes we shall overlook this still unconfirmed possibility. Electromagnetic interactions of course conserve baryon number (N) and strangeness (S)—or equivalently, hypercharge $Y = N + S$; and they conserve the third component I_3 of isotopic spin. According to the familiar formula

$$Q^{em} = I_3 + Y/2$$

it is generally supposed that j_λ^{em} contains a part which transforms like a scalar $(I = 0)$ under isotopic spin rotations and a part which transforms like the third component of an isovector ($I = 1, \Delta I_3 = 0$):

$$j_\lambda^{em} = j_\lambda^{em}(I = 0) + j_\lambda^{em}(I = 1, \Delta I_3 = 0) \quad . \tag{2.2}$$

The two pieces are separately conserved, reflecting conservation of hypercharge and third component of isotopic spin; and the corresponding "charges" measure these quantities:

$$\frac{Y}{2} = \int j_0^{em}(I = 0)\, d^3x$$

$$I_3 = \int j_0^{em}(I = 1)\, d^3x \quad . \tag{2.3}$$

Consider now an electromagnetic process of the sort $\alpha \to \beta + \gamma$, where γ is a real photon of momentum k and α and β are systems of one or more hadrons. To lowest electromagnetic order the transition amplitude is given by

$$e < \beta | j_\lambda^{em} | \alpha > \epsilon_\lambda \quad , \tag{2.4}$$

where ϵ_λ is the photon polarization vector and the states $|\alpha>$ and $|\beta>$ are determined purely by the strong interactions, with electromagnetism switched off. The photon process probes the structure of these states via the current operator j_λ^{em}. Here of course $k = P_\alpha - P_\beta$, $k^2 = 0$, $k \cdot \epsilon = 0$. Off mass shell ($k^2 \neq 0$) electromagnetic probes are provided in processes involving the interactions of electrons or muons with hadrons. For example, to lowest relevant order the process $e + \alpha \to e + \beta$ is described by the Feynman diagram shown below, where P_e and P_e' are the initial and final electron momenta and the virtual photon has momentum $k = P_e - P_e'$. The amplitude is given by

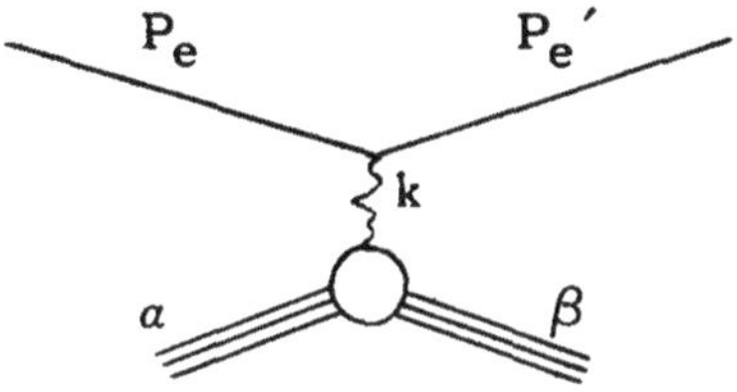

(2.5) $$e^2 <\beta|j_\lambda^{em}|\alpha> \frac{1}{k^2}\,\bar{u}\,(P_e')\gamma_\lambda u(P_e)\ .$$

The matrix element of the electron part of the overall electromagnetic current comes out here as a trivially known factor. All the complexities of the strong interactions are again isolated in the matrix element of the hadron current j_λ^{em}; but now, in general, $k^2 = (P_\beta - P_\alpha)^2$ is not zero. Other related variations have to do with processes such as $\alpha \to \beta + e^+ + e^-$ and, for colliding beam experiments, $e^+ + e^- \to$ hadrons. In all cases it is the hadronic matrix element that is of interest; and current conservation implies the relation

$$k_\lambda < \beta|j_\lambda^{em}|\alpha > = 0\ , \quad k = P_\beta - P_\alpha\ .$$

(ii) Weak Lepton Currents: –

So much for electromagnetism. Concerning the weak interactions, you recall that they group themselves phenomenologically into three classes. (1) Purely leptonic processes, of which muon decay $\mu^- \to e^- + \bar{\nu}_e + \nu_\mu$ and the conjugate μ^+ decay are the sole observed examples. (2) Semi-leptonic processes, i.e., those which involve hadrons and a lepton pair $(e\nu_e)$ or $(\mu\nu_\mu)$. There are many observed examples, e.g., $n \to p + e^- + \bar{\nu}_e$, $\bar{\nu}_\mu + p \to n + \mu^+$, $K \to e + \nu + \pi$, etc. (3) Non leptonic weak processes, e.g., $\Lambda \to p + \pi^-$, $K^+ \to \pi^+ + \pi^0$, etc.

The purely leptonic muon decay reaction is well described phenomenologically to lowest order in an interaction Hamiltonian which couples an $(e\nu_e)$ "current" with a $(\mu\nu_\mu)$ "current":

$$\mathcal{H}^{\text{leptonic}} = \frac{G}{(2)^{1/2}} \mathcal{L}_\lambda(e\nu)\mathcal{L}_\lambda^\dagger(\mu\nu) + \text{h.c.} \tag{2.6}$$

where the current operators are given by

$$\mathcal{L}_\lambda(e\nu) = i\bar{\Psi}_{\nu_e}\gamma_\lambda(1+\gamma_5)\Psi_e$$

$$\mathcal{L}_\lambda(\mu\nu) = i\bar{\Psi}_{\nu_\mu}\gamma_\lambda(1+\gamma_5)\Psi_\mu$$

The factor $(1+\gamma_5)$ expresses the presumed 2-component nature of neutrinos and gives to each current both a vector and axial vector part. The coupling constant G has the dimension $(\text{mass})^{-2}$ and can be taken generally as a characteristic (but dimensional) measure of the strength of all classes of weak interactions. From the fit to the muon decay rate, one finds that $Gm^2 \simeq 10^{-5}$, where m is the nucleon mass.

(iii) Weak Hadron Currents:—

The observed semi leptonic reactions seem to be well described to lowest order in an interaction Hamiltonian which effectively couples the lepton currents to a weak hadron current j_λ^{weak} which, in fact, contains both vector and axial vector parts:

$$\mathcal{H}^{\text{semi leptonic}} = \frac{G}{(2)^{1/2}} j_\lambda^{\text{weak}} \mathcal{L}_\lambda^\dagger + \text{h.c.} \tag{2.7}$$

where $$\mathcal{L}_\lambda = \mathcal{L}_\lambda(e\nu) + \mathcal{L}_\lambda(\mu\nu).$$

Notice that $\mathcal{L}_\lambda(e\nu)$ and $\mathcal{L}_\lambda(\mu\nu)$ couple equally to a common hadron current, reflecting the present belief in e–μ universality. In the subsequent discussion we will let the symbol ℓ denote either e or μ; and the symbol ν will stand for the appropriate neutrino ν_e or ν_μ.

Consider now a semi-leptonic process such as $\nu + \alpha \to \beta + \ell$, where α and β are hadron systems. To lowest order the transition amplitude is given by

$$\frac{G}{(2)^{1/2}} < \beta | j_\lambda^{\text{weak}} | \alpha > i\bar{u}(\ell)\gamma_\lambda(1+\gamma_5)u(\nu) \quad , \tag{2.8}$$

where the states $|\alpha>$ and $|\beta>$ are determined solely by the strong interactions. The structure here is similar to that of Eq. (2.5), which describes the process

$$e + \alpha \to e + \beta .$$

The weak lepton current matrix element again appears as a simple and known factor, all the complexities of the strong interactions appearing in the matrix element of the hadron current. The semi leptonic reactions probe the hadron states via the weak current operator j_λ^{weak}. For a decay process such as

$$\alpha \to \beta + \ell + \bar{\nu}$$

the structure is the same as above, except for the obvious change that the neutrino spinor $u(\nu)$ is replaced by the corresponding anti neutrino spinor. Notice that on our present conventions the current j_λ^{weak} *raises* the hadron electric charge by one unit:

$$Q_\beta - Q_\alpha = +1 .$$

For processes in which the hadron charge is lowered by one unit it is the matrix element of the adjoint operator $(j_\lambda^{\text{weak}})^\dagger$ that is encountered; and of course the lepton current matrix elements undergo a corresponding and obvious change.

As with the electromagnetic current, the way in which the weak current is constructed out of fundamental hadron fields cannot presently be attacked in a reliable way. The idea instead is to characterize the current in more general terms, whose implications can be extracted independently of these details. Of course detailed models can well serve here as a source of ideas, and have in fact done so in connection with the ideas of current algebra. As already said, with respect to Lorentz transformation properties, it is known that j_λ^{weak} has both a vector and an axial vector part:

$$J_\lambda^{weak} = V_\lambda + A_\lambda .$$

For some physical processes, e.g.,

$$K^- \to K^0 + e + \bar{\nu}$$

only the vector part contributes; for others, e.g. $\pi \to \mu + \nu$, only A_λ contributes; and for still others, e.g.

$$n \to p + e + \bar{\nu} ,$$

both contribute. The currents preserve baryon number. With respect to strangeness both V_λ and A_λ have pieces which conserve strangeness ($\Delta S = 0$) and pieces which change it by one unit, ΔS and ΔQ being correlated according to $\Delta S/\Delta Q = +1$. It may be that there are additional pieces, say with $\Delta S/\Delta Q = -1$, or $|\Delta S| > 1$; but in the absence of experimental confirmation we ignore these possibilities for the present. With respect to isospin it is certain that the $\Delta S = 0$ currents contain pieces which transform like the charge raising part of an isovector ($I = 1$). For the $\Delta S/\Delta Q = 1$ currents there are pieces which transform like the charge raising member of an isotopic doublet ($I = 1/2$). Other terms, with different behavior under isotopic spin, do not at present seem to be required. For the various well established pieces we may now summarize the situation by writing

$$V_\lambda = \cos\theta_c \, V_\lambda \, (Y = 0, I = 1) + \sin\theta_c \, V_\lambda \, (Y = 1, I = 1/2)$$

$$A_\lambda = \cos\theta_c \, A_\lambda \, (Y = 0, I = 1) + \sin\theta_c \, A_\lambda \, (Y = 1, I = 1/2) \tag{2.9}$$

The characterization with respect to third component of isotopic spin can be left unsymbolized, following as it does from $Q = I_3 + Y/2$. The factors $\cos\theta_C$ and $\sin\theta_C$ which appear in the above expressions are at this stage gratuitous, as is the factor $\frac{G}{(2)^{1/2}}$ displayed in Eq. (2.7). The point is that no "scale" has as yet been set for the various currents and these factors could all have been absorbed into the currents. Some ideas which serve to set the scales will however be discussed shortly and these will give to the angle θ_C an objective significance. For the present, let us merely assign a name to θ_C: it is the Cabibbo angle.

In connection with strangeness conserving currents, vector and axial vector, there is one further respect in which they can be classified under operations involving isospin. This has to do with G parity, which involves a familiar isospin rotation that changes the sign of I_3, followed by the operation of charge conjugation that again reverses the sign of I_3. The $Y = 0$ currents V_λ $(Y = 0, I = 1)$ and A_λ $(Y = 0, I = 1)$ are the charge raising members of isotopic triplets. The adjoints $V_\lambda^\dagger$ and $A_\lambda^\dagger$ are therefore the charge lowering members of isotopic triplets. In general, however, it need not be that V_λ and $V_\lambda^\dagger$ belong to the *same* triplet, and similarly for A_λ vs. $A_\lambda^\dagger$. But if a current and its adjoint do in fact belong to a common isotopic multiplet, then a definite G parity quantum number, $G = +1$ or -1, can be assigned to the multiplet. It is in fact known that the vector $Y = 0$ current has a part which is *even* under G parity, the axial vector current a part which is *odd* under G parity. There is presently no evidence for the alternative possibilities. Notice that the *isovector* part of the electromagnetic current has $G = +1$, the *isoscalar* part $G = -1$.

The isovector part of the electromagnetic current is thus the neutral member of a $G = 1$ isotopic triplet; whereas V_λ $(I = 1)$ and $V_\lambda^\dagger$ $(I = 1)$ are the charge raising and lowering members of what in general might be a different isotopic triplet. It is clearly a tempting proposition that these triplets are in fact one and the same. This is the substance of the celebrated CVC hypothesis of Zeldovitch and Feynman and Gell-Mann. It has received some experimental confirmation and represents, insofar as it is true, a considerable simplification. The electromagnetic currents, isovector and isoscalar, are of course separately conserved. The CVC hypothesis is so-called because it entails also the conservation (C) of the strangeness conserving weak vector current (VC): $\partial V_\lambda(I = 1)/\partial x_\lambda = 0$. Moreover, corresponding matrix elements for V_λ $(I = 1)$ and j^{em} $(I = 1)$ now become related through isospin symmetry considerations. For example, for pions at rest

$$< \pi^+ | j_0^{em} \, (I = 1) | \pi^+ > = 1 \; .$$

It then follows, as is relevant for pion β*–decay, that*

$$< \pi^+ | V_0(I = 1) | \pi^0 > = (2)^{1/2} \; .$$

The "scale" of the weak vector current $V_\lambda(I = 1)$ has now acquired a sharp definition.

From the point of view of behavior under the hypercharge and isospin symmetry operations of the strong interactions, the weak and electromagnetic currents have now been characterized in as much detail as is possible. The various currents do not all preserve these symmetries, but they transform in very simple ways. The more complicated options do not seem (at present) to be taken up by nature. Although we're now done with hypercharge and isospin, we're still not finished with strong interaction symmetries. There's SU_3! It is certainly not an outstandingly exact symmetry, but for the moment let's pretend that it is. It would then be natural to ask how the various currents behave under SU_3. For each of the currents that have been discussed the simplest possibility, consistent with the isospin and hypercharge quantum numbers, is that it transforms like a member of an SU_3 octet. Cabibbo proposed something even stronger, however,[7] namely, that all of the vector currents (weak and electromagnetic) belong to a *common* octet. On this view the isoscalar electromagnetic current $(Y = 0,\ I = 0)$, the triplet of electromagnetic and weak strangeness conserving currents (Y = 0, I = 1), and the weak strangeness changing currents (Y = ±1, I = 1/2) all stand in the same relation to each other under SU_3 as η^0, $(\pi^{\pm}, \pi^0)$, (K^+, K^-). Cabibbo similarly proposed that the various weak axial vector currents all belong to a *common* octet of axial vector operators. These ideas, if true, represent an enormous simplification, which leaves only two independent and unrelated sets of objects to be contemplated, the vector and axial vector octets. Within each octet the various pieces are related by symmetry operations, which means also that the relative scales within an octet are fixed. Moreover, for the vector octet the absolute scale has already been fixed, in that this octet provides operators which measure charge and hypercharge. It would seem that there only remains to fix the scale of the axial octet relative to the vector one, something which however seems difficult to imagine for objects which have different Lorentz transformation properties. On the other hand there is already the problem that SU_3 is after all not such a good strong interaction symmetry. The question arises whether anything of these SU_3 symmetry characterizations can be given a useful meaning that survives under the breakdown of SU_3. On both of these issues Gell-Mann's current algebra conjectures have suggested solutions. We will begin our review of these matters very shortly. In the meantime it should be remarked that various applications of the Cabibbo model have been carried out on the assumption of exact SU_3 symmetry for the strong interactions; and treated in this hopeful way the model has in fact been fairly successful in correlating various baryon and meson β decays, with Cabibbo angle $\theta_C \doteq 0.23$ radians. You will no doubt hear more of this from Prof. Willis.

(iv) Weak Non Leptonic Interactions:

The weak leptonic and semi leptonic Hamiltonia each have the form of a current × current interaction. What about the non leptonic interactions? Since reactions of this class involve particles which are all subject to strong interaction effects, there is no reliable way to read off from experiment anything of the inner structure of the non leptonic Hamiltonian – in connection with the possibility, say, that it factors into interesting pieces. This is in contrast with the situation, say, for semi leptonic reactions, where it can be inferred that the interaction involves the local coupling of a lepton current with a hadron current. On the other hand we can in the usual way attempt to characterize the non leptonic Hamiltonian with respect to various strong interaction symmetries. It violates C and P invariance (and perhaps also CP symmetry); and it certainly contains pieces which violate strangeness and isotopic spin conservation. Indeed, all of the most familiar weak non leptonic reactions involve a change of strangeness of one unit. However, recent evidence has been produced for a small degree of parity violation in certain $\Delta S = 0$ nuclear radiative transitions. The effect occurs at a strength level which roughly corresponds to the strength of the usual strangeness changing reactions. This suggests that the weak non leptonic interaction contains a (parity violating) $\Delta S = 0$ piece, as well as the usual $|\Delta S| = 1$ pieces. But there is no evidence at present for $|\Delta S| > 1$. With respect to isotopic spin the strangeness changing pieces appear to transform like the members of an isotopic doublet. This is the famous $\Delta I = 1/2$ rule and it continues to hold up well experimentally, with discrepancies that are perhaps attributable to electromagnetic effects – although the effects are a bit uncomfortably large on this interpretation. In the main, however, things again seem to be simple with respect to isospin and hypercharge. The SU_3 properties of the weak non leptonic interactions are much less clear. The simplest possibility would be that the $\Delta S = 0$, $\Delta S = 1$, and $\Delta S = -1$ pieces all have octet transformation properties.

Although there is no convincing way to draw from experiment any inferences about the "inner" structure of the non leptonic Hamiltonian, there does exist a very attractive *theoretical* possibility. Namely, one can imagine that it is built up out of local products of the hadron currents already encountered in the semi leptonic interactions. An even more compact possibility is that *all* the classes of weak interactions are described by a master current x current coupling. That is, let $J_\lambda = j_\lambda^{weak} + \mathcal{L}_\lambda$ be a master charge raising current composed of the previously discussed hadron and lepton pieces; and consider the Feynman, Gell-Mann[8] inclusive weak interaction Hamiltonian

$$\mathcal{H}^{weak} = \frac{G}{(2)^{1/2}} J_\lambda J_\lambda^\dagger \quad . \tag{2.10}$$

This clearly contains the purely leptonic interaction responsible for muon decay, as well as the hadron-lepton couplings responsible for weak semi leptonic processes. But there are additional pieces: (1) *Self* coupling of the lepton currents,

$$\mathcal{L}_\lambda\,(e\nu)\,\mathcal{L}_\lambda^\dagger\,(e\nu) \text{ and } \mathcal{L}_\lambda\,(\mu\nu)\,\mathcal{L}_\lambda^\dagger\,(\mu\nu)\ .$$

This leads to the prediction of processes such as $\nu_e + e \to \nu_e + e$, with conventional strength and structure. (2) Self coupling of the hadron currents, $j_\lambda j_\lambda^\dagger$. This describes non leptonic interactions and includes $\Delta S = 0$ and $|\Delta S| = 1$ pieces, as wanted. There is the difficulty however that the $|\Delta S| = 1$ couplings contain not only an $I = 1/2$ part but also an $I = 3/2$ part. And with respect to SU_3, in addition to an octet part there is also a term belonging to the representation $\underset{\sim}{27}$.

The SU_3 troubles need not be decisive, since the experimental evidence for an octet structure is scanty; and anyhow, SU_3 is not all that good a strong interaction symmetry. But the violation of the $\Delta I = 1/2$ rule on this model could well be serious, unless one wishes to invoke dynamical accidents to quantitatively suppress the violation (or, as it is sometimes said, to enhance the $\Delta I = 1/2$ contribution). It is possible to restore the $\Delta I = 1/2$ rule and, further, achieve a pure octet structure, by bringing in appropriate self couplings of the *neutral* members of the hadron current octets. These neutral currents, which have so far tagged along as mathematical objects, do not seem to figure in semi leptonic interactions with neutral lepton currents, such as $(\nu\bar{\nu})$. To this extent the introduction of self couplings of the neutral hadron currents seems to be artificial. On the other hand, it may be a good idea to find an actual physical significance for these neutral hadron currents. Later on we will discuss some current algebra tests of this scheme.

III. CURRENT COMMUTATION RELATIONS

We have by now assembled a collection of various hadron currents, vector and axial vector, which nature is supposed to employ in electromagnetic and weak semi leptonic interactions. Each current can be labeled by an index which specifies the isospin and hypercharge quantum numbers (I, I_3, Y). Between the electromagnetic and semi leptonic interactions, six different vector currents have been encountered; and four axial vector currents arise in the semi leptonic interactions. Anticipating SU_3 considerations to be taken up shortly, let us complete each set to a full octet, introducing further currents which may or may not have a physical significance but which in any case will be useful as mathematical objects. Moreover it will be convenient to introduce, in place of the quantum numbers I, $I_3 Y$, a single SU_3 index α which is appropriate to a real tensor basis. The vector and axial vector octets are denoted by the symbols V_λ^α, A_λ^α, $\alpha = 1, 2, \ldots 8$.

In this notation the electromagnetic current is

$$j_\lambda^{em} = V_\lambda^3 + \frac{1}{(3)^{1/2}} V_\lambda^8 , \tag{3.1}$$

where the first term is the isovector part of the current, the second term the isoscalar part. Similarly, the semi leptonic hadron current is now written

$$j_\lambda^{weak} = \cos\theta_c \, j_\lambda^{1+i2} + \sin\theta_c \, j_\lambda^{4+i5} \tag{3.2}$$

and, for the charge lowering adjoint,

(3.2) $(j_\lambda^{weak})^\dagger = \cos\theta_c\, j_\lambda^{1-i2} + \sin\theta_c\, j_\lambda^{4-i5}$

Here

(3.3) $j_\lambda^\alpha = V_\lambda^\alpha + A_\lambda^\alpha$

and

(3.3′) $j_\lambda^{1+i2} \equiv j_\lambda^1 + ij_\lambda^2$,   etc.

For each vector current V_λ^α let us introduce a generalized "charge" operator Q^α, according to

(3.4) $Q^\alpha(x_0) \equiv \int d^3x\, V_0^\alpha\, (\vec{x},x_0)$.

Conservation of hypercharge in the strong interactions implies that V_λ^8 is a conserved current, hence that Q^8 is in fact a time independent operator. Up to a factor, it is just the hypercharge operator:

(3.4′) $Q^8 = \dfrac{(3)^{\frac{1}{2}}}{2}\, Y$.

Similarly, on the CVC hypothesis the vector currents $V_\lambda^1, V_\lambda^2, V_\lambda^3$ are all conserved and all belong to a common isotopic triplet. The associated charges Q^i, $i = 1,2,3$, are therefore time independent and are nothing other than the three components of the isotopic spin operator $\vec{I}$:

(3.5) $Q^i = I^i$, $i = 1,2,3$.

The Q^i are the generators of isospin symmetry (SU_2) transformations and satisfy the familiar commutation relations

(3.6) $[Q^i, Q^j] = i\epsilon_{ijk} Q^k$, $i, j, k, = 1,2,3$.

Finally, let us for the moment suppose that SU_3 symmetry breaking effects are switched off in the strong interactions. There would then be a full octet of *conserved* vector currents; and on the Cabibbo hypothesis we would identify this with the octet $V_\lambda{}^\alpha$, $\alpha = 1,2,\ldots,8$. The members $V_\lambda{}^6$ and $V_\lambda{}^7$ do not enter physically in electromagnetic and semi leptonic interactions, but we are entitled to carry them along mathematically. In a world with exact SU_3 symmetry for the strong interactions, all eight charges Q^α would be time independent. They would be identified as the generators of SU_3 symmetry transformations and would satisfy the corresponding commutation relations

$$[Q^\alpha, Q^\beta] = if^{\alpha\beta\gamma}Q^\gamma \quad . \tag{3.7}$$

The SU_3 structure constants $f^{\alpha\beta\gamma}$ are totally antisymmetric in the indices. The non-vanishing $f^{\alpha\beta\gamma}$ (apart from permutations in the indices) are listed below:

$$f^{123} = 1; \;\; f^{147} = f^{246} = f^{257} = f^{345} = -f^{156} = -f^{367} = \tfrac{1}{2} \;;$$

$$f^{458} = f^{678} = (3)^{1/2}/2 \quad . \tag{3.8}$$

In the mythical world with exact SU_3 symmetry for the strong interactions we can go still farther on Cabibbo's model. For it asserts that not only the charges Q^α but also the current *densities* $V_\lambda{}^\alpha(x)$ form an SU_3 octet; similarly that the axial vector current densities $A_\lambda{}^\alpha(x)$ form an octet. These transformation properties are reflected in the more inclusive commutation relations

$$[Q^\alpha, V_\lambda{}^\beta(x)] = if^{\alpha\beta\gamma}V_\lambda{}^\gamma(x)$$

$$[Q^\alpha, A_\lambda{}^\beta(x)] = if^{\alpha\beta\gamma}A_\lambda{}^\gamma(x) \quad . \tag{3.9}$$

In the mythical world under discussion the hadrons would have well-defined SU_3 transformation properties and symmetry relations would connect various matrix elements of the different currents. The Clebsch-Gordon coefficients appearing in these relations of course reflect the structure of the SU_3 group, hence the structure of the commutation relations of Eq. (3.7). But apart from their technical usefulness in the derivation of the Clebsch-Gordon coefficients, the commutation relations would not be terribly interesting. We all pay passing tribute to the isospin commutation relations of Eq. (3.6) and thereafter look to the Condon-Shortley tables for practical isospin facts. The important thing, rather, is that SU_3 is not really a good strong interaction symmetry. With symmetry breaking terms switched on the hadrons cease having well defined SU_3 transformation properties, the charges Q^{α} (apart from $\alpha = 1,2,3,8$) now become time dependent, and the Cabibbo model loses a direct significance so far as symmetry considerations are concerned. It was Gell-Mann's idea that something of the model could nevertheless survive under SU_3 symmetry breakdown. In the first place, he proposed what at the time was a novel way of looking at things: namely, he proposed to characterize the currents in terms of their equal time commutation relations. The equal time commutation relations, formally at least, depend only on the structure of the currents when regarded as functions of the canonical field variables, the q's and p's, so to speak. If the symmetry breaking terms in the Lagrangian are reasonably decent (no derivative couplings, say), then independent of further details the currents will retain the original structure in terms of the canonical fields, and the commutation relations will correspondingly remain unchanged. It is safest to imagine all these things for the space integrated charge densities, i.e., for the charge operators. In short, Gell-Mann suggested that the *equal-time* version of Eq. (3.7) may be exact for the real world. Somewhat stronger is the conjecture that the equal-time version of Eq. (3.8) is exact. Even under the breakdown of SU_3 symmetry for the strong interactions, these conjectures suggest a precise, if abstract, meaning for the notion that the vector and axial vector currents have octet transformation properties. It also sets a relative scale for the various vector currents and similarly for the various axial vector currents. But at this stage the scale of the axial vector current octet is unrelated to that of the vector octet. The connection that is wanted here was again provided by Gell-Mann, who introduced conjectures concerning the equal-time commutators of two axial vector currents. In many ways these represent the most striking part of his scheme. To state these new hypotheses forthwith, let us introduce a set of eight "axial vector charges" $\bar{Q}^{\alpha}$ defined in analogy with Eq. (3.5)

$$\bar{Q}^{\alpha}(x_0) \approx \int d^3x \, A_0^{\alpha}(\vec{x},x_0) \quad . \tag{3.10}$$

Gell-Mann then postulates the equal time commutation relations

$$[\bar{Q}^{\alpha}(x_0), \bar{Q}^{\beta}(x_0)] = if^{\alpha\beta\gamma}Q^{\gamma}(x_0) \quad . \tag{3.11}$$

Since this is bilinear in axial vector charges, and since the right hand side contains vector charges, the relative scale becomes fixed as between the axial vector and vector quantities.

Now the equal-time version of the commutation relations of Eqs. (3.7) and (3.9) could be motivated by what is at least a roughly visible symmetry of the real world and by the attractive features of the Cabibbo model based on this symmetry. It is less easy to motivate Eq. (3.11). One could do so by contemplating a larger symmetry group, $SU_3 \times SU_3$, with generators $Q^{\alpha} + \bar{Q}^{\alpha}$ and $Q^{\alpha} - \bar{Q}^{\alpha}$. But this larger symmetry can hardly be said to be visible to the naked eye. Alternatively we might seek motivation in a model. Following Gell-Mann, suppose that the only fundamental fields are those associated with quarks. Then, just as we group, say, the proton and neutron fields in a 2 component vector in isospin space, $\psi = \binom{\psi_p}{\psi_n}$, so for SU_3 let $q = \begin{pmatrix} q_p \\ q_n \\ q_\lambda \end{pmatrix}$ represent the trio of quark fields. And in analogy with the three traceless 2 x 2 matrices, τ^i, i = 1,2,3, which operate in the isospin space of the nucleon doublet, here introduce 8 traceless 3 x 3 matrices λ^α, which operate in the SU_3 space of the quarks. A basis can be chosen such that the λ^α matrices satisfy the commutation relations

$$\left[\frac{\lambda^\alpha}{2}, \frac{\lambda^\beta}{2}\right] = if^{\alpha\beta\gamma} \frac{\lambda^\gamma}{2} \quad . \tag{3.12}$$

Let us now form vector and axial vector currents in the simplest way possible, as bilinear expressions in the quark fields:

$$V_\lambda{}^\alpha = i\bar{q}\,\gamma_\lambda \lambda^\alpha q$$

$$A_\lambda{}^\alpha = i\bar{q}\,\gamma_\lambda \gamma_5 \lambda^\alpha q \quad . \tag{3.13}$$

Of course we would be free to multiply the above expression by numerical factors, perhaps different ones for the vector and axial vector currents. But let's not! Using the canonical equal time anticommutation relations for spinor fields, together with the relations of Eq. (3.12), we can work out formally the equal-time commutation relations among the various components of the currents. I will not record the space-space commutators, but for the rest one finds the formal results

$$[V_0{}^\alpha(x), V_\lambda{}^\beta(y)]_{x_0=y_0} = if^{\alpha\beta\gamma}\, V_\lambda{}^\gamma(x)\delta(\vec{x}-\vec{y})$$

$$[V_0{}^\alpha(x), A_\lambda{}^\beta(y)]_{x_0=y_0} = if^{\alpha\beta\gamma} A_\lambda{}^\gamma(x)\delta(\vec{x}-y)$$

$$[A_0{}^\alpha(x), A_\lambda{}^\beta(y)]_{x_0=y_0} = if^{\alpha\beta\gamma} V_\lambda{}^\gamma(x)\delta(\vec{x}-\vec{y})$$

$$[A_0{}^\alpha(x), V_\lambda{}^\beta(y)]_{x_0=y_0} = if^{\alpha\beta\gamma} A_\lambda{}^\gamma(x)\delta(\vec{x}-\vec{y}) \quad . \tag{3.14}$$

These equations evidently contain the results of Eq. (3.11) and of the equal-time version of Eq. (3.9). But these local commutation relations are of course stronger and more far reaching. So far as I am aware, the local *time-time* relations of Eq. (3.14) are at least theoretically tenable. For the *space-time* commutators the local relations are too formal – they run afoul of the so-called Schwinger diseases, which presumably reflect the dangers of proceeding in a purely formal way with objects so singular as quantum fields. I will not enter into these delicacies, since all the applications to be discussed here can be based on the local time-time commutators, or on the presumably safe integrated versions of Eq. (3.14) (charge, current density commutators). So the Eqns. (3.14) will be allowed to stand, formally. As for the space-space commutators, I leave these to later speakers.

With the qualifications stated above, the idea now is to put the commutation relations of Eq. (3.14) forward as definite conjectures for the real world, independent of the quark model and other considerations that motivated them. The question is whether they happen to be true for the real world. So the problem is how are the physical consequences of these abstract propositions to be extracted, ideally with a minimum of approximation or extra assumptions? In the most direct approach, one can convert the commutation relations into sum rules. In this connection it is worthwhile to recall the famous oscillator strength sum rule for electric dipole radiation in a non-relativistic quantum mechanical system. For simplicity consider a single electron moving one-dimensionally in a potential field. Let the energy eigenvalues be E_i, the corresponding eigenstates $|i>$. The probability for electric dipole radiation between the levels i and j is proportional to the so-called oscillator strength, defined by

$$f_{ij} = \frac{2m}{\hbar^2}(E_j - E_i)\,|<j|x|i>|^2 \quad .$$

The energy levels, eigenstates, and hence the oscillator strengths all depend on the details of the potential field in which the electron moves. Nevertheless there is a constraint on the oscillator strengths that transcends these details, and which follows from the most primitive commutation relation of all, namely

$$[x,p] = ih \ .$$

It is an elementary exercise, left to the reader, to convert this commutation relation into the sum rule

$$\sum_j f_{ji} = 1 \ .$$

On the other hand, for the current algebra commutators the practical implications do not spring forward quite so easily. The art of extracting the physics from the current algebra conjectures will be one of our major occupations here. However, since some part of this art makes use of the notions of PCAC, let us drop current algebra for a while and turn to PCAC.

IV. PCAC

The pion decay reaction $\pi \to \ell + \bar{\nu}$ plays a central role for the notion of PCAC and we must turn to this first. The process is induced solely by the strangeness conserving axial vector current and has a very simple structure. For π^- decay, say, the amplitude is given by the expression

$$\text{(4.1)} \qquad \frac{G}{(2)^{1/2}} \cos\theta_c < 0|A_\lambda{}^{1+i2}|\pi^- > i\bar{u}(\ell)\gamma_\lambda(1+\gamma_5)u(\bar{\nu}) \ .$$

With p denoting the pion 4-momentum, the hadronic matrix element must clearly have the form

$$< 0|A_\lambda^{1+i2}|\pi^- > = i(2)^{1/2} f_\pi p_\lambda \ ,$$

where f_π, the so-called pion decay constant, is a parameter with dimensions of a mass. More generally, if we characterize the isospin state of the pion in a real basis, with index i ($i = 1,2,3$), and recall that

$\pi^- = \frac{1}{(2)^{1/2}} (\pi^1 - i\pi^2)$, then the above expression generalizes to

$$< 0|A_\lambda^i|\pi^j > = i f_\pi p_\lambda \, \delta_{ij} \ . \tag{4.2}$$

With the symbol μ henceforth denoting the mass of the pion and m_ℓ the mass of the charged lepton, the pion decay rate works out to be

$$\text{Rate}(\pi \to \ell + \bar{\nu}) = \frac{G^2 f_\pi^{\ 2}}{4\pi} \mu \, m_\ell^{\ 2} \left(1 - \frac{m_\ell^{\ 2}}{\mu^2}\right)^2 \cos^2\theta_c \ .$$

The experimental rate corresponds to

$$f_\pi \cos\theta_c \approx 96 \text{ MeV} \ . \tag{4.3}$$

Since the PCAC idea will focus especially on the matrix element of the divergence of the axial vector current, let us observe here that

$$< 0|\frac{\partial A_\lambda^{\ i}}{\partial x_\lambda}|\pi^j > = i p_\lambda < 0|A_\lambda^{\ i}|\pi^j > = \mu^2 f_\pi \delta_{ij} \ . \tag{4.4}$$

The divergence $\partial A_\lambda^{\ i}/\partial x_\lambda$ is a pseudoscalar operator, with odd G parity, unit isospin, zero hypercharge. It has, that is, all the quantum numbers of the ith component of the pion triplet. The essential idea of PCAC is going to be that this operator is a "good" one for use in describing the creation and destruction of pions. What will be involved here, for processes involving a pion, is the notion of continuing the amplitude off mass shell in the pion mass variable. The presumed goodness of the axial vector divergence is supposed to mean that the matrix elements of this operator are very slowly varying in the mass variable as one goes a little way off shell.

These vague statements have, of course, to be elaborated. As a beginning, let us consider the reaction $a+b \rightarrow c+d$, parameterized by the usual Mandelstam variables $s = -(p_a + p_b)^2$, $t = -(p_a - p_c)^2$. These quantities can vary continuously over some physical range. It is only over this range that the amplitude is of interest and is initially defined. However, it is known that the amplitude is in fact the boundary value of an analytic function, defined in some larger domain in the complex planes of s and t. That is, the physical amplitude admits of an analytic continuation into "unphysical" regions; and thanks to Cauchy, the continuation is unique. At the end, of course, one is interested in statements that can be made about the *physical* amplitude. But the process of discovering such statements is facilitated by contemplating the amplitude with respect to its full analytic properties. In fact, this has become a major industry.

On the other hand, the mass of a given particle cannot be varied experimentally – over any range whatsoever! So for this kind of parameter the notion of analytic continuation has no objective meaning. Cauchy is not available here. This is not to say that off mass shell amplitudes cannot be *defined* by the theoretical structures that people invent – for the ultimate purpose of describing physical processes. But it does mean that different procedures, dealing with the *same* physics, may validly give different off mass shell results. The familiar LSZ reduction formulas, which we shall be using, would seem to provide a natural basis for defining off mass shell amplitudes. But an arbitrariness is contained in the question: which operator is to be used for describing the creation and destruction of a given kind of particle. Any local operator formed from the fields of the underlying theory will do, provided that it has the appropriate quantum numbers and that it is properly normalized (see below). No doubt the purists have their additional requirements; but the point is that different choices will give the same results on mass shell, but, in general, will differ off mass shell. The "canonical" pion field, for example, even if it appears as a fundamental field of the underlying theory, need not take precedence over other choices. In fact, the PCAC enthusiasts advocate that the role of "pion" field be taken by the axial vector divergence. It can happen, as in the σ-model of Gell-Mann and Levy,[9] that the underlying theory contains a canonical pion field ϕ^i and that an axial vector current $A_\lambda{}^i$ can be formed in such a way that the relation $\partial A_\lambda{}^i/\partial x_\lambda \sim \phi^i$ follows from the equations of motion. In this case the axial vector divergence *is* the canonical pion field, up to a proportionality constant. But this special circumstance does not seem to be crucial to the idea of PCAC. The operational hypothesis of PCAC is that $\partial A_\lambda{}^i/\partial x_\lambda$ is a gentle operator for the purpose of going off shell in the pion mass variable.

Why one *wants* to go off shell will appear in the subsequent discussions. Briefly, it is because with the PCAC mode of continuation one can make interesting statements when the off shell pion mass variable goes to zero. The hope is that these statements remain true, more or less, back on the physical mass shell.

For use in standard reduction formulas, any choice of pion field operator ϕ^i is supposed to be normalized according to

$$< 0|\phi^i|\pi^j > = \delta_{ij} \quad .$$

If we are to make the PCAC choice, the identification must then be

$$\frac{1}{\mu^2 f_\pi} \frac{\partial A_\lambda{}^i}{\partial x_\lambda} \equiv \phi^i \quad . \tag{4.5}$$

To see in a more concrete way that one is free to make this identification, consider the process $\alpha \to \beta +$ lepton pair and focus on the axial vector contribution, as described by the matrix element $<\beta|\, A_\lambda{}^i\,|\alpha>$. Let

$q = p_\alpha - p_\beta$ be the momentum carried by the lepton pair. The quantity $iq_\lambda \langle \beta | A_\lambda^i | \alpha \rangle$ is the matrix element of the current divergence and let us, in fact, concentrate on this object, i.e., on

$$\langle \beta | \frac{\partial A_\lambda^i}{\partial x_\lambda} | \alpha \rangle \ .$$

Among the Feynman graphs which contribute to this, single out those in which a single virtual pion connects between the current and the hadrons, as shown. These graphs have a pole at $q^2 = -\mu^2$, the factor $(q^2+\mu^2)^{-1}$ arising from the pion propagator. The residue of the pole is a product of the factor $\langle 0 | \partial A_\lambda^i / \partial x_\lambda | \pi^i \rangle = \mu^2 f_\pi$ and a factor

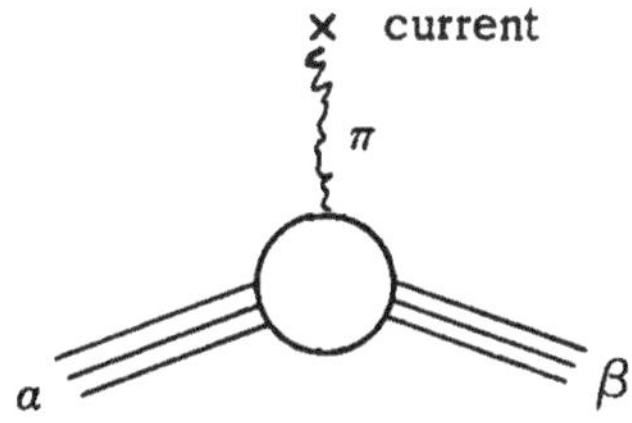

corresponding to the pion mass shell amplitude for the process $\alpha \to \beta + \pi^i$. Thus we can write

$$\langle \beta | \frac{\partial A_\lambda^i}{\partial x_\lambda} | \alpha \rangle = \frac{f_\pi \mu^2}{q^2 + \mu^2} \times \mathrm{Amp}(\alpha \to \beta + \pi^i) + \text{terms regular at } q^2 = -\mu^2.$$

In the limit $q^2 \to -\mu^2$, the pole term dominates, hence

$$\text{(4.6)} \qquad \mathrm{Amp}(\alpha \to \beta + \pi^i) = \lim_{q^2 \to -\mu^2} \frac{\mu^2 + q^2}{\mu^2 f_\pi} \langle \beta | \frac{\partial A_\lambda^i}{\partial x_\lambda} | \alpha \rangle \ .$$

But compare this with the standard LSZ formula for the same amplitude, as expressed in terms of "the pion field" ϕ^i:

$$\mathrm{Amp}(\alpha \to \beta+\pi^i) = \lim_{q^2 \to -\mu^2} (\mu^2+q^2) < \beta|\phi^i|\alpha > .$$

(i) Neutron β Decay: –

So long as we remain on mass shell, Eq. (4.6) is evidently a mere identity. Upon removing the instruction $q^2 \to -\mu^2$ we *define* an off mass shell amplitude. PCAC acquires content when we assert that the so-defined off shell amplitude is gently varying in q^2, typically over the range at least from $q^2 = -\mu^2$ to $q^2 = 0$. The notion of gentleness is not a very precise one, but let's turn to a simple application where the hypothesis takes on a rather unambiguous meaning operationally.[10] Namely, consider the β decay reaction $n \to p+e^-+\bar{\nu}_e$ and focus in particular on the axial vector matrix element. The most general structure is given by

$$(4.7) \qquad <p|A_\lambda^{1+i2}|n> = -i\bar{u}_p \{g_A(q^2)\gamma_\lambda\gamma_5+iG_A(q^2)q_\lambda\gamma_5\}u_n ,$$

where

$$q = n-p$$

is the momentum transfer between neutron and proton. The form factors $g_A(q^2)$ and $G_A(q^2)$ are real functions of the invariant momentum transfer variable q^2. In physical β decay it varies over a small range near $q^2 \approx 0$. For the matrix element of the divergence of the axial vector current we have, using the Dirac equation,

$$(4.8) \qquad <p|\frac{\partial A_\lambda^{1+i2}}{\partial x_\lambda}|n> = iq_\lambda <p|A_\lambda^{1+i2}|n> =$$

$$= i[2mg_A(q^2)-q^2G_A(q^2)]\,\bar{u}_p\gamma_5 u_n .$$

With the PCAC identification of pion field, however, we can write

$$(4.9) \qquad \langle p | \frac{\partial A_\lambda^{1+i2}}{\partial x_\lambda} | n \rangle = (2)^{1/2} \mu^2 f_\pi \langle p | \frac{\phi^{1+i2}}{(2)^{1/2}} | n \rangle \; .$$

The matrix element on the right defines the pion nucleon vertex function $g_r(q^2)$, according to

$$(4.10) \qquad \langle p | \frac{\phi^{1+i2}}{(2)^{1/2}} | n \rangle = (2)^{1/2} \frac{g_r(q^2)}{\mu^2+q^2} \, i\bar{u}_p \gamma_5 u_n \; .$$

The vertex function $g_r(q^2)$ has an objective meaning at $q^2 = -\mu^2$, where it becomes the pion-nucleon coupling constant. Off shell, i.e., for $q^2 \neq -\mu^2$, the vertex function does not have an objective significance. It depends on the choice of pion operator used to effect the off shell continuation. Here we have made the PCAC choice. In fact, so far we have done nothing with content. Physical content enters when we conjecture that with the PCAC way of going off mass-shell, $g_r(q^2)$ is a gently varying function. In particular, let us choose $q^2 = 0$ and suppose that $g_r(0) \approx g_r(-\mu^2) \equiv g_r$. From Eqns. (4.8)–(4.10) we then find

$$(4.11) \qquad f_\pi = \frac{m g_A(0)}{g_r} \; .$$

With $g_A(0) \simeq 1.22$, $g_r{}^2/4_\pi \approx 14.6$ we find $f_\pi \simeq 87$ MeV, in reasonably good agreement with the empirical value ≈ 96 MeV. More generally, Eqns. (4.8)–(4.10) lead to the relation

$$2m g_A(q^2) - q^2 G_A(q^2) = \frac{2g_r(q^2)}{\mu^2+q^2} \mu^2 f_\pi = 2g_r(q^2) f_\pi - \frac{2q^2 g_r(q^2) f_\pi}{\mu^2+q^2} \; .$$

The pole on the right hand side is associated with the form factor $G_A(q^2)$

on the left side. If $g_r(q^2)$ is slowly varying, as we assume say for $-\mu^2 \lesssim q^2 \lesssim 0$, and if also $g_A(q^2)$ is slowly varying, then

$$G_A(0) \simeq \frac{2mg_A(0)}{\mu^2} \quad . \tag{4.12}$$

We have noted this result because the term proportional to G_A in Eq. (4.7) is responsible for the so-called pseudoscalar coupling in β decay and in muon capture. This has been detected for muon capture in hydrogen and the sign and magnitude of G_A seem to be roughly right.

(ii) PCAC for Neutrino Reactions: –

Another remarkable application of the PCAC hypothesis was first noticed by Adler in connection with high energy neutrino reactions.[11] These processes are of enormous interest from many points of view and we will return to some of these other aspects later on. Here it is PCAC that is in view. Consider a neutrino induced reaction of the sort

$$\nu + p \rightarrow \ell + \beta \ ,$$

where β is some multihadron system with zero strangeness and where the target particle for definiteness is taken to be a proton. The amplitude is given by

$$M = \frac{G}{(2)^{1/2}} < \beta | j_\lambda^{\text{weak}} | p > i\bar{u}_\ell \gamma_\lambda (1+\gamma_5) u_\nu \quad . \tag{4.13}$$

Let $q = p_\nu - p_\ell$, where p_ν is the neutrino momentum, p_ℓ the outgoing lepton momentum. Suppose that the lepton mass m_ℓ can be ignored, as seems reasonable once the energy is well above threshold; and specialize now to the configurations where the neutrino and lepton have parallel 3-momenta, $\vec{p}_\ell \,||\, \vec{p}_\nu$ (we have the laboratory frame in mind, though for massless particles this characterization is an invariant one). For these configurations it is evident that $q^2 = 0$. Moreover a short calculation shows that

(4.14) $$\bar{u}_\ell \gamma_\lambda (1+\gamma_5) u_\nu = 2 \left(\frac{\epsilon_\nu \epsilon_\ell}{(\epsilon_\nu - \epsilon_\ell)^2} \right)^{1/2} q_\lambda$$

where ϵ_ν and ϵ_ℓ are the neutrino and lepton energies. But

(4.15) $$-iq_\lambda < \beta | j_\lambda^{\text{weak}} | p > \; = \; < \beta | \partial j_\lambda^{\text{weak}} / \partial x_\lambda | p > \; .$$

Thus, for the parallel configurations the amplitude M involves the divergence of the weak current. We are dealing, however, only with the strangeness conserving current and can therefore invoke the CVC hypothesis to eliminate the divergence of the *vector* part of j_λ^{weak}. The matrix element M can then be written, for $q^2 \to 0$ (i.e., for the parallel configurations),

(4.16) $$\underset{q^2 \to 0}{M \to} \; -(2)^{1/2} G \left[\frac{\epsilon_\nu \epsilon_\ell}{(\epsilon_\nu - \epsilon_\ell)^2} \right]^{1/2} < \beta | \partial A_\lambda^{\;1+i2} / \partial x_\lambda | a > \cos\theta_c \; .$$

Set this aside for the moment and consider now the *pion* reaction $\pi^+ + p \to \beta$, letting q denote the pion momentum. According to the standard reduction formula, the amplitude for the pionic process is

(4.17) $$M_\pi = \lim_{q^2 \to -\mu^2} \frac{(q^2 + \mu^2)}{(2)^{1/2}} < \beta | \phi^{1+i2} | \, p >$$

where $\phi^{1+i2}/(2)^{1/2}$ is the field that destroys the π^+ (or creates the π^-). Let us now make the PCAC identification for the pion field, according to Eq. (4.5). Moreover, in Eq. (4.17) let us remove the restriction $q^2 \to -\mu^2$ and continue M_π off mass shell in q^2, to the point $q^2 = 0$. At this point the physical neutrino reaction amplitude M becomes related to the continued pion amplitude M_π, and we find

$$(4.18)\qquad |M|^2 \underset{q^2\to 0}{\to} 4G^2 \frac{\epsilon_\nu \epsilon_\ell}{(\epsilon_\nu - \epsilon_\ell)^2} f_\pi^{\ 2} \cos^2\theta_c |M_\pi|^2 \ .$$

Again, the physics enters when we invoke the PCAC hypothesis that the pion amplitude continued to zero pion mass is not very different from the physical (i.e., on mass shell) pion amplitude. Hereby, through PCAC a weak process becomes related to a strong one! The more detailed expression of this relation, in terms of cross sections, is a matter now of routine kinematics. To illustrate, let $\sigma_\pi(W)$ be the total cross section for the reaction $\pi^+ + p \to \beta$, where W is the center of mass energy, or equivalently, the invariant mass of the hadron system β. For the neutrino reaction, let $\partial\sigma/\partial\Omega\,\partial W$ be the lab frame differential cross section for forward lepton production, with W the invariant mass of the system β. Then

$$(4.19)\qquad \frac{\partial\sigma}{\partial\Omega\partial W} = \frac{G^2}{\pi^3} f_\pi^{\ 2}\cos^2\theta_c \left(\frac{W^2}{m^2}\right) \left(\frac{2m\epsilon_\nu - W^2 + m^2}{W^2 - m^2}\right)^2 \left(\frac{W^2 - m^2}{2W}\right) \sigma_\pi(W)\ ,$$

where m is the nucleon mass, ϵ_ν the laboratory energy of the neutrino. Reliable tests of this relation are unfortunately not yet available.

(iii) The Adler Consistency Condition: –

A third application of PCAC, again devised by Adler,[12] may be illustrated on the example of pion-nucleon scattering. Let p_1 and p_2 be the initial and final nucleon momenta, k and q the initial and final meson momenta. Let the indices α and β denote the isotopic spin states of initial and final mesons. To symbolize all this, write: $k(\alpha) + p_1 \to q(\beta) + p_2$. With the PCAC choice of pion field, the amplitude is given by

$$M^{\beta\alpha} = \lim_{q^2+\mu^2 \to 0} \frac{q^2+\mu^2}{\mu^2 f_\pi} < p_2 | \frac{\partial A_\lambda{}^\beta}{\partial x_\lambda} | p_1, k(\alpha) >$$

(4.20)
$$= \lim_{q^2+\mu^2 \to 0} \frac{q^2+\mu^2}{\mu^2 f_\pi} iq_\lambda < p_2 | A_\lambda{}^\beta | p_1, k(\alpha) > .$$

The general structure of the amplitude is expressed by

(4.21)
$$M^{\beta\alpha} = \bar{u}(p_2) \left[A^{\beta\alpha} - \frac{i\gamma\cdot(k+q)}{2} B^{\beta\alpha} \right] u(p_1)$$

where $A^{\beta\alpha}$ and $B^{\beta\alpha}$ are scalar functions of the two kinematic variables of the problem, say the Mandelstam variables s and t; or more conveniently for present purposes, choose the variables

$$\nu = - \frac{(p_1 + p_2)\cdot q}{2m}$$

(4.22)
$$\nu_B = \frac{q\cdot k}{2m}$$

Physically, of course, the pion masses are fixed: $q^2 = k^2 = -\mu^2$. The function $B^{\beta\alpha}$ has nucleon pole terms arising from the Feynman graphs shown:

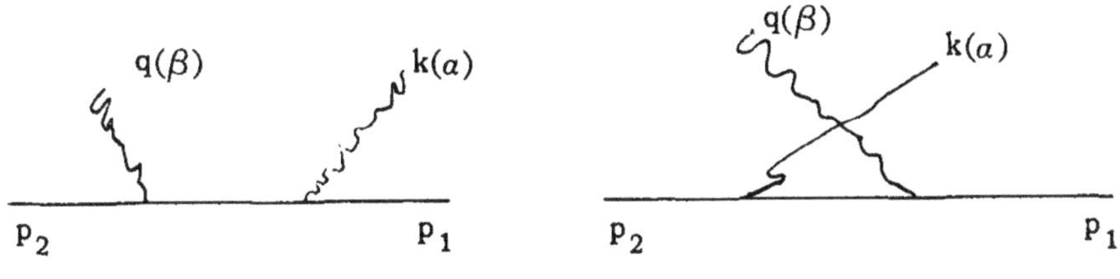

$$(4.23) \qquad B^{\beta a}\Big|_{\text{pole term}} = \frac{g_r^{\,2}}{2m}\left\{\frac{\tau_\beta \tau_a}{\nu_B - \nu} - \frac{\tau_a \tau_\beta}{\nu_B + \nu}\right\} .$$

Let us now return to Eq. (4.20), remove the instruction $q^2 \to -\mu^2$, and let the 4-vector q go to zero (so, incidently, $q^2 \to 0$). That is, we are defining an off mass shell amplitude and are continuing it to $q^2 = 0$ in the mass variable. But as $q \to 0$ we are also taking the "true" variables ν and ν_B to unphysical points: $\nu \to 0$, $\nu_B \to 0$. Because of the factor q_λ which appears on the right hand side of Eq. (4.20), it would appear that $M^{\beta a}$ must vanish as $q \to 0$. This is, of course, the case, except for terms in the matrix element $< p_2|A_\lambda{}^B|p_1, k(a) >$ which are singular in the limit $q \to 0$. It is easy to see that such terms arise only from the one-nucleon pole diagrams for this matrix element.

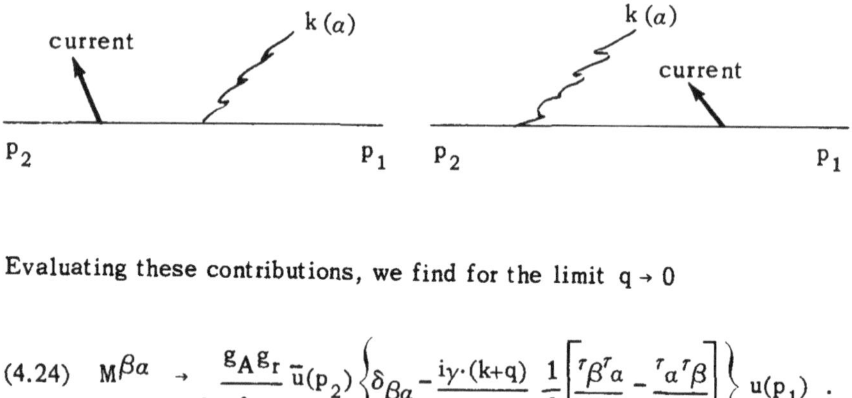

Evaluating these contributions, we find for the limit $q \to 0$

$$(4.24) \quad M^{\beta a} \underset{q \to 0}{\to} \frac{g_A g_r}{f_\pi} \bar{u}(p_2)\left\{\delta_{\beta a} - \frac{i\gamma\cdot(k+q)}{2}\,\frac{1}{2}\left[\frac{\tau_\beta \tau_a}{\nu_B - \nu} - \frac{\tau_a \tau_\beta}{\nu_B + \nu}\right]\right\} u(p_1) .$$

Comparing with Eqns. (4.21) and (4.23), we see that the continued function $B^{\beta a}$ has the poles of the mass shell function $B^{\beta a}$, provided that we set

$$\frac{g_A g_r}{2f_\pi} = \frac{g_r^{\,2}}{2m} \quad \text{or} \quad f_\pi = \frac{m g_A}{g_r}$$

But this is the relation obtained earlier from the PCAC application to β-decay, i.e., PCAC has passed this consistency test. We also obtain a new result however for the amplitude $A^{\beta\alpha}$. In the limit $q \to 0$

$$A^{\beta\alpha}(\nu=0,\ \nu_B=0,\ q^2=0) = \delta_{\alpha\beta}\frac{g_A g_r}{f_\pi} = \frac{g_r^2}{m}\delta_{\alpha\beta} \ . \tag{4.25}$$

The amplitude which appears here is evaluated at an unphysical point with respect to all three variables, ν, ν_B, q^2. Physics corresponds to $q^2 = -\mu^2$, The quantities ν and ν_B are continuous variables, but the point $\nu = \nu_B = 0$ lies outside the physical region. In principle this is no problem, however. The physical amplitude admits of a unique analytic continuation to $\nu = \nu_B = 0$; so we may regard $A^{\beta\alpha}(\nu=0,\ \nu_B=0,\ q^2=-\mu^2)$ as knowable, and indeed, Adler has carried out the extrapolation. In relating this quantity to the amplitude on the left side of Eq. (4.25) we now invoke the PCAC hypothesis that the amplitude varies gently as q^2 varies from $-\mu^2$ to zero. Thus we take PCAC to mean

$$A^{\beta\alpha}(\nu=0,\ \nu_B=0,\ q^2=0) \approx A^{\beta\alpha}(\nu=0,\ \nu_B=0,\ q^2=-\mu^2) \tag{4.26}$$

and the prediction is then

$$A^{\beta\alpha}(\nu=0,\ \nu_B=0,\ q^2=-\mu^2) \approx \frac{g_r^2}{m}\delta_{\beta\alpha} \ . \tag{4.27}$$

The experimental fit is quite good. The essential features can be grasped more readily if we consider the pi-nucleon amplitude at physical threshold, $\nu=\mu$, $\nu_B=-\mu^2/m$, which is about as close as we can come physically to $\nu=\nu_B=0$. Decompose the amplitude $M^{\beta\alpha}$ into parts which are respectively even and odd in the isotopic indices, according to

$$M^{\beta\alpha} = M^{(e)}\delta_{\beta\alpha} + \frac{1}{2}[\tau^\beta, \tau^\alpha]\, M^{(0)} \tag{4.28}$$

and similarly decompose $A^{\beta a}$ and $B^{\beta a}$. Focus on the even amplitude $M^{(e)}$. At *threshold* it is given by

$$M^{(e)} = A^{(e)} + \mu B^{(e)}.$$

For $A^{(e)}$ let us take the Adler prediction of Eq. (4.27), ignoring the small variation between the point $\nu = \nu_B = 0$ and the threshold point $\nu = \mu$, $\nu_B = -\mu^2/m$. For $B^{(e)}$ take the pole approximation of Eq. (4.23), since threshold is fairly close to the pole. Moreover drop terms which are of order μ/m. At threshold one then finds the remarkable prediction

$$M^{(e)} \approx 0. \tag{4.29}$$

In terms of the s-wave scattering lengths $a_{3/2}$ and $a_{1/2}$ (the indices refer to isotopic spin) the prediction is

$$\frac{1}{3}(2a_{3/2} + a_{1/2}) \approx 0.$$

This is well satisfied by the experimental values $a_{1/2} = 0.17\mu^{-1}$, $a_{3/2} = -0.088\mu^{-1}$.

(iv) Discussion: –

It is important to comment here (as we will do again, from time to time) on some of the ambiguities that arise for the notion of PCAC. In the discussion leading to Eq. (4.11) the PCAC choice of pion field was used to define an off mass shell pion-nucleon vertex function $g_r(q^2)$. Only at the point $q^2 = -\mu^2$ does this quantity have an independent and objective meaning. The relation $f_\pi = mg_A(0)/g_r(0)$ is therefore in itself without content. The physics entered when the PCAC assumption was made that $g_r(q^2)$, in the manner defined, is slowly varying, so that $g_r(0) \approx g_r(-\mu^2) \equiv g_r$. In the simple situation encountered here, where the continued quantity depends only on the mass variable q^2, this seems to

be the only reasonable way to interpret the PCAC idea. In other applications, however, the off shell amplitude depends not only on the mass variable q^2 but on other variables as well. Thus, in the pi-nucleon example, $A^{(e)}$ depends on two physically continuous variables. These were taken to be the quantities ν and ν_B. Our procedures then led to a statement concerning $A^{(e)}(\nu = 0, \nu_B = 0, q^2 = 0)$. As usual, this is free of content until some hypothesis is made about its connection to the mass shell amplitude $A^{(e)}(\nu = 0, \nu_B = 0, q^2 = -\mu^2)$. To be sure, this latter quantity is itself not physical with respect to the variables ν and ν_B, but that's not the issue — the point $\nu = 0$, $\nu_B = 0$ can, in principle at least, be reached by extrapolation from the physical region. The physics entered when the assumption was made that the variation with q^2 is gentle, for ν and ν_B held fixed, i.e. the operational assumption was that
$A^{(e)}(\nu = 0, \nu_B = 0, q^2 = 0) \approx A^{(e)}(\nu = 0, \nu_B = 0, q^2 = -\mu^2)$. But instead of variables ν and ν_B we could originally have chosen a different set of variables, e.g., the Mandelstam quantities s and t. On mass shell the choice is merely a matter of convenience. But off mass shell, the statement that the amplitude is slowly varying in q^2, for fixed ν and ν_B, is in general inequivalent to the same statement made for fixed s and t: the relations connecting one set with the other involve the off shell mass variable q^2. It is only when the *physical* amplitude is in fact slowly varying in the region of interest, i.e., slowly varying in the physically continuous variables, that the specification of these variables becomes irrelevant for the purposes of PCAC. This is presumably the situation for the amplitude $A^{(e)}$ in the region of interest. Here it wouldn't matter much if we had applied the PCAC hypothesis to this amplitude regarded as a function of, say, s and t rather than ν and ν_B. Strictly speaking, however, if the physical amplitude has any variation whatsoever, however slight, a sufficiently extreme alteration in choice of variables could always be arranged to accomplish any desired change in PCAC predictions. Crudely speaking, one is guided in choice of variables by the requirement that the pion pole singularity at $q^2 = -\mu^2$ shall dominate over other singularities, i.e., that

the other singularities be as far away from the point $q^2 = 0$ as possible. But the distance of a singularity and the strength of its contribution are not obviously correlated in a simple way in general.

V. *CURRENT ALGEBRA: METHODS*

A number of techniques that have proved useful for extracting the implications of current algebra are assembled in this section.

Consider the matrix element

$$M_\nu{}^\beta = i\int dx\, e^{-iq\cdot x} < b|\theta(x_0)\, [j_\nu{}^\beta(x),\, F(0)]\,|\, a> \quad , \tag{5.1}$$

where $j_\nu{}^\beta$ is some current operator, specified by the index β; and $F(0)$ is some unspecified local operator. In our subsequent discussions we will see how such matrix elements, with appropriate choices for the operators and states, arise in various physical problems. For the moment we leave interpretations open. Form the contraction

$$iq_\nu\, M_\nu{}^\beta = -i\int dx \frac{\partial}{\partial x_\nu}(e^{-iq\cdot x}) < b|\theta(x_0)\, [j_\nu{}^\beta(x),\, F(0)]\,|a>$$

and integrate by parts, ignoring surface terms (along with the anguished cries of purists). Since $\partial\theta(x_0)/\partial x_0 = \delta(x_0)$, this leads to the equation

$$iq_\nu M_\nu{}^\beta = i\int dx\, e^{-iq\cdot x} < b|\theta(x_0)\left[\frac{\partial j_\nu{}^\beta(x)}{\partial x_\nu},\, F(0)\right]|a>$$

$$+\, i\int dx\, e^{-iq\cdot x}\, \delta(x_0) < b|[j_0{}^\beta(x), F(0)]\,|\, a> . \tag{5.2}$$

The first term on the right involves the divergence of the current, which is a useful thing for PCAC purposes; the second term on the right involves an equal time commutator, the sort of thing one likes to see displayed for the subject of current algebra.

For the case where the operator F is itself a current, we consider, in analogy with Eq. (5.1), the matrix element

$$T_{\nu\mu}{}^{\beta\alpha} = i\int dx\; e^{-iq\cdot x} < b|\theta(x_0)[j_\nu{}^{\beta}(x), j_\mu{}^{\alpha}(0)|\, a > \quad , \tag{5.3}$$

and the analog of Eq. (5.2)

$$iq_\nu T_{\nu\mu}{}^{\beta\alpha} = i\int dx\; e^{-iq\cdot x} < b|\theta(x_0)\left[\frac{\partial j_\nu{}^{\beta}(x)}{\partial x_\nu},\, j_\mu{}^{\alpha}(0)\right]|\, a >$$

$$+\, i\int dx\; e^{-iq\cdot x}\, \delta(x_0) < b|\,[j_0{}^{\beta}(x), j_\mu{}^{\alpha}(0)\,]\,|\, a > \quad . \tag{5.4}$$

We may think of $T_{\nu\mu}{}^{\beta\alpha}$ as amplitude for the "process": current $(\alpha)+a \to$ current $(\beta)+b$, where q is the momentum carried by the outgoing current and

$$k \equiv p_b - p_a + q \tag{5.5}$$

is the momentum carried by the incoming current. For some later purposes it will be useful to consider a twice contracted equation

$$q_\nu T_{\nu\mu}{}^{\beta\alpha}k_\mu = k_\mu\int dx\; e^{-iq\cdot x} < b|\theta(x_0)\left[\frac{\partial j_\nu{}^{\beta}(x)}{\partial x_\nu},\, j_\mu{}^{\alpha}(0)\right]|\, a >$$

$$+\, k_\mu\int dx\; e^{-iq\cdot x}\delta(x_0) < b|\,[j_0{}^{\beta}(x), j_\mu{}^{\alpha}(0)\,]\,|\, a > \quad .$$

Using translation invariance one can express the first term on the right in the form

$$k_\mu \int dx\, e^{ik\cdot x} < b|\theta(-x_0) \left[\frac{\partial j_\nu{}^\beta(0)}{\partial x_\nu}, j_\mu{}^\alpha(x) \right] |a> .$$

This can be integrated by parts, as before. Then again using translation invariance one finds

$$q_\nu T_{\nu\mu}{}^{\beta\alpha} k_\mu = i \int dx\, e^{-iq\cdot x} < b|\theta(x_0) \left[\frac{\partial j_\nu{}^\beta(x)}{\partial x_\nu}, \frac{\partial j_\mu{}^\alpha(0)}{\partial x_\mu} \right] |a>$$

$$+ k_\mu \int dx\, e^{-iq\cdot x}\, \delta(x_0) < b|[j_0^\beta(x), j_\mu{}^\alpha(0)]\, a>$$

$$+ i \int dx\, e^{-iq\cdot x}\, \delta(x_0) < b|[j_0{}^\alpha(0), \frac{\partial j_\nu{}^\beta(x)}{\partial x_\nu}]\, a> . \tag{5.6}$$

Finally, consider an amplitude of the form

$$F_\mu{}^\alpha = i \int dx\, dy\, e^{-iq\cdot x - ik\cdot y} < b|T(j_\mu{}^\alpha(x)\, B(y)\, C(0))|a> \tag{5.7}$$

where T ( ) denotes time-ordering. Following the procedures used above one finds

$$iq_\mu F_\mu{}^\alpha = i \int dx\, dy\, e^{-iq\cdot x - ik\cdot y} < b|T(\frac{\partial j_\mu{}^\alpha(x)}{\partial x_\mu} B(y)\, C(0))|a>$$

$$+ i \int dx\, dy\, e^{-iq\cdot x - ik\cdot y}\, \delta(x_0 - y_0) <b|T([j_0^\alpha(x), B(y)]C(0)|a>$$

$$+ i \int dx\, dy\, e^{-iq\cdot x - ik\cdot y}\, \delta(x_0) <b|T([j_0{}^\alpha(x), C(0)]B(y))|a> . \tag{5.8}$$

Several important applications of current algebra ideas (the Cabibbo-Radicati sum rule for Compton scattering, the Adler sum rule for neutrino induced reactions, the Adler-Weisberger relation for pi-nucleon scattering) can be treated on the same footing, up to a point. For these cases we focus on the amplitude $T_{\nu\mu}{}^{\beta a}$ defined by Eq. (5.3) and specialize to the situation where the states $|a>$ and $|b>$ are identical states corresponding to a single hadron (proton, say) of momentum p. To bring this out, write $|a> = |b> = |p>$. From Eq. (5.4): $k = q$. Moreover, if the hadron has spin, let it be understood that $T_{\nu\mu}{}^{\beta a}$ is averaged over spin. The different applications will be distinguished by the choice of currents and by the interpretation of certain general results which are now to be discussed. For a while longer we continue with formalism, leaving these distinguishing features till later on. However, in all the cases under present discussion we take j^{a} to be a charge raising current and j^{β} to be its adjoint, a charge lowering current. For brevity, write $j^{a} = j$, $j^{\beta} = j^{\dagger}$. Then in the present situation we are studying

$$(5.9) \qquad T_{\nu\mu} = i\int dx\, e^{-iq\cdot x} < p|\theta(x_0)\,[j_\nu{}^{\dagger}(x), j_\mu(0)]|p> \quad .$$

Since $T_{\nu\mu}$ is understood to be averaged over spin of the hadron state $|p>$, it is a tensor that can depend only on the momenta p and q. It can therefore be decomposed as follows:

$$(5.10) \quad T_{\nu\mu} = A\, p_\mu p_\nu + B_1\, \delta_{\mu\nu} + B_2\, p_\mu\, q_\nu + B_3\, q_\mu p_\nu + B_4\, q_\mu q_\nu + B_5\, \epsilon_{\mu\nu a\beta}\, q_a p_\beta$$

where the coefficients $A, B_1, \ldots B_5$ are scalar functions of two covariant variables—the "mass" variable q^2 and the laboratory energy of the current

$$\nu = -\, p\cdot q/m, \quad p^2 = -m^2 \ .$$

Notice that the function B_5 arises only in a parity violating situation, from interference of vector and axial vector components of j and $j^{\dagger}$.

Refer now to Eq. (5.4) and decompose the first term on the right according to

$$(5.11)\qquad \int dx\, e^{-iq\cdot x} < p|\theta(x_0)\,[\frac{\partial j_\nu{}^\dagger(x)}{\partial x_\nu},\, j_\mu(0)]|p> = Dp_\mu + Eq_\mu \ ,$$

where D and E are scalar functions of q^2 and ν. Equation (5.4) also contains an equal time commutator term and for this we invoke the hypotheses of current algebra. For the current j_μ we will be dealing, variously, with the choices $V_\mu{}^{1+i2}$, $A_\mu{}^{1+i2}$, and $j^{weak} = \cos\theta_c$ $(V_\mu{}^{1+i2} + A^{1+i2}) + \sin\theta_c\,(V_\mu{}^{4+i5} + A_\mu{}^{4+i5})$. Let us recall then some of the relevant commutation relations of current algebra:

$$(5.12)\qquad [V_0{}^{1-i2}(x), V_\mu{}^{1+i2}(0)]_{x_0=0} = [A_0{}^{1-i2}(x), A_\mu{}^{1+i2}(0)]_{x_0=0}$$

$$= -2V_\mu{}^3(x)\delta(\vec{x}) \ ,$$

$$(5.12')\qquad [V_0{}^{4-i5}(x), V_\mu{}^{4+i5}(0)]_{x_0=0} = [A_0{}^{4-i5}(x), A_\mu{}^{4+i5}]_{x_0=0}$$

$$= -\left\{(3)^{1/2}\, V_\mu{}^8(x) + V_\mu{}^3\right\}\delta(\vec{x}) \ .$$

The following may also be noted

$$(5.13)\qquad [V_0{}^{1-i2}(x),\, A_\mu{}^{1+i2}(0)]_{x_0=0} = [A_0{}^{1-i2}(x),\, V_\mu{}^{1+i2}(0)]_{x_0=0}$$

$$= -2A_\mu{}^3(x)\delta(\vec{x}) \ ,$$

$$(5.13')\qquad [V_0{}^{4-i5}(x), A_\mu{}^{4+i5}(0)]_{x_0=0} = [A_0{}^{4-i5}(x), V_\mu{}^{4+i5}(0)]_{x_0=0}$$

$$= -\left\{(3)^{1/2} A_\mu{}^8(x) + A_\mu{}^3(x)\right\}\delta(\vec{x}) \ ;$$

but these won't figure into the present considerations since we are averaging over hadron spins and therefore $< p|A_\mu^a|p > = 0$. Finally, observe that

$$< p|V_\mu^{\ 3}|p > = \frac{p_\mu}{m} I_3 \quad , \tag{5.14}$$

$$< p|V_\mu^{\ 8}|p > = \frac{p_\mu}{m} \frac{(3)^{1/2}}{2} Y \quad , \tag{5.15}$$

where Y and I_3 are the hypercharge and isospin quantum numbers of the hadron. In all the cases considered here, therefore, the equal time commutator term in Eq. (5.4) can be written

$$\int dx\, e^{-iq\cdot x}\, \delta(x_0) < p|\,[j_0^\dagger(x),\, j_\mu(0)]\,|\,p > = Cp_\mu \quad , \tag{5.16}$$

where the commutator constant C will depend on the particular case under discussion. Now insert Eqs. (5.10), (5.11), and (5.16) into Eq. (5.4), which really constitutes two equations, one for the coefficients of p_μ, another for the coefficients of q_μ. The former yields the relation

$$(q\cdot p)\, A + q^2\, B_2 = D + C \quad . \tag{5.17}$$

Recall that A, B_2, and D all depend on the variables ν and q^2, whereas C is a constant determined by the commutation relations.

Now comes the major technical assumption. The functions A, B_2, D all satisfy dispersion relations in the variable ν, for q^2 fixed (we are always concerned here only with space-like q). Let us suppose that all three in fact satisfy unsubtracted dispersion relations. For example

$$A(\nu,q^2) = \frac{1}{\pi}\int_{-\infty}^{\infty} d\nu' \;\frac{\mathrm{Im}\ A(\nu',q^2)}{\nu'-\nu} \quad ,$$

and similarly for B_2 and D. But in Eq. (5.17) the function A appears multiplied by the factor $q \cdot p = -m\nu$. In the limit $\nu \to \infty$, the functions B_2 and D make no contribution in Eq. (5.17) and therefore

$$\lim_{\nu \to \infty} (q \cdot p)\, A = \frac{m}{\pi} \int_{-\infty}^{\infty} d\nu' \, \mathrm{Im}\, A(\nu', q^2) = C \, . \tag{5.18}$$

This is the basic result.

The absorptive part of $T_{\nu\mu}$ is given by

$$\mathrm{Abs}\, T_{\nu\mu} \equiv t_{\nu\mu} = \frac{1}{2} \int dx \, e^{-iq\cdot x} < p|[j_\nu{}^\dagger(x), j_\mu(0)]|p> \tag{5.19}$$

and its tensor decomposition is related to that of Eq. (5.10) in a simple way:

$$t_{\nu\mu} = \mathrm{Im}\, A\, p_\nu\, p_\mu + \mathrm{Im}\, B_1\, \delta_{\mu\nu} + \ldots \tag{5.20}$$

Expanding $t_{\nu\mu}$ in contributions from a complete set of states $|s>$ one has

$$t_{\nu\mu} = \frac{(2\pi)^4}{2} \sum_s \Big\{ <p|j_\nu{}^\dagger|s> <s|j_\mu|p> \delta(s-p-q)$$

$$- <p|j_\mu|s> <s|j_\nu{}^\dagger|p> \delta(s+q-p) \Big\} \; . \tag{5.21}$$

For $\nu > 0$ only the first term makes a contribution, whereas for $\nu < 0$ only the second term contributes. Let us now introduce a new matrix element $\tilde{T}_{\nu\mu}$, which differs from $T_{\nu\mu}$ only in that the charge raising and charge lowering currents are interchanged:

$$\tilde{T}_{\nu\mu} = i \int dx \, e^{-iq\cdot x} < p|\theta(x_0)\, [j_\nu(x), j_\mu{}^\dagger(0)]\, | p> \; , \tag{5.22}$$

with decomposition as in Eq. (5.10) but $A \to \tilde{A}$, $B_1 \to \tilde{B}_1$ etc. The absorptive part of $\tilde{T}_{\nu\mu}$ is

$$(5.23) \qquad \tilde{t}_{\nu\mu} = \frac{1}{2}\int dx\ e^{-iq\cdot x} < p|\,[j_\nu(x),\ j_\mu{}^\dagger(0)]\,|\,p> \ ;$$

expanding this in the manner of Eq. (5.21) one finds that

$$(5.24) \qquad t_{\nu\mu}\,(p,-q) = -\tilde{t}_{\mu\nu}\,(p,q),$$

hence in particular

$$(5.25) \qquad \text{Im}\ A(-\nu,q^2) = \text{Im}\ \tilde{A}\,(\nu,q^2)\ .$$

The sum rule of Eq. (5.18) can thus be written

$$(5.26) \qquad \frac{1}{\pi}\int_0^\infty d\nu\ [\text{Im}\ A(\nu,q^2) - \text{Im}\ \tilde{A}(\nu,q^2)\,] = \frac{C}{m}\ .$$

This result would have been obtained if we had carried out the whole discussion for the isotopically "odd" amplitude $\frac{1}{2}(T_{\nu\mu} - \tilde{T}_{\nu\mu})$. It is only for this amplitude that the no-subtraction assumption had to be made. It may be said that standard Regge model ideas support the assumption.

The Infinite Momentum Frame Method: –

The result of Eq. (5.18), or its equivalent Eq. (5.26), is sufficiently important to merit discussion from an alternative viewpoint. Start with the absorptive amplitude $t_{\nu\mu}$ defined in Eq. (5.19) and integrate over q_0, holding the three-vectors $\vec{q}$ and $\vec{p}$ fixed. Using

$$\int dq_0\ e^{iq_0x_0} = 2\pi\ \delta(x_0)\ \ ,$$

we find

$$\frac{1}{\pi p_0}\int dq_0\, t_{\nu\mu}\,(\vec{q},\vec{p},q_0) = \frac{1}{p_0}\int dx\; e^{-i\vec{q}\cdot\vec{x}}\,\delta(x_0)$$

$$< p|\,[j_\nu{}^\dagger(x),\, j_\mu(0)]\,|p> \quad ,$$

where, for convenience, we have divided through by a factor p_0. Now choose time components for both of the indices ν and μ and refer to Eq. (5.16):

$$\frac{1}{\pi p_0}\int dq_0\; t_{00}\;(\vec{q},\vec{p},q_0) = C \quad , \tag{5.27}$$

where

$$t_{00} = \text{Im } A\; p_0{}^2 - \text{Im } B_2 + \ldots \quad . \tag{5.28}$$

The scalar functions Im A, Im B_1, ... depend on the variables $q^2 = |\vec{q}|^2 - q_0{}^2$ and $\nu = -q\cdot p/m = (q_0 p_0 - \vec{q}\cdot\vec{p})/m$. Choose a frame where $\vec{q}\cdot p = 0$, so that

$$q_0 = \frac{m\nu}{p_0}\;,\; q^2 = \vec{q}^2 - \frac{m^2\nu^2}{p_0{}^2} \;. \tag{5.29}$$

Converting (5.27) to an integral over ν, we have

$$\frac{m}{\pi p_0{}^2}\int d\nu\;\{\text{Im } A\; p_0{}^2 - \text{Im } B_1 + \ldots\} = C \quad , \tag{5.30}$$

where

$$\text{Im } A = \text{Im } A\;(\nu, q^2 = \vec{q}^{\,2} - \frac{m^2\nu^2}{p_0{}^2}),\quad \text{etc.}$$

In the integration over ν, the quantity q^2 itself varies, in such a way that $q^2 \to \infty$ as $|\nu| \to \infty$. However, there is still freedom left in the choice of Lorentz frame. Suppose we adopt the "infinite momentum" frame for the target state $|p\rangle$, where $p_0 \to \infty$. Make the *assumption* now that the limiting operation can be taken under the integral in Eq. (5.30). That is, suppose it is legitimate, for each fixed ν, to first pass to the limit $p_0 \to \infty$, the integration over ν following thereafter. If this is allowed then only the Im A term survives in Eq. (5.30) and moreover the argument $q^2 \to \vec{q}^2$ becomes independent of ν. We then find the earlier result

$$\frac{1}{\pi} \int_{-\infty}^{\infty} d\nu \, \mathrm{Im}\, A(\nu, q^2) = \frac{C}{m} \, ,$$

where the integration over ν is carried out for fixed q^2.

VI. *SOME APPLICATIONS OF CURRENT ALGEBRA*

The basic result of Eq. (5.26) is exploited in this section for Compton scattering, high energy neutrino reactions, and pi-nucleon scattering. The first two applications are independent of the notions of PCAC and thus, apart from the additional technical assumptions that have gone into Eq. (5.26), they constitute rather clear tests of current algebra. The pi-nucleon discussion will culminate in the Adler-Weisberger formula and will bring in also the hypotheses of PCAC.

(i) Cabibbo-Radicati Sum Rule: –

Consider the forward elastic scattering of a photon of momentum q on a hadron target of momentum p, specializing to the case of no helicity flip for photon or hadron[13]. For definiteness take the hadron to be a proton. Let ϵ be the photon polarization vector, so that $\epsilon^2 = 1$, $q^2 = 0$, $q \cdot \epsilon = 0$. The scattering amplitude is expressed by $e^2 \epsilon_\nu T_{\nu\mu} \epsilon_\mu$, where $T_{\nu\mu}$ is given by Eq. (5.3), with $j^\alpha = j^\beta = j^{em}$. Recall that the electromagnetic current is composed of an isoscalar part and an isovector part. The absorptive part of the amplitude $T_{\nu\mu}$ is related, through the optical theorem (unitarity), to the total cross section for production of hadrons in the collision of a photon with the hadron target. In principle, by use of both proton and neutron targets and by an isotopic analysis of the final hadron states, one can separately determine the contributions to the cross section coming from the isoscalar and isovector parts of the current, and the latter in turn can be decomposed into contributions from final states with $I = \frac{1}{2}$ and $I = \frac{3}{2}$. Insofar as we have something to say about the absorptive part of the Compton amplitude, it is not unphysical then to consider Compton scattering induced solely by the isovector part of the electromagnetic current: $j_\lambda^{em} \to V_\lambda^3$. But in fact, let us consider the scattering of "charged" isovector photons, where we employ the charge raising and charge lowering members of the triplet of which $V_\lambda{}^3$ is the neutral member. At the end our results can be related through isospin symmetry considerations to quantities of direct physical interest. The amplitude to be considered is therefore

$$(6.1) \qquad T_{\nu\mu} = i \int dx\, e^{-iq\cdot x} < p|\theta(x_0)\, [V_\nu^{1-i2}(x),\, V_\mu^{1+i2}(0)]|p > .$$

Its absorptive part is

$$(6.2) \qquad t_{\nu\mu} = \frac{1}{2} \int dx\, e^{-iq\cdot x} < p|\, [V_\nu^{1-i2}(x),\, V_\mu^{1+i2}(0)]\, |\, p > .$$

These have the tensor decompositions given respectively by Eqs. (5.10) and (5.20), of course with $B_5 = 0$.

Because we are dealing here with conserved currents it follows that

$$(6.3) \qquad q_\nu\, t_{\nu\mu} = t_{\nu\mu} q_\mu = 0;\; q_0{}^2\, t_{00} = q_i q_j t_{ij},\; i,j = 1,2,3.$$

For a while let us retain q^2 at a non zero value, although of course at the end $q^2 \to 0$. The other scalar variable of the problem is $\nu = -q \cdot p/m$, the laboratory energy of the photon. From Eq. (6.3) it is easily established that

$$\text{Im } A(\nu,\ q^2 = 0) = 0; \tag{6.4}$$

but

$$\frac{\partial}{\partial q^2} \text{Im } A(\nu, q^2|_{q^2=0} = \frac{\text{Im } B_1\ (\nu,0)}{m^2\nu^2} \quad , \tag{6.5}$$

a result which will shortly be required.

The amplitude $T_{\nu\mu}$, hence the scalar quantities A, B_1, ... refer to the process $\gamma^+ + p \to \gamma^+ + p$ for $\nu \geq 0$. But these scalar quantities can be analytically continued to $\nu < 0$ and refer there to the crossed reaction $\gamma^- + p \to \gamma^- + p$. That is, let $\tilde{A}(\nu,q^2)$, $\tilde{B}$, (ν,q^2), etc. refer to the crossed reaction, with physical energy $\nu > 0$. Then Im $A(-\nu,q^2)$ = Im $\tilde{A}(\nu,q^2)$, etc. Refer now to Eqs. (5.26), (5.16), (5.14), (5.12). The commutator coefficient C is given by

$$C = \frac{-2}{m} I_3 \quad , \tag{6.6}$$

where I_3 is the isospin quantum number of target ($I_3 = \frac{1}{2}$ for proton); and Eq. (5.26) now reads

$$\frac{m^2}{\pi} \int_0^\infty d\nu\ [\text{Im } \tilde{A}(\nu,q^2) - \text{Im } A(\nu,q^2)\ ] = 1\ . \tag{6.7}$$

For Im A, which corresponds to $\gamma^+ + p \to \gamma^+ + p$, the integral begins at the continium threshold: $-(q+p)^2 = (m+\mu)^2$. For Im $\tilde{A}$, which corresponds to $\gamma^- + p \to \gamma^- + p$ the continium begins at the same threshold, but there is a "pole" contribution from the one-neutron intermediate state, i.e., $\tilde{A}$ has a pole at $-(q+p)^2 = m^2$ coming from $\gamma^- + p \to$ neutron $\to \gamma^- + p$. The residue depends on the isovector electromagnetic form factors of the nucleon, evaluated at momentum transfer q^2. Separating off this term explicitly, we find

$$(6.8) \qquad 1 = [F_1^V(q^2)]^2 - \frac{q^2}{4m^2}[F_2^V(q^2)]^2 + \frac{m^2}{\pi}\int_{\text{thresh.}}^{\infty} d\nu\, [\mathrm{Im}\tilde{A}(v,q^2) - \mathrm{Im}A(\nu,q^2)] \; ,$$

where $F_1^V(q^2)$ and $F_2^V(q^2)$ are respectively the isovector charge and anamolous magnetic moment form factors of the nucleon: $F_1^V(0) = 1$, $F_2^V(0) = \mu_p^{(a)} - \mu_n^{(a)}$, the difference of anomalous magnetic moments of protons and neutrons. Since $\mathrm{Im}\,\tilde{A} = \mathrm{Im}\,A = 0$ for $q^2 = 0$, the above equation is a trivial identity for the limit $q^2 = 0$. However, differentiate Eq. (6.8) with respect to q^2 and then set $q^2 = 0$, using Eq. (6.5). In this way one finds

$$(6.9) \qquad 2\left(\frac{dF_1^V}{dq^2}\right)_{q^2=0} - \frac{1}{4m^2}[F_2^V(0)]^2 + \frac{1}{\pi}\int_{\text{thresh.}}^{\infty} \frac{d\nu}{\nu^2}[\mathrm{Im}\,\tilde{B}_1(\nu,0) - \mathrm{Im}\,B_1(\nu,0)] = 0 \; .$$

To interpret the integrand in terms of physical quantities, recall that the amplitude for, say, $\gamma^+ + p \rightarrow \gamma^+ + p$ scattering is given by $\frac{1}{2}e^2\, \epsilon_\nu T_{\nu\mu}\epsilon_\mu$, where the factor 1/2 arises because it is $V_\mu^{1+i2}/(2)^{1/2}$ that creates γ^+, destroys γ^-, etc. In Eq. (6.9) we have passed to the limit $q^2 = 0$, so that $\epsilon \cdot q = 0$, $\epsilon^2 = 1$. Let us work in the laboratory frame ($\vec{p} = 0$), where ν represents the photon energy, and let us choose the gauge where $\epsilon \cdot p = 0$. The absorptive amplitude then reduces to

$$(6.10) \qquad \mathrm{Im}F = \frac{1}{2}e^2\, \epsilon_\nu\, t_{\nu\mu}\, \epsilon_\mu = \frac{1}{2}e^2\, \mathrm{Im}\, B_1\, (\nu,0) \; .$$

For $\gamma^- + p \to \gamma^- + p$ scattering $B_1 \to \tilde{B}_1$. The expression in Eq. (6.10), and its analog for $\gamma^- + p$ scattering, is the imaginary part of the Feynman amplitude F for forward scattering. The ordinary scattering amplitude f(the thing whose absolute square gives $d\sigma/d\Omega$) differs from this by a factor of 4π. By the optical theorem, however,

$$\mathrm{Im}\ f = \frac{\mathrm{Im}\ F}{4\pi} = \frac{\nu\sigma(\nu)}{4\pi} \ .$$

Thus, with $\sigma^+(\nu)$ and $\sigma^-(\nu)$ the total cross sections for hadron production in $\gamma^+ + p$ and $\gamma^- + p$ collisions, the integral in Eq. (6.9) can be written

$$\frac{2}{e^2\pi} \int_{\mathrm{Thresh.}}^{\infty} \frac{d\nu}{\nu} [\sigma^-(\nu) - \sigma^+(\nu)] \ .$$

It remains only to relate the quantities σ^- and σ^+ to cross sections for processes induced by neutral isovector photons corresponding to the isovector current $V_\lambda{}^3$. For the latter one can in principle isolate the cross sections for hadron production in states with $I = \frac{1}{2}$ and $I = \frac{3}{2}$. Let $\sigma^V_{1/2}$ and $\sigma^V_{3/2}$ be the corresponding cross sections. These can be related to σ^- and σ^+ by a simple isotopic analysis. In this way one finds the Cabibbo-Radicati formula

$$(6.11) \quad \frac{1}{2\pi^2\alpha} \int_{\mathrm{Thresh.}}^{\infty} \frac{d\nu}{\nu} [\sigma^V_{3/2}(\nu) - 2\sigma^V_{1/2}(\nu)] =$$

$$= \left(\frac{F_2{}^V(0)}{2m}\right)^2 - 2\left(\frac{dF_1{}^V}{dq^2}\right)_{q^2=0} \ .$$

Although the cross sections which appear here are in principle accessible to experiment, the practical demands are clearly very great. Gilman and

Schnitzer[14] have attempted to test the formula by saturating the integral with contributions from low lying baryon resonances. The trends are reasonably satisfactory.

(ii) The Adler Sum Rule For Neutrino Reactions: –

The procedures employed above can be adapted with little change to a discussion of high energy neutrino reactions. In Eq. (5.9) j_μ is however now taken to be the total weak semi leptonic current

$$j_\mu = j_\mu^{\text{weak}} = \cos\theta_C\,(V_\mu^{1+i2} + A_\mu^{1-i2}) + \sin\theta_C\,(V_\mu^{4+i5} + A_\mu^{4+i5})\,, \tag{6.12}$$

and $j_\mu^\dagger$ is the charge lowering adjoint. With this choice of currents $T_{\nu\mu}$ describes the improbable reaction lepton pair + target → lepton pair + target. The absorptive part $t_{\nu\mu}$ is more physically practical: it enters into the description of the reactions ν + target → ℓ + hadrons. Again, in Eq. (5.9) an average over target spin is understood. Otherwise, in this equation we leave open the specification of the state $|p\rangle$. Again, the tensor decomposition of $T_{\nu\mu}$ is given in Eq. (5.10). But for present purposes it is conventional to rearrange the terms and to introduce a special notation. We write out the decomposition as follows:

$$T_{\nu\mu} = T_1\,(\delta_{\nu\mu} - \frac{q_\nu q_\mu}{q^2}) + \frac{T_2}{m^2}\,(p_\nu - \frac{p\cdot q}{q^2}\,q_\nu)(p_\mu - \frac{p\cdot q}{q^2})$$

$$- \frac{1}{2}\,\frac{T_3}{m^2}\,\epsilon_{\nu\mu\alpha}\,q_\alpha\,p_\beta + \frac{T_4}{m^2}\,q_\nu\,q_\mu \tag{6.13}$$

$$+ \frac{T_5}{m^2}\,q_\nu\,p_\mu + \frac{T_6}{m^2}\,p_\nu q_\mu\,.$$

The absorptive part $t_{\nu\mu}$ has the same structure, with $T_i \to \text{Im}\,T_i$. But it is now customary to write

(6.14) $$W_i\,(q^2,\nu) = \frac{1}{\pi}\,\text{Im}\;T_i\,(q^2,\nu)\;.$$

It is only the "structure functions" W_1, W_2, W_3 that will enter into the present discussion. The W_i are scalar functions of the variables q^2 and $\nu = -q\cdot p/m$, where m is the mass of the hadron target. Notice that in taking over results from our earlier discussion, we must make the identification

(6.15) $$\text{Im}\;A \to \frac{\pi W_2}{m^2}\;.$$

In the present notation, and with $\tilde{W}_2(\nu,q^2) = -W_2(-\nu,q^2)$, Eq. (5.26) can be written

(6.16) $$\int_0^\infty d\nu\,[W_2(\nu,q^2) - \tilde{W}_2\,(\nu,q^2)] \equiv mC\;,$$

where the commutator coefficient C is defined by Eq. (5.16). With the present choice of currents [see Eq. (6.12)], and from Eqs. (5.12), (5.13), (5.14), (5.15), we obtain the coefficient C and hence the Adler sum rule

(6.17) $$\int_0^\infty d\nu\,[\tilde{W}_2\,(\nu,q^2) - W_2\,(\nu,q^2)] = 2[2I_3\cos^2\theta_c + (I_3 + \tfrac{3}{2}\,Y)\sin^2\theta_c]\;.$$

Here Y and I_3 are the hypercharge and isospin quantum numbers of the hadron target. In the above integral we have not bothered to separate off the contributions arising from the one-baryon intermediate states. Notice on the right hand side of Eq. (6.17) that the term proportional to $\cos^2\theta_c$ arises from the strangeness conserving currents; the other term from the strangeness changing currents. Notice also that the right hand side of this equation is independent of the variable q^2.

Having derived the Adler sum rule of Eq. (6.17), let us now interpret the structure functions which appear in it. For this purpose consider the reaction

$$\nu + \text{hadron target} \rightarrow \ell + \text{hadrons} \quad . \tag{6.18}$$

For the moment suppose that some definite system, s, of final hadrons is being contemplated, where the symbol s names the system and also stands for its total 4-momentum. Let $q(\nu)$, $q(\ell)$, and p be respectively the momenta of neutrino, outgoing lepton, and hadron target, so that $q(\nu) + p = q(\ell) + s$. Let ϵ be the laboratory energy of the neutrino, ϵ' the energy of the lepton, and θ the laboratory angle between lepton and neutrino directions of motion. In the following discussion the lepton mass is systematically neglected. Now define

$$q = q(\nu) - q(\ell), \quad \nu = -q \cdot p/m \quad , \tag{6.19}$$

and observe that

$$\begin{aligned} q^2 &= 4\epsilon\epsilon' \sin^2 \frac{\theta}{2} \geq 0 \\ \nu &= \epsilon - \epsilon' \geq 0 \quad . \end{aligned} \tag{6.20}$$

The invariant mass M of the hadron system s is related to these quantities by

$$M^2 = -(q + p)^2 = 2m\nu - q^2 + m^2 \ . \tag{6.21}$$

So much for kinematics!

The amplitude for the reaction (6.18) is given by

$$\text{Amp} = \frac{G}{(2)^{1/2}} < s|j_\mu^{\text{weak}}|p > i\, \bar{u}_\ell\, \gamma_\mu\, (1 + \gamma_5)\, u_\nu \ . \tag{6.22}$$

Let us now compute the differential cross section in its dependence on energy ϵ' and angle θ of the outgoing lepton, but summed over *all* final hadron states which are kinematically accessible. An average over target spin is also taken (unpolarized target). In computing this differential cross section we encounter a quantity

$$\frac{(2\pi)^4}{2} \sum_s < p|j_\nu{}^+|s > < s|j_\mu|p > \delta(s - p - q) \ .$$

Recall that the variable ν satisfies $\nu \geq 0$; and observe that the above quantity is just the absorptive part $t_{\nu\mu}$ of the amplitude $T_{\nu\mu}$(for $\nu > 0$ only the first sum in Eq. (5.21) contributes to $t_{\nu\mu}$). But from Eqns. (6.13) and (6.14)

$$\frac{1}{\pi} t_{\nu\mu} = W_1 (\delta_{\nu\mu} - \frac{q_\nu q_\mu}{q^2}) + \frac{W_2}{m^2} (p_\nu - \frac{p \cdot q}{q^2} q_\nu)(p_\mu - \frac{p \cdot q}{q^2} q_\mu)$$

$$- \frac{1}{2} \frac{W_3}{m^2} \epsilon_{\nu\mu\alpha\beta} q_\alpha p_\beta + \cdots \ . \tag{6.23}$$

In the approximation where we neglect the lepton mass, $q_\mu \bar{u}_\ell \gamma_\mu (1+\gamma_5) u_\nu = 0$; hence the remaining structure functions W_4, W_5, W_6 make no contribution to the cross section. Carrying out the remaining arithmetic, we find

$$\frac{\partial \sigma^{(\nu)}}{\partial \Omega \partial \epsilon'} = \frac{G^2}{2\pi^2} \epsilon'^2 \left\{ 2W_1 \sin^2 \frac{\theta}{2} + W_2 \cos^2 \frac{\theta}{2} - (\frac{\epsilon + \epsilon'}{m}) W_3 \sin^2 \frac{\theta}{2} \right\} . \tag{6.24}$$

The superscript (ν) on $d\sigma^{(\nu)}$ reminds that we are dealing with the neutrino reaction (6.18). The structure functions obey the following positivity conditions

$$(1 + \frac{\nu^2}{q^2}) W_2 \geq W_1 \geq \frac{1}{2m} (\nu^2 + q^2)^{1/2} |W_3| \geq 0 \ . \tag{6.25}$$

For antineutrino reactions

$$\bar{\nu} + \text{hadrons} \to \ell^+ + \text{hadrons}$$

the whole kinematic analysis is as above. But here one encounters the structure functions $\tilde{W}_i\ (\nu,q^2)$. The other change is in the overall sign of the last term in Eq. (6.24). The cross section formula for the anti-neutrinos process is

$$\frac{\partial\sigma^{(\bar{\nu})}}{\partial\Omega\partial\epsilon'} = \frac{G^2}{2\pi^2}\epsilon'^2 \left\{ 2\tilde{W}_1 \sin^2\frac{\theta}{2} + \tilde{W}_2 \cos^2\frac{\theta}{2} \right.$$

$$\left. + \left(\frac{\epsilon+\epsilon'}{m}\right)\tilde{W}_3 \sin^2\frac{\theta}{2} \right\} . \tag{6.26}$$

Whether for the neutrino or antineutrino case, the differential cross section depends on three variables, $\epsilon,\epsilon',\theta$, or equivalently ϵ,ν,q^2. The structure functions depend on only two variables ν and q^2. In principle the three structure functions can thus be disentangled; and Eq. (6.17) represents an important sum rule relating to one type of structure function.

As yet there is very little experimental evidence that bears on this Adler sum rule. The importance of the sum rule should be evident. It brings in parts of current algebra involving commutators of strangeness changing as well as strangeness concerning currents, vector and axial; and for $q^2 \neq 0$ one is testing the *local* commutation relations.

(iii) The Adler-Weisberger Formula: –

In the next sections we discuss a number of problems in which current algebra and PCAC are used jointly. The first, and still the most striking application culminates in the Adler-Weisberger formula. It merits discussion from different angles. One approach can be based closely on the methods of the present section and it is natural to take this up here. In the next section we return to the Adler-Weisberger formula from another point of view.

Return again to the amplitude $T_{\mu\nu}$ of Eq. (5.9), defined this time for axial vector currents

$$j_\mu = A_\mu^{1+i2}, \quad j_\mu^\dagger = A_\mu^{1-i2} \ .$$

The tensor decomposition of $T_{\nu\mu}$, the absorptive part $t_{\nu\mu}$ etc. are as discussed before; and the basic result of the current algebra hypotheses is still expressed in Eq. (5.26). For definiteness take the hadron target to be a proton. Then the commutator coefficient which appears in Eq. (5.26) has, according to Eq. (6.6), the value $C = -1/m$. Now in Eq. (6.26) the continuum part of the integral starts at $\nu = \mu + (\mu^2 + q^2)/2m$. But there is an isolated contribution to $\mathrm{Im}\tilde{A}$ arising from the one-neutron intermediate state. In the remaining discussion we specialize to $q^2 = 0$; and we separate off the one one-nucleon contribution explicitly. This evidently brings in the axial vector form factor $g_A(q^2 = 0) \equiv g_A$ of β-decay. One finds

$$(6.27) \qquad g_A{}^2 + \frac{m^2}{\pi} \int_{\text{Thresh.}}^{\infty} d\nu \ [\mathrm{Im}\ \tilde{A}(\nu,0) - \mathrm{Im}\ A(\nu,0)] = 1.$$

From the tensor decomposition of the absorptive amplitude $t_{\nu\mu}$ (see Eq. (5.20)) it is easily verified that

$$(6.28) \qquad q_\nu \, t_{\nu\mu} \, q_\mu \xrightarrow[q^2 \to 0]{} m^2 \nu^2 \ \mathrm{Im}\ A\,(\nu,0) \ .$$

But from Eq. (5.21), for $\underline{\nu > \mu}$ (when only the first sum contributes), and $q^2 \to 0$, one has

$$m^2 \, \nu^2 \ \mathrm{Im}\ A(\nu,0) = q_\nu q_\mu \frac{(2\pi)^4}{2} \sum_s \langle p | A_\nu{}^{1-i2} | s \rangle$$

$$(6.29) \qquad \langle s | A_\mu{}^{1+i2} | p \rangle \ \delta(s-p-q)$$

Notice however that

$$q_\nu < p|A_\nu|s >_{q=s-p} = -i < p|\frac{\partial A_\nu}{\partial x_\nu}| s >$$

$$q_\mu < s|A_\mu|p >_{q=s-p} = i < s|\frac{\partial A_\mu}{\partial x_\mu}| p > \; .$$

With the PCAC definition of pion field, we can write

$$\frac{\partial A_\mu^{\;1+i2}}{\partial x_\mu} = (2)^{1/2}\,\mu^2 f_\pi \phi; \quad \frac{\partial A_\mu^{\;1-i2}}{\partial x_\mu} = (2)^{1/2}\,\mu^2\, f_\pi \phi^\dagger \; ,$$

where ϕ creates π^+, destroys π^-, and conversely for $\phi^\dagger$. Thus, for $\nu > \mu$ and $q^2 = 0$

$$\text{(6.30)} \qquad m^2\nu^2 \text{ Im } A(\nu,0) = 2\mu^4\, f_\pi{}^2\, \frac{(2\pi)^4}{2} \sum_s < p|\phi^\dagger|s >$$

$$< s|\phi|p > \delta(s-p-q) \; .$$

But apart from the factor $2f_\pi{}^2$, the quantity on the right can now be recognized to be the imaginary (absorptive) part of the forward amplitude for $\pi^+ + p$ scattering, except of course that $q^2 = 0$, hence the pion has zero mass. That is, for $\pi^+ + p \to \pi^+ + p$ forward scattering the amplitude is

$$\text{(6.31)} \quad M^+(\nu) = \lim_{q^2 \to -\mu^2} (q^2 + \mu^2)^2\; i \int dx e^{-iq\cdot x}$$

$$< p|\theta(x_0)\; [\phi^\dagger(x),\phi(0)]|p > \; .$$

The absorptive part is

$$(6.32) \qquad \text{Im}\ M^{(+)} = \frac{1}{2} \lim_{q^2 \to -\mu^2} (q^2 + \mu^2)^2 \int dx\ e^{-iq\cdot x} \langle p|[\phi^+(x), \phi(0)]|p\rangle .$$

When expanded in a sum of contributions over a complete set of states, $\text{Im}M^{(+)}$ for $\nu > \mu$ and $q^2 = 0$ is given by

$$\text{Im}M^{(+)} = \mu^4 \frac{(2\pi)^4}{2} \sum_s \langle p|\phi^+|s\rangle \langle s|\phi|p\rangle\ \delta(s-p-q)\ ;$$

so indeed, for $\nu > \mu$ and $q^2 = 0$,

$$(6.33) \qquad \text{Im}\ A(\nu, 0) = \frac{2}{m^2\nu^2} f_\pi^2\ \text{Im}\ M^{(+)}(\nu) \quad .$$

In the same way one finds that $\text{Im}\ \tilde{A}(\nu, 0)$, for $\nu > \mu$, is related to the imaginary part of the forward amplitude for $\pi^- + p$ scattering (with pions of zero mass):

$$(6.34) \qquad \text{Im}\ \tilde{A}(\nu, 0) = \frac{2}{m^2\nu^2} f_\pi^2\ \text{Im}\ M^{(-)}(\nu) \quad .$$

The sum rule therefore can be written in the form

$$(6.35) \qquad g_A{}^2 + \frac{2}{\pi} f_\pi{}^2 \int_{\text{Thresh.}}^{\infty} \frac{d\nu}{\nu^2} [\text{Im}\ M^{(-)}(\nu) - \text{Im}\ M^{(+)}(\nu)] = 1.$$

As always with PCAC affairs this result for off mass shell amplitudes is without content until the assumption is made that the amplitudes for $q^2 = 0$ are not so different from the physical ones, $q^2 = -\mu^2$. Let's make this PCAC assumption and then relate the imaginary part of the amplitude to the total pi-nucleon cross section, via the optical theorem

$$\text{Im}\ M^{(\pm)}(\nu) = |\vec{q}|\ \sigma^{\pm}(\nu)$$

where

$$|\vec{q}| = (\nu^2 - \mu^2)^{1/2}$$

Then

$$(6.36) \qquad g_A{}^2 + \frac{2}{\pi} f_\pi{}^2 \int_\mu^\infty d\nu \frac{|\vec{q}|}{\nu^2} [\sigma^{(-)}(\nu) - \sigma^{(+)}(\nu)] = 1$$

This is the Adler-Weisberger formula. It can be regarded as an equation which determines g_A in terms of f_π and cross section data. This yields $|g_A| = 1.21$. Or one can employ the earlier PCAC result to express f_π itself in terms of g_A: $f_\pi = mg_A/g_r$. Then g_A is determined solely by strong interaction quantities. This yields $|g_A| = 1.15$. On either view the results are remarkable when one recalls $g_A{}^{exp} \approx 1.23$.

Concerning the current algebra input to the Adler-Weisberger formula, note that the pion momentum q has been set equal to zero. So what is involved here is the equal time commutator of axial vector *charges:* $[\bar{Q}^i, \bar{Q}^j] = i\epsilon^{ijk} Q^k$.

VII. THE A-W FORMULA AND GENERALIZATIONS

The Adler-Weisberger formula, as it is written in Eq. (6.35), has the appearance of a sum rule. The content can however be displayed in the form of a low energy theorem on pi-nucleon scattering, and it is this way of looking at things that we wish to elaborate now. Consider the isotopically antisymmetric amplitude

$$(7.1) \qquad M^{(0)}(\nu) = \frac{M^{(-)}(\nu) - M^{(+)}(\nu)}{2} ,$$

where $M^{(-)}$ and $M^{(+)}$ are the amplitudes for forward $\pi^- p$ and $\pi^+ p$ scattering. We suppose that $M^{(0)}(\nu)$ has been continued off pion mass shell in the PCAC manner already described. In the end interesting results will be obtained for the limit $q^2 = 0$, where q^2 is the pion mass variable. For the moment, let us only require that q^2 be real and $\geq -\mu^2$. For fixed q^2, $M^{(0)}(\nu)$ is an analytic function of the complex variable ν, with nucleon poles at $\nu = \pm q^2/2m$ and with a cut running along the real ν axis except for the gap

$$-\mu - \left(\frac{q^2+\mu^2}{2m}\right) < \nu < \mu + \left(\frac{q^2+\mu^2}{2m}\right) .$$

Moreover $M^{(0)}(\nu)$ is an odd function of ν, $M^{0}(\nu) = -M^{(0)}(-\nu)$; and $M^{(0)}(\nu)/\nu$ presumably satisfies an unsubtracted dispersion relation. In general, because it is an odd function, $M^{(0)}$ must vanish at the origin $\nu = 0$. But for the special case that concerns us, $q^2 \to 0$, there is the delicacy that the poles move to the origin. These poles arise from the one-neutron intermediate state and the delicacy can be circumvented by restoring the so-far neglected mass difference $\Delta m = m_n - m_p$ between neutron and proton. Extracting the pole terms explicitly, we then have the dispersion relation for $q^2 = 0$,

$$\text{(7.2)} \qquad \frac{M^{(0)}(\nu)}{\nu} = \frac{g_r^2}{2m^2}\,\frac{(\Delta m)^2}{(\Delta m)^2 - \nu^2} + \frac{2}{\pi}\int_{\text{Thresh}}^{\infty} d\nu'\,\frac{\operatorname{Im} M^{(0)}(\nu')}{\nu'^2 - \nu^2}$$

where the continum threshold is at $\nu' = \mu + \mu^2/2m$. It now follows that

$$\text{(7.3)} \qquad \lim_{\nu \to 0} \frac{M^{(0)}(\nu)}{\nu} = \frac{g_r^2}{2m^2} + \frac{1}{\pi}\int_{\text{Thresh.}}^{\infty} \frac{d\nu'}{\nu'^2}\left[\operatorname{Im} M^{(-)}(\nu') - \operatorname{Im} M^{(+)}(\nu')\right] .$$

Compare this with Eq. (6.35), employing for f_π the PCAC result $f_\pi = m g_A/g_r$. Then the sum rule of Eq. (6.35) becomes the low energy theorem

$$\text{(7.4)} \qquad \frac{M^{(0)}(\nu)}{\nu} \underset{\nu \to 0}{\to} \frac{g_r^2}{2m^2 g_A}$$

Let us now see how to get this result more directly. For this purpose turn to Eq. (5.6), with $|a> = |b> = |p>$ and $k = q$. Take for the currents $j_\mu^{\,a} = A_\mu^{\,a}$, $j_\nu^{\,\beta} = A_\nu^{\,\beta}$, $a,\beta = 1,2,3$. For practice we now work with isotopic indices, in a real basis. Via the PCAC identification of pion field, the first term on the right side of Eq. (5.6) becomes related to the amplitude for forward pi-nucleon scattering, with a and β the isotopic indices of initial and final pions, q the pion momentum:

$$(7.5) \qquad M^{\beta a}(\nu) = \lim_{q^2 \to -\mu^2} \frac{i(q^2+\mu^2)^2}{\mu^4 f_\pi^{\,2}} \int dx\, e^{-iq\cdot x} < p|\theta(x_0)[\frac{\partial A_\nu^{\,\beta}(x)}{\partial x_\nu}, \frac{\partial A_\mu^{\,a}(0)}{\partial x_\mu}]|p > \ .$$

We go off mass shell by dropping the instruction $q^2 \to -\mu^2$, and we pass to the limit $q \to 0$, retaining terms only up to first order in the four vector q (hence $q^2 = 0$, and ν is retained up to first order). Thus the first term on the right of Eq. (5.6) is replaced by $f_\pi^{\,2}\, M^{\beta a}(\nu)$. The left hand side of Eq. (5.6) is second order in q, $(k = q)$ on our artifice (?) of retaining the finite neutron proton mass difference, so to our order the left hand term is set equal to zero. The second term on the right is a familiar equal time commutator

$$(7.6) \qquad q_\mu \int dx\, \delta(x_0) < p|\, [A_0^{\,\beta}(x),\ A_\mu^{\,a}(0)]|p > = i\epsilon^{\beta a \gamma}\, q_\mu < p|V_\mu^{\,\gamma}|\, p > = -i\epsilon_{\beta a \gamma} \frac{\tau^\gamma}{2} \nu \ .$$

Finally, the third term on the right side of Eq. (5.6) involves the equal time commutator of a current with the divergence of a current. It is not part of the standard apparatus of current algebra and must be regarded, in its details, as model dependent. Nevertheless some useful things can be said about this so-called "σ-term". It is evidently even in the vector $\vec{q}$ and so, to first order at least, it is independent of $\vec{q}$:

$$\sigma^{\beta a} \equiv i \int d^3x < p| \, [A_0^{\beta} (\vec{x},x_0), \frac{\partial A_{\mu}^{a}}{\partial x_{\mu}} (0,x_0)] \,| \, p >$$

$$= i \int d^3x < p| \, [A_0^{\beta} (0,x_0), \frac{\partial A_{\mu}^{a}}{\partial x_{\mu}} (\vec{x},x_0)] \,| \, p >$$

(7.7) $$\approx i \int d^3x < p| \, [A_0^{\beta} (\vec{x},x_0), \frac{\partial A_0^{a}}{\partial x_0} (0,x_0)] \,| \, p > \; .$$

The first equality employs translation invariance and a change of integration variable: $\vec{x} \to -\vec{x}$. The second is Gauss' theorem. The third again employs translation invariance. But

$$\sigma^{\beta a} = i \frac{\partial}{\partial x_0} \int d^3x < p|[A_0^{\beta}(\vec{x},x_0), \; A_0^{a} (0,x_0)] \,| \, p >$$

$$-i \int d^3x < p| \, [\frac{\partial A_0^{\beta}(\vec{x},x_0)}{\partial x_0}, \; A^{a}(0,x_0)] \,| \, p > \; .$$

The second term on the right is however equal to $\sigma^{a\beta}$; hence

$$\sigma^{\beta a} - \sigma^{a\beta} = i \frac{\partial}{\partial x_0} \int d^3x < p|[A_0^{\beta}(\vec{x},x_0), \; A_0^{a} (0,x_0)] \,| \, p > \; .$$

But the right hand matrix element is evidently time independent. Hence

(7.8) $$\sigma^{\beta a} = \sigma^{a\beta} \; .$$

Let us now write for the pi-nucleon amplitude as it enters Eq. (5.6)

(7.9) $$M^{\beta a} = M^{(e)}\,\delta_{\beta a} + i\epsilon^{\beta a \gamma}\,\frac{\tau^{\gamma}}{2}\,M^{(0)} \ .$$

Evidently the isotopically antisymmetric part $M^{(0)}$ receives no contribution from the $\sigma^{\beta a}$ term. One then finds

$$\frac{M^{(0)}(\nu)}{\nu} \xrightarrow[\nu \to 0]{} \frac{1}{2f_{\pi}^{2}} ,$$

which agrees with Eq. (7.4) when the relation $f_{\pi} = mg_A/g_r$ is employed. The point $\nu = 0$ is of course unphysical for real pions. So in applying the low energy theorem in the real world one continues to $\nu = 0$ precisely by using the dispersion relation that converts the low energy theorem back into the A-W sum rule.

But there's another approach. Take the $\nu = 0$ prediction for $M^{(0)}(\nu)/\nu$ and believe it at threshold for physical pi-nucleon scattering, i.e., at $\nu = \mu$:

(7.10) $$\frac{M^{(0)}(\mu)}{\mu} \approx \frac{g_r^{2}}{2m^{2}g^{2}}$$

After all, the exact mass shell dispersion relation reads

$$\frac{M^{(0)}(\nu)}{\nu} = \frac{g_r^{2}}{2m^{2}}\,\frac{\mu^{2}}{\nu^{2}-(\mu^{2}/2m)^{2}} + $$

$$+ \frac{1}{\pi}\int_{\mu}^{\infty} \frac{d\nu'}{\nu'^{2}-\nu^{2}}\,[M^{(-)}(\nu') - M^{(+)}(\nu')]$$

Set $\nu = \mu$, drop corrections of order μ^2/m^2, and, in the dispersion denominator $\nu'^2-\mu^2$, drop the meson mass. Then

$$\frac{M^{(0)}(\mu)}{(\mu)} \approx \frac{g_r^{\,2}}{2m^2} + \frac{1}{\pi}\int_\mu^\infty \frac{d\nu'}{\nu'^2}\,[M^{(-)}(\nu') - M^{(+)}(\nu')] \ .$$

Compare with Eq. (6.35), setting $f_\pi = mg_A/g_r$, and observe that the result agrees with Eq. (7.10).

The threshold formula of Eq. (7.10) can be related to a statement about the s-wave scattering length, according to

$$(7.11) \qquad M^{(0)}(\mu) = 4\pi\,(1 + \frac{\mu}{m})a^{(0)}, \quad a^{(0)} = \frac{1}{2}\,[a^{(-)} - a^{(+)}]$$

$$= \frac{1}{3}\,(a_{1/2} - a_{3/2}) \ ;$$

where the subscripts in the last equality refer to isotopic spin. On the present interpretation of the Adler-Weisberger formula we have for the pi-nucleon scattering length

$$(7.12) \qquad a^{(0)} = \frac{L}{1+\frac{\mu}{m}}\ , \quad L \equiv \left(\frac{\mu}{2m^2 g_A^{\,2}}\right)\left(\frac{g_r^{\,2}}{4\pi}\right) \approx 0.11\mu^{-1} \ .$$

Experimentally $a^{(0)} \approx 0.086\mu^{-1}$. The fit is quite satisfactory.

On the present procedure current algebra and PCAC have furnished a prediction for the isotopically antisymmetric amplitude. Concerning the isotopically even amplitude, $M^{(e)}$, we run up against the so-called σ term of Eq. (7.7). On the other hand, recall Adler's PCAC consistency result, Eq. (4.27). We in fact supposed this result could be applied at physical threshold and were led to the prediction that the isotopically even amplitude vanishes there [see Eq. (4.29)]. All these small shifts in venue presumably involve only small errors, of order, say $(\mu/m)^2$, where m is the target (proton) mass. The success which was encountered suggests that the σ term must be similarly small. We are then tempted to generalize these results to threshold behavior for pions scattering on an arbitrary massive hadron target.[16] Systematically neglecting the σ terms, one finds that Eqs. (7.9), (7.10), and (4.29) generalize to

$$M^{\beta a} = \frac{-\mu}{f_\pi^2} (I_\pi^\gamma)_{\beta a} < t|I^\gamma|t > . \tag{7.13}$$

Here I_π^γ is the isotopic spin operator in the SU_2 space of pions; I^γ is the same in the isospin space of the hadron target t. Let I_t be the isotopic spin of the target. Then the threshold amplitude M^I for scattering in a state of total isotopic spin I is

$$M^I(\mu) = \frac{\mu}{f_\pi^2} \left\{ \frac{I_t(I_t+1) + 2 - I(I+1)}{2} \right\} . \tag{7.14}$$

Equivalently, for scattering lengths:

$$a_I = \frac{L}{1 + \frac{\mu}{m_t}} \left\{ I_t(I_t+1) - 2 - I(I+1) \right\} . \tag{7.14'}$$

Weinberg's Treatment of π–π Scattering: –

On the interpretation of the Adler-Weisberger formula described above, we simply took over the prediction for $M^{(0)}(\nu)/\nu$ at $\nu = 0$ and decided to believe it at threshold, $\nu = \mu$, which is as near to $\nu = 0$ as one can get physically. On this view it can be said that we expanded $M^{(0)}$ in powers of the momentum q, fixed the linear term on the basis of current algebra and PCAC, and then accepted that the expansion to first order is sufficiently good at threshold. Weinberg has argued that in general this is a proper way to operate with PCAC.[16] Namely, interesting results are obtained typically only in the limit where a pion momentum goes to zero. This not only puts the mass into an unphysical region but other quantities also become unphysical (e.g. $\nu = -q \cdot p/m$). A systematic way to cope with the fact that all components of q have gone to zero is then to imagine expanding an amplitude in powers of q, using current algebra and PCAC to fix the coefficients of the low order terms insofar as possible, then supposing that the first few terms in fact provide an adequate approximation near "threshold" (where the various invariants involving q are "as small" as physically possible). In the pi-nucleon problem the expansion was carried out to first order in q.

An especially interesting application of these ideas was made by Weinberg for the problem of π–π scattering. Consider this reaction, symbolized by

$$k(a) + p(c) \to q(b) + \ell(d) \; ,$$

where k,p,q,ℓ represent the momenta of the pions and a,b,c,d are their isotopic labels. Here if all the pions are to be treated on an equal footing the amplitude has to be expanded in powers of all the momenta; and it is clear that only even powers can occur. Suppose that we are interested in the behavior of the amplitude near threshold; how far must the expansion be carried? The operational answer of course is: only so far that all the coefficients are determined, and no farther! Here this means an expansion to second order in the momenta.

Introduce the usual Mandelstam variables

$$s = -(p+k)^2, \; t = -(k-q)^2, \; u = -(p-q)^2$$

$$s + t + u = -(p^2 + k^2 + q^2 + \ell^2). \tag{7.15}$$

Bose statistics, crossing symmetry, and isospin invariance dictate the following structure for the amplitude, to second order in momenta,

$$M^{ba}_{cd} = \delta_{ab}\,\delta_{cd}\,[A + B(s+u) + Ct] + \delta_{ad}\,\delta_{bc}\,[A + B(s+t) + Cu]$$

$$+ \delta_{ac}\,\delta_{bd}\,[A + B(u+t) + Cs] \; , \tag{7.16}$$

where A,B,C are constants, independent of the pion momenta. To check the Bose requirement observe that the amplitude is indeed even, e.g., under $a \leftrightarrow c$ and $k \leftrightarrow p$ (hence $s \to s$, $t \leftrightarrow u$). Similarly, crossing symmetry requires that the amplitude be even under $c \leftrightarrow b$, $p \leftrightarrow -q$ (hence $s \leftrightarrow t$, $u \leftrightarrow u$); and we see that this is satisfied too. And so on. What is remarkable about Eq. (7.16) is that the amplitude does not depend explicitly on the "mass" variables k^2, p^2, q^2, ℓ^2, except for their appearance in the relation $s + t + u = -(k^2 + p^2 + q^2 + \ell^2)$. Expressed in terms of the parameters A,B,C, the physical s-wave scattering lengths a_I (I is the total isotopic spin) are given by

$$a_0 = \frac{1}{32\pi\mu}\,[5A + 8\mu^2\,B + 12\mu^2\,C]$$

$$a_2 = \frac{1}{32\pi\mu}\,[2A + 8\mu^2\,B]\ . \tag{7.17}$$

Let's now proceed to determine the coefficients A, B, C. Introduce the amplitude.

$$T_{\nu\mu} = i\int dx\; e^{-iq\cdot x} < \ell(d)\,|\,\theta\,(x_0)\,[A_\nu{}^b\,(x),\ A_\mu{}^a\,(0)\,]\,|\,p\,(c) >$$

and refer to Eq. (5.6). The first term on the right is, up to factors, the $\pi-\pi$ amplitude M^{ba}_{cd}. The third term on the right, the "σ-term" σ^{ba}_{cd}, is symmetric in the indices b and a, as we have already discussed, when $k = q \to 0$. Let $q = k \to 0$, but keep $p = \ell$ on mass shell. To first order in q and k the left side of Eq. (5.6) can be neglected – there are no pole term delicacies in the $\pi-\pi$ problem. The second term on the right is a familiar commutator

$$k_\mu \int dx\; e^{-iq\cdot x}\;\delta(x_0) < \ell(d)\,|\,[A_0{}^b(x)\;A_\mu{}^a(0)]\,|\,p\;(c) >$$

$$= i\epsilon^{ba\gamma} < \ell(d)\,|\;V_\mu{}^\gamma\,|p(c) > k_\mu$$

$$\approx (i\epsilon^{ba\gamma})\;(i\epsilon^{d\gamma})^c\;2p\cdot k.$$

Thus, for $k = q \to 0$, to first order in $k = q$, we have

(7.18) $$M_{cd}^{ba} = \frac{1}{f_\pi^2} [\delta_{bc}\, \delta_{da} - \delta_{ba}\, \delta_{dc}]\, 2p\cdot k + \sigma_{dc}^{ba} \ .$$

But to this order

(7.19) $$2p\cdot k \to \frac{u - s}{2} \ .$$

The first term on the right side of Eq. (7.18) is antisymmetric in the indices a and b, whereas σ_{dc}^{ba} is symmetric. Comparing the antisymmetric parts of Eqs. (7.16) and (7.18), and using Eq. (7.19), one finds

$$C - B = \frac{1}{f_\pi^2} \simeq \frac{g_r^2}{m^2 g_A^2} \ .$$

From Eqs. (7.17) it then follows that

(7.20) $$2a_0 - 5a_2 = 6L, \quad L = \left(\frac{\mu}{2m^2 g_A^2}\right)\left(\frac{g_r^2}{4\pi}\right) \simeq 0.11\mu^{-1} \ .$$

Next, consider the implications of the Adler PCAC consistency condition. It asserts that the amplitude must vanish when the momentum of any one pion goes to zero, all the other pions being held on mass shell. Kinematically this corresponds to the point $s = t = u = \mu^2$; and the PCAC condition yields the result

(7.21) $$A + \mu^2\,(2B + C) = 0 \ .$$

To complete the analysis one more relation is needed, and here Weinberg introduces an extra physical assumption concerning the σ term in Eq. (7.18). With $p = \ell$ on mass shell, and to zeroeth order in $k = q$, the σ term is symmetric in the indices a and b. Weinberg now supposes that it is in fact proportional to $\delta_{ab}\,\delta_{cd}$. This property is at any rate true in the σ model of Gell-Mann and Levy. To zeroeth order in $k = q$ we then have

$$M_{cd}^{ba} \propto \delta_{ab}\,\delta_{cd}, \quad s = u = \mu^2, \quad t = 0.$$

From Eq. (7.16) it then follows that

$$A + \mu^2\,(B + C) = 0. \tag{7.22}$$

Altogether one finds

$$B = 0, \quad A = -\mu^2 C, \quad C \approx \frac{g_r^{\,2}}{m^2 g_A^{\,2}}, \tag{7.23}$$

and for the scattering lengths

$$a_0 = \frac{7}{4}\,L, \quad a_2 = -\frac{1}{2}\,L\ . \tag{7.24}$$

The definition of the length L, as given in Eq. (7.20), is based on the relation $f_\pi = mg_A/g_r$. In fact $(f_\pi)_{exp}$ is larger than mg_A/g_r by about 10%, so that if $(f_\pi)_{exp}$ were used L would decrease by ~20%. However it doesn't seem profitable to worry about such fine points, PCAC being the rough notion that it is. The important thing about Weinberg's treatment of $\pi-\pi$ scattering is that it leads to what intuitively seem to be surprisingly small scattering lengths. Unfortunately, direct tests still lie in the remote future. Various modes of indirect access to $\pi-\pi$ scattering have been widely discussed, based for example on Chew-Low extropolation techniques for π + nucleon → 2π + nucleon. But these methods seem least reliable for low energy $\pi-\pi$ scattering, which is what concerns us here. In principle a rather clean attack can be based on analysis of the spectrum structure for K_{e4} decay; and the difficult experiments are now underway in several laboratories.

VIII. SEMI LEPTONIC DECAY PROCESSES

In the application of current algebra to Compton scattering and to neutrino reactions, the currents in question arose in their direct role as interaction currents. In the derivation of the Adler-Weisberger formula, the currents entered the analysis solely through PCAC. In various other situations one encounters a mixture of these elements: one or more currents enter the analysis as "proper" interaction currents. Others enter via PCAC to represent soft pions. The photo production reaction $\gamma + N \to N + \pi$, and the closely related electroproduction reaction $e + N \to e + N + \pi$, are obvious examples when the pion is to be treated as soft. These processes have in fact been much discussed, but they do not lend themselves to brief review. Instead we shall concentrate on two applications to weak decay processes: $K \to \pi + \ell + \nu$ and $K \to \pi + \pi + \ell + \nu$.

(i) K_{ℓ_3} Decay: –

For definiteness consider the reaction

$$K^- \to \pi^0 + \ell + \bar{\nu} .$$

Let the symbols K and q represent the kaon and pion momenta. Observe that this reaction is induced solely by the vector part of the strangeness changing weak current, V_μ^{4+i5}; so the amplitude is given by

$$\text{Amp} = \frac{G}{(2)^{1/2}} \sin\theta_c < q|V_\mu^{4+i5}|K > i\, \bar{u}_\ell\, \gamma_\mu(1+\gamma_5)\, u_{\bar{\nu}} .$$

The hadronic matrix element has the structure

$$(8.1) \qquad M_\mu \equiv < q|V_\mu^{4+i5}|K > = \frac{1}{(2)^{1/2}} \{f_+ (K+q)_\mu + f_- (K-q)_\mu\}$$

where the form factors $f_\pm$ are scalar functions of the momentum transfer variable $Q^2 \equiv -(K - q)^2$. When we define an off mass shell continuation, as we will shortly do, the form factors will come to depend also on the pion mass variable q^2. Indeed, using the PCAC *definition* of pion field and the standard LSZ reduction procedure one writes

$$M_\mu = \lim_{q^2 \to -\mu^2} \frac{i(q^2+\mu^2)}{\mu^2 f_\pi} \int dx\; e^{-iq\cdot x} < 0|\theta(x_0) \left[\frac{\partial A_\nu{}^3(x)}{\partial x_\nu}\, ,\; V_\mu{}^{4+i5}(0)\right] | K > . \tag{8.2}$$

Let us now drop the instruction $q^2 \to -\mu^2$ and use the above expression to define an off shell amplitude. In the usual way an interesting result will be found for the unphysical limit $q \to 0$. Namely, define

$$T_{\nu\mu} = i\int dx\; e^{-iq\cdot x} < 0|\theta(x_0)\,[A^3{}_\nu(x),\, V_\mu{}^{4+i5}(0)\,]|K > \;,$$

as in Eq. (5.3); and from Eq. (5.4) observe that

$$iq_\nu\, T_{\nu\mu} = \frac{\mu^2 f_\pi}{q^2+\mu^2} M_\mu + i\int dx\; e^{-iq\cdot x}\,\delta(x_0) < 0|[A_0{}^3(x),\, V_\mu{}^{4+i5}(0)]|K >$$

In the limit $q \to 0$ the left side vanishes (there are no pole term delicacies here) and

$$M_\mu(K,\, q \to 0) = \frac{-i}{f_\pi}\int dx\; \delta(x_0) < 0|[A_0{}^3(x),\; V_\mu{}^{4+i5}(0)]|K >$$

$$= \frac{-i}{2f_\pi} < 0|A_\mu{}^{4+i5}\,|\,K > \;. \tag{8.3}$$

But

$$<0|A_\mu{}^{4+i5}|\,K> = i(2)^{1/2}\, f_K\, K_\mu \tag{8.4}$$

is evidently the hadron matrix element that arises for $K \to \ell + \nu$ decay; the parameter f_K plays the same role here as does the parameter f_π for $\pi \to \ell + \nu$ decay. Thus, regarding the $K_{\ell 3}$ form factors as (scalar) functions of the kaon and pion momenta, K and q, we have

(8.5) $$f_+ (K, q = 0) + f_- (K, q = 0) = f_K/f_\pi \quad .$$

This is the basic result of a standard PCAC approach to $K_{\ell 3}$ decay. For the rest the question is how to interpret it. The $K_{\ell 3}$ problem has received a great deal of attention, both experimental and theoretical. From the present point of view the $K_{\ell 3}$ problem is perhaps a decisive one, because the steps leading to Eq. (8.5) are so straightforward and seemingly in the spirit of current algebra and PCAC. What is especially interesting then is that, on its most natural interpretation, the result expressed in Eq. (8.5) seems to disagree seriously with the present experimental trend.

The form factors are scalar functions of K and q. But physically, $K^2 = -m_K{}^2$ and $q^2 = -\mu^2$ are fixed; so there is only one physically variable quantity, customarily chosen to be the invariant momentum transfer

$$Q^2 \equiv -(K - q)^2 \ .$$

The form factors have however been continued off mass shell and have come to depend also on the mass variable q^2. Regarded as functions of Q^2 and q^2, the form factors $f_\pm$ are to be evaluated at $Q^2 = m_K{}^2$, $q^2 = 0$ for purposes of Eq. (8.5). This point is unphysical with respect to both variables, since physically $q^2 = -\mu^2$ and $m_\ell{}^2 < Q^2 < (m_K - \mu)^2$. However, the physical form factors are analytic in Q^2 and they can in principle be extrapolated to $Q^2 = m_K{}^2$; i.e., we may regard $f_\pm(Q^2 = m_K{}^2,\ q^2 = -\mu^2)$ as knowable. For Eq. (8.5), however, we need $f_\pm\ (Q^2 = m_K{}^2,\ q^2 = 0)$. As usual the equation has no content until it is said how the off mass shell quantities are related to on shell quantities. For this we invoke the PCAC hypothesis, which presumably says that the form factors are gently varying in q^2; so we are tempted to set $f_\pm(Q^2,\ q^2 = 0) \approx f_\pm(Q^2,\ q^2 = -\mu^2)$. But is it indeed so that the q^2 dependence is gentle when Q^2 is held fixed; or does gentleness obtain when some other quantity is held fixed, e.g., $2K\cdot q = Q^2 + q^2 - m_K{}^2$? If it were to happen that the physical form factors are slowly varying in Q^2 (over the region of interest) they would of course also be slowly varying in $K\cdot q$, or any other variable. In this situation the choice of variables would make no difference for the purposes of PCAC. If the form factors were rapidly varying, the choice of variables would matter and PCAC would lose meaning (in the absence of further instructions, not yet available).

Regarding the physical form factors as functions of Q^2, one customarily employs a linear parameterization of the data, according to

$$f_\pm(Q^2, q^2 = -\mu^2) = f_\pm(0, -\mu^2)\,\{1 + \lambda_\pm \frac{Q^2}{\mu^2}\} \;. \tag{8.6}$$

There is no compelling evidence to suggest that a linear fit won't do, although the last word may well not yet be in on this question. For f_+ the slope parameter seems to be in the vicinity of $\lambda_+ \approx 0.03$. This is small; nevertheless it implies a variation of about 40% in f_+, between $Q^2 = 0$ and $Q^2 = m_K{}^2$. For f_- the slope parameter is less well determined; the evidence seems to be compatible with $|\lambda_-| << \lambda_+$.

For the moment let us accept the linear parameterization of Eq. (8.6) – this is an experimentally resolvable matter. More dubious, let us for the moment guess that the form factors vary gently in q^2 when Q^2 is held fixed: i.e., let's take PCAC to apply when the variables for the off shell form factors are taken to be q^2 and Q^2. Then Eq. (8.5) becomes

$$(1 + \lambda_+ \frac{m_K{}^2}{\mu^2}) + \xi(0, -\mu^2)\,(1 + \lambda_- \frac{m_K{}^2}{\mu^2}) = \frac{f_K}{f_\pi f_+(0, -\mu^2)} \tag{8.7}$$

where

$$\xi(0, -\mu^2) = f_-(0,-\mu^2)/f_+(0,-\mu^2) \;. \tag{8.8}$$

We may recall that the overall $K_{\ell 3}$ decay amplitude contains a Cabibbo factor $\sin\theta_C$. But the same factor occurs for $K_{\ell 2}$ decay. Hence the ratio $f_K/f_+(0, -\mu^2)$ is experimentally measurable, up to algebraic sign, independently of the Cabibbo factor. On the other hand, from $\pi_{\ell 2}$ decay one measures $f_\pi \cos\theta_C$. The Cabibbo angle is rather small; and it is well enough known so that $\cos\theta_C$ is not sensitive to the residual uncertainties. Thus the right hand side of Eq. (8.7) is reasonably well known, up to algebraic sign. Now if SU_3 were an exact strong interaction symmetry, the Cabibbo model would predict $f_- = 0$, hence $\xi = 0$; also $f_+(0,-\mu^2) = 1$. Moreover, it would follow that $f_K = f_\pi$. So the right side of Eq. (8.7) would be equal to unity. Empirically, the *magnitude* is about 1.28; and we may believe enough of SU_3 to accept that the algebraic sign is positive. Thus experimentally

$$\frac{f_K}{f_\pi\, f_+(0, -\mu^2)} \approx 1.28 \quad .$$

If now we also accept that $\lambda_+ \approx 0.03$ and that $\lambda_- \approx 0$, then Eq. (8.7) leads to the prediction $\xi(0,-\mu^2) \approx -0.1$. The experiments,[18] however, seem to be converging on a much more negative value, $\xi(0, -\mu^2) \approx -1$!

It may happen that this large discrepancy will go away: the ξ parameter as determined from Eq. (8.7) is fairly sensitive to the slope parameters and, for that matter, to the assumption of linearity in Eq. (8.6). Should a substantial discrepancy nevertheless survive, one could of course call current algebra into question – at least that part of the doctrine that figures in $K_{\ell 3}$ decay. But it is PCAC that would most likely be suspect. As repeatedly emphasized, PCAC is not a notion that can be stated with any precision in universal terms. Indeed, with no change in its basic spirit we could have gotten a different prediction for $K_{\ell 3}$ decay. Suppose that the presumed gentleness in off shell mass extrapolation applies when some quantity other than Q^2 is held fixed, e.g., a quantity $X = C_1\,(Q^2 + \rho q^2 + C_2)$, where C_1 and C_2 are irrelevant constants, ρ a relevant one. For example, with K·q chosen as the "natural" variable X, one has $\rho = 1$. A little arithmetic shows that the left hand side of Eq. (8.7) would acquire a correction term $\rho[\lambda_+ + \xi\lambda_-]$. Once given that λ_+ and/or λ_- are non-vanishing, by a suitable choice of "natural" variable one can get any answer he wishes. To be sure, some reasonable guidelines have to be followed. In order to favor pion pole dominance one wants to choose variables in such a way that other singularities in the q^2 plane are as unimportant—"far away"—as possible. But this is hard to formulate in a general way. Empirically successful procedures discovered for one problem do not, a priori, indicate what procedures are to be followed in another problem.

Choice of variables is one ambiguity. But there are other and perhaps deeper ambiguities. Let us return for a moment to the case of forward pi-nucleon scattering: $q(\alpha) + p \to q(\beta) + p$. Using the PCAC choice of pion field, we had earlier represented the amplitude according to

$$M^{\beta\alpha} = \frac{(q^2+\mu)^2}{\mu^4 f_\pi{}^2}\, i \int dx\, e^{-iq\cdot x} < p|\theta(x_0) \left[\frac{\partial A^\beta_\nu(x)}{\partial x_\nu}, \frac{\partial A^\alpha_\mu(0)}{\partial x_\mu}\right]|p> . \tag{8.9}$$

This is of course exact on mass shell, $q^2 \to -\mu^2$, but we used this expression to define the amplitude off mass shell, supposing that the amplitude changes slowly between $q^2 = -\mu^2$ and $q^2 = 0$. Alternatively, we could have proceeded in two steps, as follows. Define

$$\bar{M}^{\beta\alpha} = \frac{(q^2+\mu^2)}{\mu^2 f_\pi} < p, q(\beta)|\frac{\partial A^\alpha_\mu}{\partial x_\mu}|p> = -i\frac{(q^2+\mu^2)}{\mu^2 f_\pi} q_\mu < p, q(\beta)|A^\alpha_\mu|p>$$

$$= \frac{(q^2+\mu^2)^2}{\mu^4 f_\pi{}^2} q_\mu \int dx\, e^{-iq\cdot x} < p|\theta(x_0) \left[\frac{\partial A^\beta_\nu(x)}{\partial x_\nu}, A^\alpha_\mu(0)\right]|p> \tag{8.10}$$

On mass shell this is again exact, and $\bar{M}^{\beta\alpha} = M^{\beta\alpha}$ there. But off mass shell $M^{\beta\alpha}$ and $\bar{M}^{\beta\alpha}$ represent two different kinds of continuations, although each seems to be within the spirit of PCAC. At $q^2 = 0$ the two expressions differ precisely by the σ-term:

$$\bar{M}^{\beta\alpha} - M^{\beta\alpha} \underset{q^2 \to 0}{\to} \frac{i}{f_\pi^{\,2}} \int dx\, e^{-iq\cdot x}\, \delta(x_0) < p|[A_0^\alpha(x), \frac{\partial A_\nu^\beta(0)}{\partial x_\nu}]|p > . \tag{8.11}$$

To be sure this difference, empirically, did not seem to amount to much for pion-nucleon scattering (i.e., the σ term seems small). But if the difference *had* been significant the problem would have arisen as to which PCAC path has to be followed. Theorists of course can always be relied on to find arguments that favor the empirically successful choices when choices have to be made. Notice that in Weinberg's treatment of $\pi - \pi$ scattering a non-vanishing σ term was essential.

Similar ambiguities show up for $K_{\ell 3}$ decay. We defined the off mass shell amplitude on the basis of Eq. (8.2) and were then led to an interesting result for the combination $f_+ + f_-$ in the limit $q \to 0$. Suppose instead that we now focus on the scalar amplitude

$$D \equiv i < q| \frac{\partial V_\mu^{\,4+i5}}{\partial x_\mu} |K > = - (K - q)_\mu < q| V_\mu^{\,4+i5}| K >$$

$$= \frac{1}{(2)^{1/2}} [(m_K^{\,2} - \mu^2) f_+ + Q^2 f_-] . \tag{8.12}$$

According to the standard reduction formula

$$D(K, q) = - \frac{(q^2 + \mu^2)}{\mu^2 f_\pi} \int dx\, e^{-iq\cdot x} < 0|\theta(x_0)$$

$$[\frac{\partial A_\nu^{\,3}(x)}{\partial x_\nu}, \frac{\partial V_\mu^{\,4+i5}(0)}{\partial x_\mu}]| K > , \tag{8.13}$$

where the limit $q^2 \to -\mu^2$ is understood for the physical amplitude. However, let us drop this instruction and base an off shell continuation on this expression. From Eq. (5.2), for the limit $q \to 0$ (where the left hand side vanishes) we then find

$$\text{(8.14)} \qquad D(K, q = 0) = \frac{1}{f_\pi} \int dx\, \delta(x_0) < 0|[A_0^3(x), \frac{\partial V_\mu^{4+i5}(0)}{\partial x_\mu}]|K> .$$

In contrast to the earlier procedure, we are now continuing a different combination of f_+ and f_- off mass shell and in a different way. Here we obtain a result for the quantity

$$f_+^D\,(K, q = 0) + \frac{m_K^{\;2}}{m_K^{\;2} - \mu^2} f_-^D\,(K, q = 0) \;,$$

where the superscript D is to remind that off shell form factors defined here need not be the same as the off shell quantities defined by the earlier procedure—although *on* mass shell they must agree. Since $m_K^{\;2} >> \mu^2$ the above combination is very nearly $f_+^D + f_-^D$ and we have

$$f_+^D\,(K, q = 0) + f_-^D\,(K, q = 0) \simeq \frac{(2)^{1/2}}{m_K^{\;2} f_\pi} \int dx\, \delta(x_0)$$

$$\text{(8.15)} \qquad < 0|[A_0^3(x), \frac{\partial V_\mu^{4+i5}(0)}{\partial x_\mu}]|K> .$$

Quite apart from the question of choice of variables, which shall we believe extrapolates gently of mass shell, $f_+ + f_-$ or $f_+^D + f_-^D$?

The procedure which leads to Eq. (8.5) has the advantage that the right hand side is known, from the standard hypotheses of current algebra. The right hand side of Eq. (8.15) on the other hand is not specified by current algebra or any other general doctrine. This of course does not preclude new hypotheses being advanced. Still, wishing to have a definite prediction one is tempted to shape the meaning of PCAC in that way which leads to Eq. (8.5). Since Eq. (8.15) nevertheless seems also to be in the spirit of PCAC, one hopes that the right hand side has the "right" value f_K/f_π. If there were some independent way to determine the right side, and if it gave the "wrong" value, one would have to choose between the alternatives, or let experiment choose, or devise other alternatives—or, what could well happen, discover that experiment doesn't accord with any reasonable alternative.

As said, the commutator term in Eq. (8.15) is not part of the usual current algebra apparatus. But some ideas which bear on this commutator are much under discussion these days. It would carry us too far afield to discuss these matters in detail. But briefly, recall that the algebra of the vector and axial vector charges, $Q^\alpha = \int d^3x V_0{}^\alpha$ and $\bar{Q}^\alpha = \int d^3x A_0{}^\alpha$ is that associated with the group $SU_3 \times SU_3$: the quantities $Q_R{}^\alpha \equiv Q^\alpha - \bar{Q}^\alpha$ commute among themselves like the generators of SU_3; similarly the $Q_L{}^\alpha \equiv Q^\alpha + \bar{Q}^\alpha$ commute among themselves like SU_3 generators; but the $Q_R{}^\alpha$ and $Q_L{}^\alpha$ have vanishing equal time commutators. If the strong interactions were invariant under $SU_3 \times SU_3$, the charges would be conserved, and the symmetry would reflect itself either through the occurrence of parity doublings of all particles or through the existence of an octet of massless (Goldstone) bosons. Although the real world is surely far from being $SU_3 \times SU_3$ invariant, perhaps some remnants can be perceived. The charges, at any rate, are supposed to have the $SU_3 \times SU_3$ algebra; and maybe the pseudoscalar octet particles $(\pi, K, \bar{K}, \eta)$ are the Goldstone bosons.

It has been suggested by Gell-Mann[20] that the strong interaction Hamiltonian has a part which is $SU_3 \times SU_3$ invariant and a part, $\mathcal{H}'$, which breaks the symmetry, badly perhaps, but in a simple way. He conjectured, namely, that

$$\mathcal{H}' = \alpha_0 u_0 + \alpha_8 u_8 , \tag{8.16}$$

where a_0 and a_8 are strength parameters and u_0 and u_8 belong to a common multiplet of scalar operators which transform under $SU_3 \times SU_3$ according to the representation $(3,\bar{3}) + (\bar{3},3)$. The operator u_0 preserves ordinary SU_3 symmetry, whereas u_8 transforms like the eighth component of an ordinary SU_3 octet. The multiplet $(3,\bar{3})+(\bar{3},3)$ contains 18 members in all: nine scalar operators u_α ($\alpha = 0, 1, 2 ...8$), and nine pseudoscalar operators v_α. The stated transformation properties specify the equal time commutators of the charges with the u's and v's; e.g.,

$$[Q^a(x_0), u_b (\vec{x}, x_0)] = if_{abc} u_c \delta(\vec{x}) \ .$$

We do not write out all the commutation relations here. The important thing is that, in the presumed absence of derivative couplings, the current divergences can now be specified:

$$(8.17) \qquad \frac{\partial V_\mu{}^a(x)}{\partial x_\mu} = i [\mathcal{H}'(x), Q^a(x_0)] \ .$$

$$(8.18) \qquad \frac{\partial A_\mu{}^a(x)}{\partial x_\mu} = i [\mathcal{H}'(x), \bar{Q}^a(x_0)] \ .$$

From the commutation relations one finds, in particular,

$$(8.19) \qquad \frac{\partial V_\mu^{4+i5}}{\partial x_\mu} = i\frac{(3)^{\frac{1}{2}}}{2} a_8 u^{4+i5}$$

and

$$(8.20) \qquad [A_0{}^3 (\vec{x},0), u^{4+i5}(0)] = \frac{i}{2} v^{4+i5}(0)\delta(\vec{x}) \ .$$

But also

$$(8.21) \qquad \frac{\partial A_\mu^{4+i5}}{\partial x_\mu} = ((\tfrac{2}{3})^{\frac{1}{2}} a_0 - \frac{1}{2(3)^{\frac{1}{2}}} a_8)v^{4+i5}(x) \ .$$

Putting these results together, and recalling that

$$< 0|\frac{\partial A_\mu^{4+i5}}{\partial x_\mu}|K > = (2)^{1/2}\, m_K^2 f_K$$

we find for the right hand side of Eq. (8.15) the expression

$$(8.22) \qquad (\frac{f_K}{f_\pi}) \cdot (\frac{3\alpha_8}{\alpha_8 - 2(2)^{1/2}\alpha_0})$$

To achieve agreement with Eq. (8.5) one requires

$$(8.23) \qquad \alpha_8 \approx -(2)^{1/2}\alpha_0$$

But with this choice of parameters Eq. (8.18) and the basic commutation relations lead to

$$\frac{\partial A_\mu^{i}}{\partial x_\mu} = 0, \ i = 1, 2, 3.$$

That is, the ratio α_8/α_0 specified by Eq. (8.23) corresponds to invariance under the subgroup $SU_2 \times SU_2$, the strangeness conserving axial vector currents being divergenceless. The real world is of course not quite like this. But consider the familiar relations

$$< 0|\frac{\partial A_\mu^{i}}{\partial x_\mu}|\,\pi^j > = \mu^2\, f_\pi\, \delta_{ij} \qquad\qquad i,j = 1,2,3$$

where the square of the pion mass appears on the right hand side. The pion is by a substantial margin the lightest of all hadrons; and in this limited sense it may be said that the axial vector divergence is indeed small, if not quite vanishing. So the relation of Eq. (8.23) is generally regarded as being a reasonably good approximation—approximate $SU_2 \times SU_2$ symmetry (or PCAC—partially conserved axial vector current). On the other hand, suppose all of this is wrong and that, at the opposite extreme, $SU_2 \times SU_2$ is much more badly broken than SU_3: $|a_8| << |a_0|$. In this extreme the right hand side of Eq. (8.15) would approach zero! And the prediction based on Eq. (8.15) would be $\xi \approx -1$! That is, with $SU_2 \times SU_2$ badly broken (relative to SU_3) the two approaches to $K_{\ell 3}$ decay described here lead to very different results. Of course the result expressed in Eq. (8.22) is built on a particular model of $SU_3 \times SU_3$ breaking. And within the model the *quantitative* prediction depends on a parameter a_8/a_0 which can, a priori, be freely adjusted. Independent evidence on both matters is clearly needed. It is not our purpose here to select between Eqs. (8.5) and (8.15), or to fix the parameters in Eq. (8.15). What is brought out, however, is again the lack of precision of the PCAC notion.

(ii) $K_{\ell 4}$ Decay: –

Consider the $K_{\ell 4}$ decay reaction

$$K^- \to \pi^+ + \pi^- + \ell + \bar{\nu} \ .$$

Here contributions arise both from the axial vector and vector parts of the strangeness changing current. The structure of the corresponding hadron matrix elements is

$$(8.24) \qquad \langle q^{(+)}, q^{(-)} | A_\mu^{4+i5} | K \rangle = \frac{i}{m_K} \{ F_1 \, (q^{(-)} + q^{(+)})_\mu + \\ + F_2 \, (q^{(-)} - q^{(+)})_\mu + F_3 \, (k - q^{(-)} - q^{(+)})_\mu \} \ ,$$

$$(8.25) \qquad \langle q^{(+)}, q^{(-)} | V_\mu^{4+i5} | K \rangle = \frac{i}{m_K} F_4 \, \epsilon_{\mu\nu\rho\sigma} K_\nu \, q_\rho^{(+)} q_\sigma^{(-)} \ .$$

The form factors F_i are scalar functions of the momenta K, $q^{(+)}$, $q^{(-)}$, so they depend on three physical variables, say $K\cdot q^{(+)}$, $K\cdot q^{(-)}$, $(q^{(+)}+q^{(-)})^2$. However we will soon be going off shell in the pion masses, so the F_i will come to depend on the mass variables $q^{(+)2}$, $q^{(-)2}$.

All of the PCAC ambiguities which have been belabored in connection with $K_{\ell 3}$ decay will arise also for $K_{\ell 4}$ decay—with respect to choice of variables (there are more of them here) and mode of off mass shell continuation. All the warnings and branchings need not again be stated. But because two pions are involved in $K_{\ell 4}$ decay a new kind of feature arises. We have the option to treat one pion or the other, or both, as soft; and it is this aspect of the problem that we want to especially focus on. The vector current matrix element of Eq. (8.25) vanishes when either pion has zero momentum and we shall not learn anything about it on present methods. So consider the axial vector matrix element. As before, the question arises whether PCAC is to be applied to this matrix element, or to the one involving the divergence of the current. Here let us follow the traditional path in order to get definite results,[17] but this is without prejudice as to the relative merit of the alternative procedure.

Define the amplitude

$$(8.26)\qquad M_\mu^{(-)} = i\,\frac{q^{(-)^2}+\mu^2}{(2)^{1/2}\mu^2 f_\pi}\int dx\; e^{-iq^{(-)}\cdot x} < q^{(+)}|\,\theta(x_0)\left[\frac{\partial A_\nu^{\;1+i2}(x)}{\partial x_\nu}, A_\mu^{\;4+i5}(0)\right]|\,K>$$

and observe that *on* mass shell, $q^{(-)^2} \to -\mu^2$, $M_\mu^{(-)} = \langle q^{(+)}q^{(-)}|A_\mu^{\;4+i5}|K\rangle$. We use Eq. (8.26) to define a continuation off mass shell for the π^- meson, with π^+ remaining on mass shell. In the standard way, one finds for the limit $q^{(-)} \to 0$

$$(8.27)\qquad M_\mu^{(-)} \underset{q^{(-)}\to 0}{\to} \frac{-i}{(2)^{1/2}f_\pi}\int dx < q^{(+)}|[A_0^{1+i2}(x), A_\mu^{\;4+i5}(0)]|\,K > \delta(x_0)$$

But the commutator on the right side of the equation is zero! So we immediately learn that

(8.28) $$F_1(K,q^{(+)},\ q^{(-)}=0) = F_2(K,q^{(+)},\ q^{(-)}=0).$$

(8.29) $$F_3(K,q^{(+)},\ q^{(-)}=0) = 0\ .$$

Next define

(8.30) $$M_\mu^{(+)} = i\,\frac{q^{(+)2}+\mu^2}{(2)^{1/2}\mu^2 f_\pi}\int dx\ e^{-iq^{(+)}\cdot x} < q^{(-)}|\theta(x_0)\left[\frac{\partial A_\nu^{1+i2}(x)}{\partial x_\nu},\ A_\mu^{4+i5}(0)\right]|K> \ ,$$

which provides a basis for continuing off shell in the π^+ mass. In the limit $q^{(+)}\to 0$ one finds

$$M_\mu^{(+)}\ \underset{q^{(+)}\to 0}{\to}\ \frac{-i}{(2)^{1/2}f_\pi}\int dx\ \delta(x_0) < q^{(-)}|[A_0^{1-i2}(x),A_\mu^{4+i5}(0)]|K>$$

$$= \frac{-i}{(2)^{1/2}f_\pi} < q^{(-)}|\,V_\mu^{6+i7}|\,K>$$

(8.31) $$= \frac{i}{f_\pi} < q^{(-)}|V_\mu^{4+i5}|K> \ .$$

The last step follows from isospin invariance and produces the matrix element for $K^-_{\ell 3}$ decay, with $q^{(-)}$ the π^0 momentum there. Thus

(8.32) $$F_3(K,\ q^{(+)}=0,\ q^{(-)}) = \frac{m_K}{(2)^{1/2}f_\pi}[f_+(K,q^{(-)}) + f_-(K,q^{(-)})]\ ,$$

$$(8.33) \qquad F_1(K,q^{(+)}=0,\ q^{(-)}) + F_2(K,q^{(+)}=0,\ q^{(-)}) = = (2)^{1/2} \frac{m_K}{f_\pi} f_+(K,q^{(-)}) \ .$$

What is striking here is that, unless $f_+ + f_- \approx 0$, F_3 has very different values for the two limits $q^{(-)} \to 0$ and $q^{(+)} \to 0$. On the other hand the present results are at least compatible with F_1 and F_2 being slowly varying functions of their arguments. On contracting the hadron matrix element with the lepton current matrix element, one finds that the form factor F_3 gets to be multiplied by the lepton mass. So F_3 is essentially undetectable in K_{e4} decay and nothing is known about it experimentally. For F_1 and F_2, in crudest approximation we are at present entitled to treat them as essentially constant (this neglects the rather small variation of f_+ in the variable $(K-q^{(-)})^2$). Hence

$$(8.34) \qquad F_1 = F_2 \approx \frac{m_K}{(2)^{1/2} f_\pi} f_+ \ .$$

The presently available experimental evidence is in rough accord with these predictions, at the 50% level, say. Although nothing is known about the form factor F_3, the very different values obtained theoretically for the two limits discussed above calls for some explanation, even though no paradox is involved. Such an explanation has been provided by Weinberg, again within the framework of current algebra and PCAC. He considered the additional information that becomes available when one allows both pions to become simultaneously soft. Rather than develop the rather lengthy techniques needed for dealing with such an analysis, let us see the essence of Weinberg's argument in a more intuitive way. In effect the rapid dependence on the momenta $q^{(+)}$ and $q^{(-)}$ which is displayed (theoretically) by F_3 arises from the Feynman diagram shown. At the vertex which couples the virtual meson to the current A_μ^{4+i5} there occurs

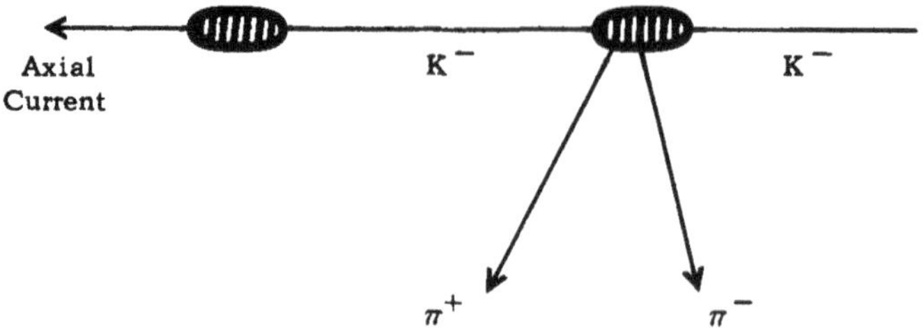

the factor $(2)^{1/2}\, f_K(K-q^{(-)}-q^{(+)})_\mu$, where $K-q^{(-)}-q^{(+)}$ is the momentum of the virtual K meson. This already shows that the above diagram contributes exclusively to the form factor F_3. The K meson propagator contributes a factor $\{(K-q^{(-)}-q^{(+)})^2+m_K{}^2\}^{-1}$, and in the limit where both $q^{(+)}$ and $q^{(-)}$ are small this is proportional to $[2K\cdot(q^{(-)}+q^{(+)})]^{-1}$. The $(KK\pi\pi)$ vertex can itself be discussed by the methods of current algebra and PCAC. For present purposes, however, it is enough to notice, on the basis of the Adler PCAC consistency condition, that in the soft pion limit the pions come out in a relative p-wave state. Hence this vertex contributes a factor proportional to $K\cdot(q^{(-)}-q^{(+)})$.

Altogether then, in the limit $q^{(-)}\to 0$, $q^{(+)}\to 0$, the diagram makes a contribution to F_3 which is of the form

$$C_1\,\frac{K\cdot(q^{(-)}-q^{(+)})}{K\cdot(q^{(-)}+q^{(+)})}\,,$$

where C_1 is some constant. Notice that this function is formally of zeroeth order in the momenta and yet is momentum dependent. All the other contributions to F_3, from non-pole diagrams, can be lumped into an additive constant, so long as we stick to zeroeth order in the momenta. Thus

$$F_3 = C_1\,\frac{K\cdot(q^{(-)}-q^{(+)})}{K\cdot(q^{(-)}+q^{(+)})} + C_2$$

The constants C_1 and C_2 can now be adjusted by matching this expression to the results obtained earlier for the two limits $q^{(-)} \to 0$ and $q^{(+)} \to 0$. In this way we find an expression for F_3 which "interpolates" between these limits, namely, the result[21]

$$F_3 = \frac{m_K}{(2)^{1/2} f_\pi} [f_+ + f_-] \frac{K \cdot q^{(-)}}{K \cdot (q^{(-)} + q^{(+)})} . \tag{8.34}$$

IX. NON LEPTONIC DECAYS

The soft pion methods described in the last section for semi leptonic processes can be transferred, formally, to non leptonic reactions which involve pions. But one encounters here the commutator of an axial vector charge with the non-leptonic Hamiltonian. So it becomes necessary here to take a position as to the structure of this Hamiltonian, or at any rate as to the commutators which enter the analysis. As we discussed earlier, the most popular and attractive model for the non-leptonic Hamiltonian is one which ascribes to it a current x current structure. In order to build in the $\Delta I = 1/2$ rule for the strangeness changing interactions, it is necessary however to invoke the coupling of neutral as well as charged currents. And if the Hamiltonian is to have an octet character, so far as SU_3 is concerned, the structure becomes fully specified. Namely, the strangeness changing Hamiltonian becomes the sixth component $\mathcal{H}^6$ of an octet of operators $\mathcal{H}^\gamma$ $\gamma = 1,2 \ldots 8$, where

$$\mathcal{H}^\gamma = \text{const. } d_{\gamma\alpha\beta}\; j_\mu{}^\alpha\, j_\mu^\beta \tag{9.1}$$

$$j_\mu{}^\alpha = V_\mu{}^\alpha + A_\mu{}^\alpha \; .$$

The coefficients $d_{\gamma\alpha\beta}$ are totally symmetric in their indices and are defined by the anticommutation relations among the SU_3 matrices λ^α:

$$\left\{\frac{\lambda^\alpha}{2}, \frac{\lambda^\beta}{2}\right\} = \frac{1}{3}\delta_{\alpha\beta} + d_{\gamma\alpha\beta}\frac{\lambda^\gamma}{2} \quad . \tag{9.2}$$

The weak non leptonic Hamiltonian $\mathcal{H}^6$ is of course charge conserving and hermitian; and it contains both a $\Delta S = 1$ and a $\Delta S = -1$ piece.

Consider now a process

$$A \to B + \pi^\alpha \quad ,$$

where the index α describes the isospin state of the pion. Let q denote the momentum of the pion. To lowest order in the weak interactions, the amplitude for this process is given by

$$\text{Amp}\,(A \to B + \pi^\alpha) = i\,\frac{q^2 + \mu^2}{\mu^2 f_\pi}\int dx\; e^{-iq\cdot x} < B|\theta(x_0)\left[\frac{\partial A_\nu^{\ \alpha}(x)}{\partial x_\nu}, \mathcal{H}^6\right]|\,A> \quad , \tag{9.3}$$

where the instruction $q^2 \to -\mu^2$ is understood. In the usual way we remove this instruction and employ the above expression to effect an off mass shell continuation; and as usual we find an interesting result for the limit $q \to 0$. Namely

$$\text{Amp}\,(A \to B + \pi^\alpha) \underset{q\to 0}{\to} \frac{-i}{f_\pi}\int dx\;\delta(x_0) < B|[A_0^{\ \alpha}(x), \mathcal{H}^6(0)]|\,A> \tag{9.4}$$

On the model which has been adopted here for the weak Hamiltonian the above commutator can be worked out explicitly. One finds

$$\int d^3x\,[A_0^{\ \alpha}(\vec{x},0), \mathcal{H}^6(0)] = i\,f_{\alpha 6\beta}\,\mathcal{H}^\beta \quad ; \tag{9.5}$$

and therefore

$$\text{Amp}\,(A \to B + \pi^\alpha) \underset{q\to 0}{\to} \frac{1}{f_\pi}\,f_{\alpha 6\beta} < B|\mathcal{H}^\beta|\,A> \quad . \tag{9.6}$$

This is the central result, and a useful one whenever the matrix element on the right is itself relateable to a physical process or is otherwise discussable.

(i) $K \to 3\pi$ Decays: –

Let us begin by reviewing the standard phenomenology of these reactions. To a reasonably good approximation the $K \to 3\pi$ decays appear to fit the $\Delta I = 1/2$ rule; and to an even better approximation they are in accord with CP invariance. Our model Hamiltonian of course has these features built into it from the start. Accordingly, if we ignore the slight CP impurities in the states $K_S{}^0$ and $K_L{}^0$, the reaction $K_S{}^0 \to 3\pi^0$ is supposed to be forbidden; and for $K_S{}^0 \to \pi^+\pi^-\pi^0$ the pions emerge in an $I = 0$ state, which is totally antisymmetric and therefore centrifugally inhibited. Neither of these processes has as yet been seen. For the remaining processes, $K^+ \to \pi^+ + \pi^+ + \pi^-$, $K^+ \to \pi^+\pi^0\pi^0$, $K_L{}^0 \to \pi^+\pi^-\pi^0$, $K_L{}^0 \to 3\pi^0$, the final pions are supposed to emerge in states of unit isotopic spin. Kinematically, the the reaction $K \to 3\pi$ is fully specified by any two of the pion energies $\omega_1, \omega_2, \omega_3$, as measured in the K meson rest frame. Of course, $m_K = \omega_1 + \omega_2 + \omega_3$. To a good approximation the maximum possible energy of any one pion is

$$\omega_{max} \approx \frac{2}{3} Q + \mu \ ; \quad Q \equiv m_K - 3\mu \ .$$

One usually supposes that the amplitude for a particular $K \to 3\pi$ reaction can be well represented by the first few terms in a power series expansion in the energies ω_i ; and indeed, phenomenologically an expansion up to linear terms only seems to work remarkably well. To this order the $\Delta I = \frac{1}{2}$ rule and Bose statistics impose a very simple structure, the amplitudes for all four processes under discussion being expressible in terms of two parameters. Namely, one finds

$$\mathrm{Amp}(K^+ \to 2\pi^+ + \pi^-) = 2\rho(1 + \tfrac{1}{2}\lambda_{++-}t_-)$$

$$\mathrm{Amp}(K^+ \to 2\pi^0 + \pi^+) = \rho(1 + \tfrac{1}{2}\lambda_{00+}t_+)$$

$$\mathrm{Amp}(K_L^{\,0} \to \pi^+\pi^-\pi^0) = -\rho(1 + \tfrac{1}{2}\lambda_{+-0}t_0)$$

(9.7)
$$\mathrm{Amp}(K_L^{\,0} \to 3\pi^0) = 3\rho \quad ,$$

where

(9.8)
$$t_i \equiv \frac{3}{Q}(\omega_i - \mu) - 1$$

and

(9.9)
$$\lambda_{00+} = \lambda_{+-0} = -2\lambda_{++-} \quad .$$

Insofar as our theoretical approximations enforce this phenomenology, it is enough then to consider the single process $K^+ \to \pi^+\pi^+\pi^-$ in order to get at the two independent parameters ρ and λ_{++-}.

From Eq. (9.6) we get two pieces of information by considering in turn the limit where the π^- momentum $q^{(-)}$ goes to zero or a π^+ momentum $q^{(+)}$ goes to zero. In either case it is only one pion at a time that is taken off mass shell.

Taking $q^{(-)} \to 0$, we have

(9.10)
$$\mathrm{Amp}(K^+ \to \pi^+ + \pi^+ + \pi^-) \xrightarrow[q^{(-)}\to 0]{} -\frac{i}{2(2)^{1/2} f_\pi}$$

$$< \pi^+\pi^+ | \mathcal{H}^{4+i5} | K^+ > = 0 \quad .$$

The matrix element on the right vanishes, simply from consideration of strangeness conservation. With $q^{(+)} \to 0$, we have

$$\mathrm{Amp}(K^+ \to \pi^+ + \pi^+ + \pi^-) \xrightarrow[q^{(+)} \to 0]{} \frac{i}{2(2)^{1/2} f_\pi} < \pi^+\pi^- | \mathcal{H}^{4-i5} | K^+ >$$

(9.11) $$= \frac{1}{2f_\pi} \mathrm{Amp}(K_S{}^0 \to \pi^+\pi^-) \ .$$

The second step follows purely from isospin considerations.

As usual with PCAC, these results for off mass shell amplitudes are purely formal until we say how they are to be related to the on shell $K \to 3\pi$ amplitude. Which variables are to be held fixed in order to produce a "gentle" dependence on off shell pion mass? Because it will lead to happy results, let's suppose that the mass dependence is gentle when the pion energies ω_i are held fixed. In effect, suppose that the parameters ρ and λ_{++-} do not depend on the pion mass variable. From Eq. (9.10) we then find

$$\mathrm{Amp}(K^+ \to 2\pi^+ + \pi^-) = 0 \quad \text{for } \omega_- = 0 \ .$$

Hence

(9.12) $$\lambda_{++-} = \frac{2Q}{m_K} \ .$$

In connection with Eq. (9.11), notice that $q^{(+)} = 0$ corresponds to $\omega_- = \frac{m_K}{2}$. Hence

(9.13) $$\rho = \frac{1}{6f_\pi} \mathrm{Amp}(K_S{}^0 \to \pi^+ + \pi^-) \ .$$

The results expressed by Eqs. (9.12) and (9.13) are in remarkably good agreement with experiment. Of course, our choice of variables in the interpretation of PCAC was somewhat arbitrary, but let us not belabor this point!

(ii) Hyperon Decay: –

Consider the reaction

$$\Sigma^+ \rightarrow n + \pi^+ \quad ,$$

and in particular its s-wave part. Take the pion off mass shell in the usual way and pass to the limit $q = 0$. In this limit we deal with the s-wave amplitude A_s and find

$$(9.14) \qquad A_s(\Sigma^+ \rightarrow n+\pi^+) \xrightarrow[q \rightarrow 0]{} \frac{i}{2(2)^{1/2} f_\pi} < n|\mathcal{H}^{4-i5}|\Sigma^+ > = 0 \ ,$$

where the vanishing of the matrix element on the right side follows purely from consideration of strangeness conservation. Making the usual PCAC assumption that the off shell variation is gentle, we are hereby led to the prediction that $\Sigma^+ \rightarrow n + \pi^+$ is essentially pure p-wave. This is in remarkably good agreement with experiment.

Current algebra and PCAC are not always so successful as in the above applications. But let us not dwell on troubles . . .

Joseph Henry Laboratories
Princeton University

REFERENCES

[1] M. Gell-Mann, *Physics* 1, 63 (1964); and references therein.

[2] S. Fubini and G. Furlan, *Physics* 1, 229 (1965).

[3] S. Adler, *Phys. Rev.* 140, B736 (1965).

[4] W. Weisberger, *Phys. Rev.* 143, 1302 (1966).

[5] B. Renner, "Current Algebras and Their Applications", Pergamon Press (Oxford 1968).

[6] Y. Nambu, *Phys. Rev. Letters* 4, 380 (1966); J. Bernstein, S. Fubini, M. Gell-Mann and W. Thirring, *Nuovo Cimento* 17, 757 (1969); M. Gell-Mann, *Phys. Rev.* 125, 1067 (1962).

[7] N. Cabibbo, *Phys. Rev. Letters* 10, 531 (1963).

[8] R. P. Feynman and M. Gell-Mann, *Phys. Rev.* 109, 193 (1958).

[9] M. Gell-Mann and M. Lévy, *Nuovo Cimento* 16, 705 (1960).

[10] M. L. Goldberger and S. B. Treiman, *Phys. Rev.* 110, 1178 (1958).

[11] S. L. Adler, *Phys. Rev.* 135, B963 (1964).

[12] S. L. Adler, *Phys. Rev.* 137, B1022 (1965).

[13] N. Cabibbo and L. A. Radicati, *Phys. Letters* 19, 697 (1966).

[14] F. Gilman and H. Schnitzer, *Phys. Rev.* 150, 1562 (1966).

[15] S. L. Adler, *Phys. Rev.* 143, 1144 (1966).

[16] S. Weinberg, *Phys. Rev. Letters* 17, 616 (1966).

[17] C. G. Callan and S. B. Treiman, *Phys. Rev. Letters* 16, 153 (1966).

[18] M. K. Gaillard and L. M. Chounet, *CERN Report* 70-14 (1970). These authors present a comprehensive survey of the $K_{\ell 3}$ situation.

[19] M. K. Gaillard, *Nuovo Cimento* 61, 499 (1969); R. A. Brandt and G. Preparata, to be published.

[20] M. Gell-Mann, *Phys. Rev.* 125, 1067 (1962); see also Ref. 1.

[21] S. Weinberg, *Phys. Rev. Letters* 17, 336 (1966).

[22] Ref. 17. Also see Y. Hara and Y. Nambu, *Phys. Rev. Letters* 16, 875 (1966); D. K. Elias and J. C. Taylor, *Nuovo Cimento* 44A, 518 (1966); H. D. I. Abarbanel, *Phys. Rev.* 153, 1547 (1967).

[23] M. Suzuki, *Phys. Rev. Letters* 15, 986 (1965).

FIELD THEORETIC INVESTIGATIONS IN CURRENT ALGEBRA

Roman Jackiw

I. INTRODUCTION

The techniques of current algebra have been developed to circumvent two difficulties which hamper progress in particle physics. These are (1) a lack of knowledge of the precise laws which govern elementary processes, other than electromagnetism; (2) an inability of solving any of the realistic models which have been proposed to explain dynamics. It was in this context that Gell-Mann,[1] in a brilliant induction from non-relativistic quantum mechanics, proposed his now famous charge algebra, which subsequently has been extended to the local algebra of charge and current densities. Just as the canonical, non-relativistic Heisenberg commutator between the momentum $p \equiv \frac{\delta L}{\delta \dot{q}}$ and position q,[2] $i[p,q] = 1$, is independent of the specific form of the Lagrangian L, and can be used to derive useful, interaction independent results such as the Thomas-Reiche-Kuhn sum rule[3]; so also it should be possible to exhibit interaction independent relations in relativistic dynamics, which can be exploited to yield physical information without solving the theory. This program has led to the algebra of current commutation relations. It has had successes in two broad categories of application: low energy theorems and high energy sum rules.

It is my purpose in these lectures to discuss research of the last two years which has elucidated the form of commutators in those model field theories which had served to derive and motivate the algebra of currents.

The remarkable results that have been obtained indicate that current commutators are *not* independent of interactions, and that some of the current algebra predictions can be circumvented. It is of course possible to *postulate* the relevant current commutators in their minimal, Gell-Mann form, and to dispense with the theoretical structure, which in the first instance led to their derivation. I shall not be taking this point of view, since such a postulational approach provides no basis for the validity of these relations. Moreover, and more importantly, the minimal current algebra has led to certain predictions which are in conflict with experiment; while the anomalies that have been found offer the possibility of eliminating some difficulties.

Examples of results which we shall be discussing in great detail are the following two: (1) The Sutherland-Veltman theorem[4] predicts that the effective coupling constant for $\pi^0 \to 2\gamma$ should vanish for zero pion mass in any field theory which exhibits current algebra, PCAC and electromagnetism, minimally and gauge invariantly coupled. However, in the σ model, which possesses all these properties, this object does not vanish.[5] (2) The Callan-Gross sum rule[6] predicts that in a large class of quark models the longitudinal cross-section, for total electroproduction off protons, vanishes in a certain high energy domain, the so-called deep inelastic limit. Explicit calculation of this cross section in the relevant models yields a non-vanishing result.[7]

It must be emphasized that the above calculations, which provide evidence for conflict between the formal, canonical reasoning of current algebra and explicit computation, are in no way ambiguous. The calculations, performed of course by perturbative techniques, are well defined consequences of the dynamics of the theory. Evidently the formal, canonical properties of the theory which lead to the erroneous predictions, are not maintained.

Our program is the following. We shall need to make frequent use of some technical results concerning Green's functions, commutators and

Ward-Takahashi identities. Thus first we shall explore the canonical and the space-time constraints which limit the possible structure of these objects, and we shall discuss the Bjorken-Johnson-Low definition of the commutator. Next we shall perform the calculations relevant to the two examples quoted above. It will then be shown how the minimal current algebra must be modified to avoid "false" theorems like those of Sutherland-Veltman and Callan-Gross. The theoretical and experimental consequences of these modifications will be examined. Finally we shall study in the context of our discoveries the recently popularized topic of broken scale invariance.[8]

REFERENCES

[1] An excellent survey of current algebra is to be found in the book by S. L. Adler and R. Dashen, *Current Algebra*, W. A. Benjamin, New York (1968).

[2] Throughout these lectures, we set $\hbar$ and c to unity. The metric we use is

$$g^{\mu\nu} = g_{\mu\nu} = \begin{pmatrix} 1 & 0 & 0 & 0 \\ 0 & -1 & 0 & 0 \\ 0 & 0 & -1 & 0 \\ 0 & 0 & 0 & -1 \end{pmatrix}$$

Greek indices are space-time while Latin indices are space.

[3] A modern treatment of these "classical" sum rules is given by H. A. Bethe and R. Jackiw, "Intermediate Quantum Mechanics", W. A. Benjamin, New York (1968).

[4] D. G. Sutherland, *Nucl. Phys.* B2, 433 (1967); M. Veltman, *Proc. Roy. Soc.* A301, 107 (1967).

[5] J. S. Bell and R. Jackiw, *Nuovo Cimento* 60, 47 (1969).

[6] C. G. Callan, Jr. and D. J. Gross, *Phys. Rev. Letters* 22, 156 (1969).

[7] R. Jackiw and G. Preparata, *Phys. Rev. Letters* 22, 975 (1969), (E) 22, 1162 (1969); S. L. Adler and Wu-Ki Tung, *Phys. Rev. Letters* 22, 978 (1969).

[8] A brief summary of the material covered here appears in R. Jackiw, "Non-Canonical Behaviour in Canonical Theories", CERN preprint, TH-1065.

II. CANONICAL AND SPACE-TIME CONSTRAINTS IN CURRENT ALGEBRA

A. Canonical Theory of Currents

An arbitrary field theory is described by a Lagrange density $\mathcal{L}$ which we take to depend on a set of independent fields ϕ and on their derivatives $\partial^\mu\phi \equiv \phi^\mu$. The canonical formalism rests on the following equal time commutators (ETC).[1]

$$i[\pi^0(t,\vec{x}),\ \phi(t,\vec{y})] = \delta(\vec{x}-\vec{y})$$

$$i[\pi^0(t,\vec{x}),\ \pi^0(t,\vec{y})] = i[\phi(t,\vec{x}),\phi(t,\vec{y})] = 0 \tag{2.1}$$

Here π^0 is the time component of the canonical 4-momentum.

$$\pi^\mu = \frac{\delta\mathcal{L}}{\delta\phi_\mu} \tag{2.2}$$

The Euler-Langrange equation of the theory is

$$\partial_\mu\pi^\mu = \frac{\delta\mathcal{L}}{\delta\phi} \ . \tag{2.3}$$

Consider now an infinitesmal transformation which changes $\phi(x)$ to $\phi(x)+\delta\phi(x)$. The explicit form for $\delta\phi(x)$ is assumed known; we have in mind a definite, though unspecified transformation. It is interesting to inquire what conditions on $\mathcal{L}$ insure this transformation to be a *symmetry* operation for the theory. This can be decided by examining what happens to $\mathcal{L}$ under the transformation.

$$\delta\mathcal{L} = \frac{\delta\mathcal{L}}{\delta\phi}\,\delta\phi + \frac{\delta\mathcal{L}}{\delta\phi^{\mu}}\,\delta\phi^{\mu} = \frac{\delta\mathcal{L}}{\delta\phi}\,\delta\phi + \pi_{\mu}\partial^{\mu}\delta\phi \tag{2.4}$$

If *without the use of equations of motion* we can show that $\delta\mathcal{L}$ is a total divergence of some object Λ^{μ},

$$\delta\mathcal{L} = \frac{\delta\mathcal{L}}{\delta\phi}\,\delta\phi + \pi_{\mu}\partial^{\mu}\delta\phi = \partial_{\mu}\Lambda^{\mu}, \tag{2.5}$$

then the action, $I = \int d^4x\mathcal{L}$, is not affected by the transformation, and the transformation is a symmetry operation of the theory. The conserved current can now be constructed in the following fashion. *With the help of the equations of motion* (2.3), an alternate formula for $\delta\mathcal{L}$ can be given which is always true, regardless whether or not we are dealing with a symmetry operation. We have from (2.3) and (2.4).

$$\delta\mathcal{L} = \partial_{\mu}\pi^{\mu}\delta\phi + \pi^{\mu}\partial_{\mu}\delta\phi = \partial_{\mu}(\pi^{\mu}\delta\phi) \tag{2.6a}$$

Equating this with (2.5) yields

$$0 = \partial_{\mu}[\pi^{\mu}\delta\phi - \Lambda^{\mu}]\ . \tag{2.6b}$$

Hence the conserved current is

$$J_{\mu} = \pi_{\mu}\delta\phi - \Lambda_{\mu}. \tag{2.7}$$

Two situations are now distinguished. If $\Lambda^\mu = 0$, we say that we are dealing with an *internal* symmetry; otherwise we speak of a *space-time* symmetry.[2] Examples of the former are the SU(3) x SU(3) currents of Gell-Mann.

$$\delta^a \phi = T^a \phi \tag{2.8}$$

T^a is a representation matrix of the group; it is assumed that the fields transform under a definite representation. The internal group index a labels the different matrices. The internal symmetry current is

$$J_\mu{}^a = \pi_\mu T^a \phi \quad . \tag{2.9}$$

Space-time symmetries are exemplified by translations.

$$\delta^\alpha \phi = \partial^\alpha \phi \tag{2.10a}$$

$$\delta^\alpha \mathcal{L} = \partial^\alpha \mathcal{L}$$

$$\Lambda_\mu{}^\alpha = g_\mu{}^\alpha \mathcal{L} \tag{2.10b}$$

Now the transformations are labeled by the space-time index α. The conserved quantity is the canonical energy-momentum tensor $\theta_c{}^{\mu\alpha}$.

$$\theta_c{}^{\mu\alpha} = \pi^\mu \phi^\alpha - g^{\mu\alpha} \mathcal{L} \tag{2.11}$$

In the subsequent we shall reserve the symbol J_μ and the term "current" for *internal* symmetries.

It is clear that when the internal transformation (2.8) is not a symmetry operation, i.e., $\delta\mathcal{L} \neq 0$, it is still possible to define the current (2.9), which is not conserved. By virtue of the canonical formalism, the charge density satisfies a model independent ETC, regardless whether or not the current is conserved.[3]

$$[J_0{}^a(t,\vec{x}), J_0{}^b(t,\vec{y})]$$
$$= [\pi_0(t,\vec{x})T^a\phi(t,\vec{x}), \pi_0(t,\vec{y})T^b\phi(t,\vec{y})]$$
$$= i\,\pi_0(t,\vec{x})\,[T^a,T^b]\,\phi(t,\vec{x})\delta(\vec{x}-\vec{y})$$
$$= -f_{abc}\,J_0{}^c(t,\vec{x})\delta(\vec{x}-\vec{y}) \tag{2.12}$$

We have used the group property of the representation matrices

$$[T^a,T^b] = if_{abc}T^c \tag{2.13}$$

Similarly the charge,

$$Q^a(t) = \int d^3x\, J_0{}^a(t,\vec{x}) \quad , \tag{2.14}$$

which for conserved currents is a time independent Lorentz scalar, generates the proper transformation on the fields, even in the non-conserved case.

$$i[Q^a(t), \phi(t,\vec{x})]$$
$$= i\int d^3y\, [\pi_0(t,\vec{y})T^a\phi(t,\vec{y}), \phi(t,\vec{x})]$$
$$= T^a\phi(x) = \delta^a\phi(x) \tag{2.15}$$

It should be remarked here that although conserved and non-conserved internal symmetry currents and charges satisfy the ETC (2.12) and (2.15), the space-time currents do not, in general, satisfy commutation relations which are insensitive to conservation, or lack thereof of the appropriate quantity; see Exercise 2.5.

The importance of relations (2.12) and (2.15) is that they have been derived without reference to the specific form of $\mathcal{L}$; i.e., without any commitment to dynamics. Thus it appears that they are *always* valid, and that any consequence that can be derived from (2.12) and (2.15) will necessarily be true. But it must be remembered that the Eqs. (2.12) and

(2.15) have been obtained in a very formal way; all the difficulties of local quantum theory have been ignored. Thus we have not worried about multiplying together two operators at the same space-time point, as in (2.9); nor have we inquired whether or not the equal time limit of an unequal time commutator really exists as in (2.1), (2.12) and (2.15). It will eventually be seen that the failure of current algebra predictions can be traced to precisely these problems.

A word about non-conserved currents. It turns out that in applications of the algebra of non-conserved currents, it is necessary to make assumptions about the divergence of the current. The assumption that is most frequently made is that $\partial^\mu J_\mu$ is a "gentle" operator, though the precise definition of "gentle" depends on the context. We shall spell out in detail what we mean by "gentle"; however for the moment the following concept of "gentleness" will delimit the non-conserved currents which we shall consider. The dimension of a current, in mass units, is 3. This follows from the fact that the charge, which is dimensionless, is a space integral of a current component. Therefore $\partial^\mu J_\mu$ has dimension 4. However if the dynamics of the theory is such that all *operators* which occur in $\partial^\mu J_\mu$ carry dimension less than 4, then we say that $\partial^\mu J_\mu$ is "partially conserved"

As an explicit example, consider the axial current constructed from Fermion fields, $J_5^\mu = i\bar{\psi}\gamma^\mu\gamma^5\psi$; and assume that the Fermions satisfy the equation of motion

$$i\gamma^\mu\partial_\mu\psi = -m\psi + e\gamma^\mu A_\mu + g\phi\gamma^5\psi \quad .$$

Here A^μ and ϕ are vector and pseudo-scalar Boson fields respectively. Recall that the dimension of a Fermion field is 3/2 while that of a Boson field is 1. (This is seen from the Lagrangian, which necessarily has dimension 4, so that the action, $I = \int d^4x\mathcal{L}$, is dimensionless. The Fermion Langrangian contains $i\bar{\psi}\gamma^\mu\partial_\mu\psi$; the derivative carries one unit of dimension, this leaves 3 for $\bar{\psi}\psi$ hence ψ has dimension 3/2. The Boson Lagrangian contains $\partial^\mu\phi\partial_\mu\phi$; the two derivatives use up 2 units of dimension; hence ϕ has dimension 1.) Evidently J_5^μ possesses in this model the divergence $\partial_\mu J_5^\mu = 2m\,\bar\psi\gamma^5\psi + 2g\,\bar\psi\psi\phi$. The operator $\bar\psi\gamma^5\psi$ has dimension 3, while $\bar\psi\psi\phi$ has dimension 4. Hence we say that J_5^μ is partially conserved only in the absence of the pseudo-scalar coupling.

Although model independent commutators for current components have been derived from canonical transformation theory, the use of these results for physical predictions requires a tacit dynamical assumption which we must expose here. The point is that in the context of transformation theory it is always possible to add to the canonical current a divergence of an antisymmetric tensor.

$$J^\mu \to J^\mu + \partial_\lambda X^{\lambda\mu}$$

$$X^{\lambda\mu} = -X^{\mu\lambda} \tag{2.16}$$

Such additions, called "super-potentials", do not change the charges nor the divergence properties of the current. (Conservation of $\partial_\lambda X^{\lambda\mu}$ is assured by the anti-symmetry of the super-potentials. The fact that the super-potential does not contribute to the charges is seen as follows: $\int d^3x\,\partial_\lambda X^{\lambda 0} = \int d^3x\,\partial_i X^{i0} = 0$.) It may be that the modified current possesses a physical significance, greater than that of the canonical expression. Indeed this state of affairs occurs with the energy momentum tensor. For reasons which I shall discuss presently, the canonical expression (2.11), is usually replaced in physical discussions by the symmetric, Belinfante form; see Exercise 2.3.

$$\theta_B^{\mu a} = \theta_c^{\mu a} + 1/2\,\partial_\lambda X^{\lambda\mu a} \tag{2.17a}$$

$$X^{\lambda\mu a} = -X^{\mu\lambda a} = \pi^\lambda\Sigma^{\mu a}\phi - \pi^\mu\Sigma^{\lambda a}\phi - \pi^a\Sigma^{\lambda\mu}\phi \tag{2.17b}$$

Here $\Sigma^{a\beta}$ is the spin matrix appropriate to the field ϕ.

The modified expressions for currents will in general possess commutators which differ from the canonical ones given above, (2.12). Thus our insistence on the canonical commutators, rather than some others, requires an assumption that the canonical currents have a unique physical significance. This significance can be derived from the seemingly well established fact that the electromagnetic and weak interactions are governed by the *canonical* electromagnetic and SU(3) x SU(3) currents respectively. The physical significance of the Belinfante tensor follows from the belief that gravitational interactions are described by Einstein's general relativity. In that theory gravitons couple to $\theta_B^{\mu\nu}$, and not to $\theta_C^{\mu\nu}$. (In our discussion of scale transformations, Chapter VII, we shall argue that a new improved energy-momentum tensor should be introduced; and correspondingly gravity theory should be modified.) It is possible to develop a general formalism based directly on the dynamical role of currents. In this context one can derive current commutators without reference to canonical transformation theory. The results are of course the same, and we shall not discuss this approach here.[4]

We conclude this section by recording another commutator which can be established by canonical reasoning; see Exercise 2.4.

$$(2.18)\qquad i[\theta^{00}(t,\vec{x}),J_0^{\ a}(t,\vec{y})] = \partial^\mu J_\mu^{\ a}(x)\delta(\vec{x}-\vec{y})+J_i^{\ a}(x)\partial^i\delta(\vec{x}-\vec{y})$$

This has the important consequence that the divergence of a current can be expressed as a commutator.

$$(2.19)\qquad i[\theta^{00}(t,\vec{x}),\ Q^a(t)] = \partial^\mu J_\mu^{\ a}(x)$$

Formulas (2.18) and (2.19) are insensitive to the choice of θ^{00}; both the canonical and the Belinfante tensor lead to the same result. Other ETC between selected components of $\theta^{\alpha\beta}$ and $J_\mu^{\ a}$ can also be derived, in a model independent fashion. We do not pursue this topic here; one can read about it in the literature.[5]

B. Space-Time Constraints on Commutators

Although interesting physical results can be obtained from the charge density algebra, (2.12), the applications that we shall study require commutation relations between other components of the currents. These cannot be derived canonically in a model independent form. For example, $J_k^{\,a}$ involves π_k, see (2.9); but the dependence of π_k on the canonical variables π^0 and ϕ is not known in general, and one cannot compute commutators involving π_k in an abstract fashion.

It is possible to determine the $[J_a^{\,0}, J_b^{\,k}]$ ETC by investigating the space-time constraints which follow from the fact (2.12) is supposed to hold in all Lorentz frames. As an example, consider the once integrated version of (2.12), for the case of conserved currents.

$$[Q^a, J_0^{\,b}(0)] = -f_{abc} J_0^{\,c}(0) \tag{2.20}$$

An infinitesimal Lorentz transformation can be effected on (2.20) by commuting both sides with M^{0i}, the generator of these transformations.

$$\begin{aligned}[M^{0i},[Q^a, J_0^{\,b}(0)]] &= [[M^{0i}, Q^a], J_0^{\,b}(0)] + [Q^a,[M^{0i}, J_0^{\,b}(0)]] \\ &= -f_{abc}[M^{0i}, J_0^{\,c}(0)]\end{aligned} \tag{2.21a}$$

The second equality in (2.21a) follows from the first by use of the Jacobi identity. All the commutators with M^{0i} may be evaluated, since the commutator of $M^{\alpha\beta}$ with $J_\mu^{\,a}$ is known from the fact that the current transforms as a vector.

$$i[M^{\alpha\beta}, J_\mu^{\,a}(x)] = (x^\alpha \partial^\beta - x^\beta \partial^\alpha) J_\mu^{\,a}(x) + (g_\mu^{\,\alpha} g^{\beta\nu} - g^{\alpha\nu} g_\mu^{\,\beta}) J_\nu^{\,a}(x) \tag{2.21b}$$

It now follows that (remember the current is assumed conserved)

$$[Q^a, J_i^{\,b}(0)] = -f_{abc} J_i^{\,c}(0) \; . \tag{2.22a}$$

The local version of (2.22a) is

(2.22b) $$[J_0{}^a(t,\vec{x}),J_i{}^b(t,\vec{y})] = -f_{abc}\, J_i{}^c(x)\delta(\vec{x}-\vec{y}) + S_{ij}{}^{ab}(y)\partial^j\delta(\vec{x}-\vec{y}) + \ldots$$

In (2.22b) we have inserted a gradient of a δ function; the dots indicate the possible higher derivatives of δ functions which may be present. Of course, all these gradients must disappear upon integration over $\vec{x}$, so that (2.22a) is regained. Such gradient terms in the ETC are called Schwinger terms (ST).[6]

Further constraints can be obtained by commuting the *local* commutator (2.12) with P^0 and M^{0i}. However, the strongest results are arrived at by commuting (2.12) with θ^{00}, rather than with once integrated moments of θ^{00} which is what P^0 and M^{0i} are. ($P^0 = \int d^3x\ \theta^{00}\ (0,\vec{x})$; $M^{0i} = -\int d^3x\ x^i\theta^{00}\ (0,\vec{x})$.) Thus we are led to consider

(2.23a) $$i[\theta^{00}\ (0,\vec{z}),\ [J_0{}^a(0,\vec{x}),J_0{}^b(0,\vec{y})\ ]] = -f_{abc}\delta(\vec{x}-\vec{y})i[\theta^{00}(0,\vec{z}),J_0{}^c(0,\vec{x})\ ]\ .$$

The left hand side is rewritten in terms of the Jacobi identity, then (2.18) is used to evaluate the $[\theta^{00},J_0{}^a]$ ETC. The result, for conserved currents, is

$$[J_0{}^b(0,\vec{y}),J_k{}^a(0,\vec{z})]\ \partial^k\delta(\vec{x}-\vec{z}) + [J_0{}^a(0,\vec{x}),J_k{}^b(0,\vec{z})]\ \partial^k\delta(\vec{z}-\vec{y}) =$$

(2.23b) $$= -f_{abc}\ \delta(\vec{x}-\vec{y})J_k{}^c(0,\vec{z})\partial^k\delta(\vec{z}-\vec{y})\ .$$

The most general form for the $[J_0{}^a,J_i{}^b]$ ETC consistent with the constraint (2.23b) is; see Exercise 2.6,

(2.24a) $$[J_0{}^a(0,\vec{x}),J_i{}^b(0,\vec{y})\ ] = -f_{abc}J_i{}^c(0,\vec{x})\delta(\vec{x}-\vec{y}) + S_{ij}{}^{ab}(0,\vec{y})\partial^i\delta(\vec{x}-\vec{y})$$

(2.24b) $$S_{ij}{}^{ab}(0,\vec{y}) = S_{ji}{}^{ba}(0,\vec{y}) \quad .$$

Thus we have determined the $[J_0{}^a, J_i{}^b]$ ETC up to *one* derivative of the δ function; all higher derivatives should vanish. The surviving ST possess the symmetry (2.24b). It will be shown later that the ST cannot vanish. The same conclusions can be obtained when the current is partially conserved, as long as the divergence of the current is sufficiently gentle so that no ST is produced when it is commuted with $J_a{}^0$.

The above methods can be used to obtain additional constraints on various current commutators. One exploits the Jacobi identity, and the model independent commutators between selected components of $\theta^{\alpha\beta}$ and J^{μ}. We do not present these results here, since they are only of limited interest. However one result is sufficiently elegant to deserve explicit mention. If $S_{ij}{}^{ab} = \delta_{ij} S^{ab}$, where S_{ab} is a Lorentz scalar, then the $[J_i{}^a, J_j{}^b]$ ETC does not have any derivatives of δ functions.[7]

C. Space-Time Constraints on Green's Functions

We must also discuss the space-time structure of Green's functions and Ward identities. The reason for emphasizing this topic here is that the theorems of current algebra concern themselves with Green's functions: scattering amplitudes, decay amplitudes and the like; while the most felicitous way of obtaining these results is by use of Ward identities.

Consider the T product of two operators A and B.

(2.25) $$\begin{aligned} T(x) &= TA(x)B(0) \\ &= \theta(x_0)A(x)B(0) + \theta(-x_0)B(0)A(x) \end{aligned}$$

Matrix elements of T(x) are related to Green's functions. However, a Green's function must be Lorentz covariant, while T(x) need not have this property because of the time ordering. It is necessary, in the general case, to add to T(x) another non-covariant term, called a seagull, $\tau(x)$, so that the sum is covariant. The sum of a term ordered product with the covariantizing seagull is called a T* product.

$$T^*(x) = T(x) + \tau(x) \tag{2.26}$$

It is required that T*(x) and T(x) coincide for $x_0 \neq 0$; hence $\tau(x)$ has support only at $x_0 = 0$; i.e., $\tau(x)$ will involve δ functions of x_0 and derivatives thereof.

We now investigate under what conditions T(x) is not covariant. We also show how to construct the covariantizing seagull. Finally we examine under what conditions Feynman's conjecture concerning the cancellation of Schwinger terms against divergences of seagulls is valid. (Feynman's conjecture will be explained, when we come to it.) To effect this analysis it is necessary to assume that the [A,B] ETC is known.

$$[A(0,\vec{x}),B(0)] = C(0)\delta(\vec{x}) + S^i(0)\partial_i\delta(\vec{x}) \tag{2.27}$$

In offering (2.27) we have assumed, for simplicity, one ST; higher derivatives can easily be accommodated by the present technique.

Our analysis[8] makes use of the device of writing non-covariant expressions in a manifestly covariant, but frame dependent notation. A unit time like vector n^μ, and a space like projection $P^{\mu\nu}$ are introduced.

$$n^0 > 0, \quad n^2 = 1$$

$$P^{\mu\nu} = g^{\mu\nu} - n^\mu n^\nu \tag{2.28}$$

In terms of n, the n dependent T product has the form

$$T(x;n) = \theta(x\cdot n)A(x)B(0) + \theta(-x\cdot n)B(0)A(x) \quad , \tag{2.29}$$

while the ETC is

$$[A(x),B(0)]\delta(x\cdot n) = C(n)\delta^4(x) + S^\alpha(n)P_{\alpha\beta}\partial^\beta\delta^4(x) \, . \tag{2.30}$$

The T* product is n independent.

$$T^*(x) = T(x;n) + \tau(x;n) \tag{2.31}$$

We now vary n. However, since n is constrained to be timelike, only space like variations of n are permitted. Hence we operate on (2.31) by $P^{\alpha\beta}\frac{\delta}{\delta n^\beta}$.

$$0 = P^{\alpha\beta}\frac{\delta}{\delta n^\beta}T(x;n) + P^{\alpha\beta}\frac{\delta}{\delta n^\beta}\tau(x;n) \tag{2.32}$$

The first term in (2.32) is evaluated from (2.29).

$$P^{\alpha\beta}\frac{\delta}{\delta n^\beta}T(x;n) = P^{\alpha\beta}x_\beta\,\delta(x\cdot n)[A(x),B(0)] \tag{2.33a}$$

Inserting the commutator from (2.30), we find

$$P^{\alpha\beta}\frac{\delta}{\delta n^\beta}T(x;n) = -P^{\alpha\beta}\,S_\beta(n)\delta^4(x) \quad . \tag{2.33b}$$

The above equation shows that the T product is not covariant (n independent) whenever the ETC of the relevant operators contains a ST.

To proceed with the construction of the seagull, (2.33b) is inserted in (2.32), and a differential equation for τ is obtained.

$$P^{\alpha\beta}S_\beta(n)\,\delta^4(x) = P^{\alpha\beta}\frac{\delta}{\delta n^\beta}\tau(x;n) \tag{2.34a}$$

The space like projection $P^{\alpha\beta}$ may be cancelled from (2.34a), if an arbitrary expression proportional to n^β is introduced. However, since $S_\beta(n)$ is defined by (2.30) only up to terms proportional to n^β, such arbitrary contributions can be absorbed into the definition of $S_\beta(n)$. Hence we have

$$S_\beta(n)\delta^4(x) = \frac{\delta}{\delta n^\beta}\tau(x;n) \quad . \tag{2.34b}$$

The solution of this differential equation is

$$\tau(x;n) = \int^{n} dn_\beta{}' \, S^\beta(n')\delta^4(x) + \tau_0(x) \quad . \tag{2.35}$$

We have thus constructed the covariantizing seagull from the Schwinger term. Of course the seagull is not uniquely determined; an arbitrary Lorentz covariant term $\tau_0(x)$ may be added. We shall see below how $\tau_0(x)$ is specified further by Ward identities. (It can be shown that Lorentz covariance insures that the line integral in (2.35) is line independent.[8])

D. Space-Time Constraints on Ward Identities

Let us now consider the case that A and B are vector operators; for example currents. We are interested in the T and the T* products of these quantities.

$$T^{\mu\nu}(x,y;n) = \theta([x-y]\cdot n)A^\mu(x)B^\nu(y) + \theta([y-x]\cdot n)B^\nu(y)A^\mu(x) \tag{2.36a}$$

$$T^{*\mu\nu}(x,y) = T^{\mu\nu}(x,y;n) + \tau^{\mu\nu}(x,y;n) \tag{2.36b}$$

(In contrast to our previous notation, we indicate here explicitly the coordinate dependence of B^ν, which earlier was set to zero.) In applications of current algebra one frequently desires to obtain the Ward identities which are satisfied by $T^{*\mu\nu}(x,y)$; i.e., one wants to know the formula for

$$\begin{aligned}\frac{\partial}{\partial x^\mu} T^{*\mu\nu}(x,y) &= \theta([x-y]\cdot n)\partial_\mu A^\mu(x)B^\nu(y) \\ &+ \theta([y-x]\cdot n)B^\nu(y)\partial_\mu A^\mu(x) \\ &+ \delta([x-y]\cdot n)[n_\mu A^\mu(x),B^\nu(y)] \\ &+ \frac{\partial}{\partial x^\mu} \tau^{\mu\nu}(x,y;n) \quad ,\end{aligned} \tag{2.37a}$$

$$\frac{\partial}{\partial y^\nu} T^{*\mu\nu}(x,y) = \theta([x-y]\cdot n)A^\mu(x)\partial_\nu B^\nu(y)$$
$$+ \theta([y-x]\cdot n)\partial_\nu B^\nu(y)A^\mu(x)$$
$$- \delta([x-y]\cdot n)[A^\mu(x), n_\nu B^\nu(y)]$$
$$+ \frac{\partial}{\partial y^\nu} \tau^{\mu\nu}(x,y;n). \tag{2.37b}$$

To evaluate such expressions, it is necessary to know the $[A^0,B^\nu]$ and the $[B^0,A^\nu]$ ETC's completely, since these objects occur in the divergence of the T product. Also one needs to know the divergence of the seagull, since that object is also present in the T* product. In the specific case of SU(3) x SU(3) currents: $A^\mu = J_a{}^\mu$, $B^\nu = J_b{}^\nu$; we have shown above that the term proportional to the δ function in the $[J_a{}^0, J_b{}^\nu]$ ETC is known explicitly; however we have not determined the ST proportional to the gradient of the δ function; see (2.12) and (2.24). Consequently the covariantizing seagull also cannot be determined. An obstacle thus has arisen in our program of calculating the Ward identity.

This obstacle is in practice overcome with the help of Feynman's conjecture.[9] Observe that if, for the moment, we pretend that there are no gradient terms in the ETC, then we may also set the seagull to zero. In this fictitious case, the Ward identity *can* be derived, since all we need to use is the known coefficient of the δ function. Feynman's conjecture is the statement that this "naive" procedure gives the correct answer. That is, one hopes that whatever the form of the ST, the associated seagull has the property that its divergence always cancels against the ST in the ETC, and one is left only with the term proportional to the δ function.

It is clear that Feynman's conjecture is crucial for progress in current algebraic analysis. Hence we now present an analysis[8] of it. Our conclusion will be a criterion for the general conditions which assure the validity of the conjecture. Later when we study anomalies, we shall see that one of the reasons for the conflict between current algebraic predictions and explicit calculations is the failure of Feynman's conjecture.

We return now to the general discussion of the T* product of the two vector operators A^μ and B^ν. We record here their commutator in abstract form.

$$[A^\mu(x),B^\nu(y)]\delta([x-y]\cdot n) = C^{\mu\nu}(y;n)\delta^4(x-y) + S^{\mu\nu\,\alpha}(y;n)P_{\alpha\beta}\partial^\beta\delta^\mu(x-y) \tag{2.38}$$

It is clear that the previous formulae for the seagull still holds.

$$\tau^{\mu\nu}(x,y;n) = \tau^{\mu\nu}(y;n)\delta^4(x-y)+\tau_0^{\mu\nu}(x,y)$$

$$\tau^{\mu\nu}(y;n) = \int^n dn_\beta{}'\, S^{\mu\nu\,\beta}(y;n') \tag{2.39}$$

$$\frac{\delta}{\delta n^\beta}\,\tau^{\mu\nu}(y;n) = S^{\mu\nu\,\beta}(y;n) \tag{2.40}$$

Now let us insert (2.38) into (2.37). We find

$$\begin{aligned}\frac{\partial}{\partial x^\mu}\,T^{*\mu\nu}(x,y) &= \theta([x-y]\cdot n)\partial_\mu A^\mu(x)B^\alpha(y) \\ &+ \theta([y-x]\cdot n)B^\nu(y)\partial_\mu A^\mu(x) \\ &+ n_\mu C^{\mu\nu}(y;n)\delta^4(x-y) + n_\mu S^{\mu\nu\,\alpha}(y;n)P_{\alpha\beta}\partial^\beta\delta^4(x-y) \\ &+ \tau^{\mu\nu}(y;n)\frac{\partial}{\partial x^\mu}\delta^4(x-y) + \frac{\partial}{\partial x^\mu}\tau_0^{\mu\nu}(x,y)\ .\end{aligned} \tag{2.41}$$

For conserved currents, the T product on the right hand side of (2.41) is absent. The remaining terms must be covariant, since the T* product on the left hand side possesses this property. Therefore, it must be true that

$$n_\mu C^{\mu\nu}(y;n) = I_1{}^\nu(y) \tag{2.42}$$

(2.43) $$\tau^{\mu\nu}(y;n) + n_\beta S^{\beta\nu\,\alpha}(y;n)P_\alpha{}^\mu = I_1{}^{\mu\nu}(y) \quad .$$

In the above $I_1{}^\nu(y)$ and $I_1{}^{\mu\nu}(y)$ are Lorentz covariant, n independent quantities. For partially conserved currents, we again obtain the same result when we interpret "partial conservation" to mean that $\partial_\mu A^\mu$ is a sufficiently gentle operator, so that it does not give rise to a ST when commuted with B^ν. In that case the T product in (2.41) is already covariant, and the remaining terms again satisfy (2.42) and (2.43). Similar results are obtained by diverging with $\frac{\partial}{\partial y^\mu}$.

(2.44) $$n_\nu C^{\mu\nu}(y;n) - \frac{\partial}{\partial y^\nu}\tau^{\mu\nu}(y;n) = I_2{}^\nu(y)$$

(2.45) $$\tau^{\mu\nu}(y;n) + n_\beta S^{\mu\beta\,\alpha}(y;n)P_\alpha{}^\nu = I_2{}^{\mu\nu}(y)$$

$I_2{}^\nu$ and $I_2{}^{\mu\nu}$ are also covariant, not necessarily equal to $I_1{}^\nu$ and $I_1{}^{\mu\nu}$. The solution to (2.42) is

(2.46) $$C^{\mu\nu}(y;n) = n^\mu I_1{}^\nu + P_\alpha{}^\mu C^{\alpha\nu}(y;n) \quad ,$$

while (2.43) and (2.45) require

(2.47a) $$n_\mu \tau^{\mu\nu}(y;n) = n_\mu I_1{}^{\mu\nu}(y)$$

(2.47b) $$n_\nu \tau^{\mu\nu}(y;n) = n_\nu I_2{}^{\mu\nu}(y) \quad .$$

Equations (2.46) and (2.47) represent constraints which must be satisfied by $C^{\mu\nu}(y;n)$ and $\tau^{\mu\nu}(y;n)$ which follow from Lorentz covariance.

Next we inquire under what conditions on the ETC can Feynman's conjecture be established; i.e., we require that divergences of seagulls cancel against ST, so that all gradient terms are absent from the Ward identity. Evidently for this to be true in the μ Ward identity, we must have, according to (2.41) and (2.43), that the following combination be free of gradients of δ functions.

$$I_1^{\mu\nu}(y)\partial_\mu\delta^4(x-y) + \frac{\partial}{\partial x^\mu}\tau_0^{\mu\nu}(x,y) \tag{2.48a}$$

Similarly, the ν Ward identity sets the requirement that no gradients of δ functions occur in

$$-I_2^{\mu\nu}(y)\partial_\nu\delta^4(x-y) + \frac{\partial}{\partial y^\nu}\tau_0^{\mu\nu}(x,y) \quad . \tag{2.48b}$$

Equations (2.48) show that *it is always possible to satisfy Feynman's conjecture in one of the two Ward identities*. For example, to satisfy (2.48a) we may set

$$\tau_0^{\mu\nu}(x,y) = -I_1^{\mu\nu}(y)\delta^4(x-y) \quad . \tag{2.49a}$$

Alternatively, to satisfy (2.48b) $\tau_0^{\mu\nu}(x,y)$ can be chosen to be

$$\tau_0^{\mu\nu}(x,y) = -I_2^{\mu\nu}(y)\delta^4(x-y) \quad . \tag{2.49b}$$

However, in the general case, it need not be possible to satisfy *both* requirements since (2.48) may overdetermine the equation for $\tau_0^{\mu\nu}$, and no solution need exist.

A sufficient condition for the solution to both Eqs. (2.48) is that there be no ST in the time-time component of the commutators. The reason for this can be see from (2.43) and (2.47). We have

$$\begin{aligned} n_\nu n_\beta S^{\beta\nu\ \alpha}(y;n)P_\alpha^{\ \mu} &= n_\nu I_1^{\mu\nu}(y) - n_\nu \tau^{\mu\nu}(y;n) \\ &= n_\nu[I_1^{\mu\nu}(y) - I_2^{\mu\nu}(y)] \quad . \end{aligned} \tag{2.50}$$

If there are no ST in the time-time ETC, the left hand side vanishes. Consequently (2.50) implies that $I_1^{\mu\nu}(y) = I_2^{\mu\nu}(y)$, and the two solutions

(2.49) become identical; thus *both* Ward identities can satisfy Feynman's conjecture. (It is also possible to derive a necessary and sufficient condition for the existence of a solution to (2.48); see reference 8.)

We conclude that if the *time* component algebra, which we derived canonically, and which does not possess a ST, survives in the complete theory, then Feynman's conjecture can be satisfied. On the other hand, if the time component algebra acquires a ST, it may not be possible to effect Feynman's conjecture. It will be seen that this indeed happens for the $\pi^0 \to 2\gamma$ problem.

E. Schwinger Terms

As a final example of space time constraints on current commutators, we show that the ST which we allowed for in the $[J_a{}^0, J_b{}^i]$ ETC cannot be zero. For simplicity we study only the conserved, electromagnetic currents; hence the group indices a and b are suppressed.

Consider the vacuum expectation value of the $[J^0, J^i]$ ETC.

$$<0|[J^0(0,\vec{x}), J^i(0,\vec{y})]|0> = <0|C^i(\vec{x},\vec{y})|0> \tag{2.51}$$

$C^i(\vec{x},\vec{y})$ is, by definition, the $[J^0, J^i]$ ETC. It has support only at $\vec{x} = \vec{y}$; thus it is composed of the δ function and derivatives thereof. However, from (2.22) we see that the δ function contribution is absent. since $a = b$. Therefore $C^i(\vec{x},\vec{y})$ is non-vanishing only to the extent that ST are non-vanishing. We now prove that $C^i(\vec{x},\vec{y})$ has non-zero vacuum expectation value; hence the ST is present. (In order to keep the discussion as general as possible, we do not use the specific form for $C^i(\vec{x},\vec{y})$, derived in (2.24).)

To effect the analysis, we begin by differentiating (2.51) with respect to $\vec{y}$, and use current conservation.

$$<0|[J^0(0,\vec{x}), -\partial_0 J^0(0,\vec{y})]|0> = <0|\frac{\partial}{\partial y^i} C^i(\vec{x},\vec{y})|0> \tag{2.52a}$$

The time derivative may be expressed as a commutator with the Hamiltonian.

$$< 0|[J^0(0,\vec{x}),\partial_0 J^0(0,\vec{y})]|0 > = < 0|[J^0(0,\vec{x}),i[H,J^0(0,\vec{y})]]|0 >$$

$$= i < 0|J^0(0,\vec{x})\ H\ J^0(0,\vec{y}) + J^0(0,\vec{y})\ H\ J^0(0,\vec{x})|0 > \tag{2.52b}$$

In presenting (2.52b) we have used the fact that the vacuum has zero energy. Finally we multiply the above by $f(\vec{x})\ f(\vec{y})$, where $f(\vec{x})$ is an arbitrary real function; and integrate over $\vec{x}$ and $\vec{y}$. One is left with

$$i\int d^3x\ d^3y < 0|C^i(\vec{x},\vec{y})|0 > f(\vec{x}) \frac{\partial}{dy^i} f(\vec{y}) = 2 < 0|FHF|0 >$$

$$F = \int d^3x\ f(\vec{x}) J^0(0,\vec{x}) \quad . \tag{2.53}$$

The right hand side is non-zero. The reason for this is that the operator F possesses in general non-vanishing matrix elements between the vacuum and other states which necessarily carry positive energy; i.e.,

$$< 0|FHF|0 > = \sum_{nm} < 0|F|m > < m|H|n > < n|F|0 >$$

$$= \sum_n E_n |< 0|F|n >|^2 > 0 \quad . \tag{2.54}$$

(The vacuum cannot be an eigenstate of F, since J^0 has zero vacuum expectation value.) It therefore follows that $C^i(\vec{x},\vec{y})$ is non-zero, and the ST cannot vanish.[10] The proof can be extended to non-conserved currents; hence one knows that there must be a ST in the $[J_a{}^0, J_a{}^i]$ ETC as well, where a is any internal group index.

When attention was first drawn to the existence of this ST, great interest was aroused because canonical computation in many instances leads to a vanishing result which, as we have seen, is inconsistent with other properties of local quantum theory: positivity and Lorentz covariance.

In spinor electrodynamics for example one believes the formula for J^μ to be $\bar{\psi}\gamma^\mu\psi$; one further believes that the ETC $[J^0,\psi]$ is given by

$$[J^0(0,\vec{x}),\psi(0)] = -\psi(0)\delta(\vec{x}) \quad . \tag{2.55}$$

From this it then follows that the $[J^0,J^i]$ ETC vanishes. (In scalar electrodynamics, on the other hand, a non-vanishing result is obtained.) Similar comments apply to SU(3) x SU(3) quark currents; there too canonical evaluation leads to a vanishing result.

This then is an example of a commutator anomaly; historically it predates the current emphasis of this topic. The reasons which were advanced at one time for ignoring these anomalies, were based on the belief that physical predictions are insensitive to the ST; essentially by virtue of the Feynman conjecture discussed earlier. We now know that this optimism was unjustified.

The present argument for the existence of the ST does not indicate whether this object is a c number or an operator – only the vacuum expectation value has been shown to be non-zero. Model calculations are also inconclusive. In scalar electrodynamics and in the σ model it comes out canonically to be a q number; in the algebra of fields it is a c number. In theories where canonical evaluation gives zero, more careful computation yields a quadratically divergent c number result in spinor electrocynamics, a q number in quark models with spin zero gluons and a c number in quark models with vector gluons.[11]

To settle the question of the nature of this object one must turn to experiment. Unfortunately only very limited experimental probes have been discovered so far. The vacuum expectation value of the ST in the electromagnetic current commutator can be expressed in terms of an integral over the total cross section for lepton annihilation.[12] Recently a sum rule has been derived relating the single proton connected matrix element of the same object with the total electro-production cross section.[13]

The latter result tests the q number nature of the electromagnetic ST. Preliminary conclusions indicate that this ST *is* a c number, although very likely it is a quadratically divergent object.

The experimental evidence for the nature of the SU(3) x SU(3) ST is even more scanty. The only thing one can say, at the present time, is that the apparent validity of the SU(2) x SU(2) Weinberg first sum rule[14] strongly indicates that $S_{ab}{}^{ij}$ possesses no $I = 1$ part; i.e., $S_{ab}{}^{ij} = S_{ba}{}^{ij}$; $a,b = 1,2,3$, for vector and axial vector currents.

The vacuum expectation value sum rule is given in the literature; we shall not examine it here; see Exercise 3.2. We shall discuss the proton matrix element relation in connection with the Callan-Gross sum rule in Chapter V. Finally, we mention that the existence of non-canonical structures analogous to the ST has also been demonstrated for ETC between selected components of $\theta^{\mu\nu}$; see the first two papers in reference 4.

F. Discussion

The constraints which we have obtained in this Chapter follow from very general model independent considerations. However, a cautionary reminder must be inserted here concerning the validity of our results. Whenever very *detailed* and formal properties of field theory are used to arrive at a conclusion, we must remember that the troubles of local quantum field theory, previously alluded to, may invalidate the argument. On the other hand, when *general* properties of the theory are exploited, greater confidence may be placed in the result. Sections A and B above are examples of the former; thus we must not be surprised when the canonical constraints are evaded. Sections C, D and E rely merely on Lorentz covariance, positivity and the existence of equal time commutators. So far no one has found counter examples to the conclusions given there.

REFERENCES

[1] We ignore the complications which arise with Fermion operators: *anti*-commutation relations, etc. One can verify that these complications do not modify our final conclusions.

[2] Our distinction between *internal* and *space-time* symmetries must be refined to account for the possibility of adding total divergences to a Lagrangian without affecting dynamics. Thus even when

$$\delta\mathcal{L} = \partial_\mu \Lambda^\mu \neq 0 \quad ,$$

we would still call this an *internal* symmetry if it is possible to find a dynamically equivalent Lagrangian $\mathcal{L}'$ such that

$$\mathcal{L}' = \mathcal{L} + \partial_\mu X^\mu$$

$$\delta\mathcal{L}' = \delta\mathcal{L} + \partial_\mu \delta X^\mu = \partial_\mu [\Lambda^\mu + \delta X^\mu] = 0 \quad .$$

When it is impossible to remove $\partial_\mu \Lambda^\mu$ by this method, we are dealing with an internal symmetry.

[3] A more conventional form for the current commutator is

$$[J_0{}^a(t,\vec{x}), J_0{}^b(t,\vec{y})] = i f_{abc} J_0{}^c(t,\vec{x}) \delta(\vec{x}-\vec{y}) \quad .$$

This form is equivalent to (2.12), once one replaces $J_0{}^a$ by $iJ_0{}^a$.

[4] This formalism has been given by J. Schwinger, *Phys. Rev.* 130, 800 (1963). Various applications of this point of view are found in the work of Schwinger, and D. G. Boulware and S. Deser, *J. Math. Phys.* 8, 1468 (1967); D. J. Gross and R. Jackiw, *Phys. Rev.* 163, 1688 (1967).

[5] D. J. Gross and R. Jackiw, *Phys. Rev.* 163, 1688 (1967); R. Jackiw, *Phys. Rev.* 175, 2058 (1968); S. Deser and L. K. Morrison, *J. Math. Phys.* 11, 596 (1970).

[6] J. Schwinger, *Phys. Rev. Letters* 3, 296 (1959). These gradient terms were first discovered by T. Goto and I. Imamura, *Prog. Theoret. Phys.* 14, 196 (1955).

[7] This argument is detailed in the papers by D. J. Gross and R. Jackiw, Ref. 5.

[8] This analysis is presented by D. J. Gross and R. Jackiw, *Nucl. Phys.* B14, 269 (1969).

[9] R. P. Feynman, unpublished; see also M. A. B. Bég, *Phys. Rev. Letters* 17, 333 (1966).

[10] The derivation presented here is due to Schwinger, Ref. 6; Goto and Imamura, Ref. 6, previously had presented a different argument toward the same end.

[11] D. G. Boulware and R. Jackiw, *Phys. Rev.* 186, 1442 (1969); D. G. Boulware and J. Herbert, *Phys. Rev.* D2, 1055 (1970); see also Chapter VI.

[12] J. D. Bjorken, *Phys. Rev.* 148, 1467 (1966); V. N. Girbov, B. L. Ioffe and I. Ya. Pomeranchuk, *Phys. Letters* 24B, 554 (1967); R. Jackiw and G. Preparata, *Phys. Rev. Letters* 22, 975 (1969); (E) 22, 1162 (1969).

[13] R. Jackiw, R. Van Rogen and G. B. West, *Phys. Rev.* D2, 2473 (1970). See also L. S. Brown lectures delivered at Summer Institute for Theoretical Physics, University of Colorado (1969), to be published in *Lectures in Theoretical Physics*, ed. W. E. Brittin, B. W. Downs and J. Downs, *Interscience*, (New York); J. M. Cornwall, D. Corrigan and R. E. Norton, *Phys. Rev. Letters* 24, 1141 (1970).

[14] S. Weinberg, *Phys. Rev. Letters* 18, 507 (1967).

III. THE BJORKEN-JOHNSON-LOW LIMIT

In order to understand the failure of formal, current algebraic predictions, we shall need to calculate the ETC in order to ascertain whether or not the canonical value is maintained. The first difficulty that is encountered in this program is that the ETC frequently is ambiguous, and the result one obtains may depend on the rules one adopts towards the handling of ambiguous expressions. For example, how one proceeds to equal times from an unequal time commutator can affect the answer. A related, further problem is that in calculating commutators of operators, which themselves are products of other operators, ambiguities and infinities arise in forming these products. I have in mind, for example, the construction of currents which are bilinear in Fermion fields. There have appeared in the literature many evaluations of commutators which yield different results; this variety is traceable to the various ways one can choose to handle the attendant ambiguities.

Fortunately, for our purposes, we can prescribe a unique method for calculating commutators. The reason for this is that we are comparing the current algebraic predictions with explicit dynamical predictions of the theory. For a consistent check, the ETC must be evaluated by the same techniques as the solutions of the theory. The only known tool for calculating physical consequences of a field theory is renormalized perturbation theory, which provides one with finite, well-defined Green's functions. Thus the ETC must be computed from the (known) Green's function. This is achieved by the method of Bjorken, Johnson and Low (BJL),[1,2] which we now discuss.

Consider a matrix element of the T product of two operators A and B.

$$T(q) \equiv \int d^4x\ e^{iqx} < \alpha | T A(x) B(0) | \beta > \tag{3.1}$$

The BJL definition of the matrix elements of the [A,B] ETC is

$$\lim_{q_0 \to \infty} q_0 T(q) = i \int d^3x\ e^{-i\vec{q}\cdot\vec{x}} < \alpha | [A(0,\vec{x}), B(0)] | \beta > \ . \tag{3.2}$$

Alternatively we can say that the $1/q_0$ term in T(q), at large q_0, determines the commutator. In all our subsequent calculations, we shall determine the commutator from the (known) T products by Eq. (3.2). If the above limit diverges, this is interpreted as the statement that this particular matrix element of the ETC is divergent.

As a justification of the BJL formula, one can present "derivations" of it, which are valid if various mathematical manipulations are permitted. These derivations serve merely to *motivate* the result, and to insure that in non-singular situations the BJL definition corresponds to the usual ones for the ETC. Three derivations of increasing amount of rigor are offered.

The first method[3] begins by rewriting (3.1) as

$$T(q) = -i \frac{1}{q_0} \int d^4x \left(\frac{\partial}{\partial x_0} e^{iqx} \right) < \alpha | T A(x) B(0) | \beta > \ . \tag{3.3a}$$

The time integration is performed by parts, the surface terms are dropped, and one has

$$\begin{aligned} q_0 T(q) &= i \int d^4x\ e^{iqx} < \alpha | T \dot{A}(x) B(0) | \beta > \\ &+ i \int d^3x\ e^{-i\vec{q}\cdot\vec{x}} < \alpha | [A(0,\vec{x}), B(0)] | \beta > \ . \end{aligned} \tag{3.3b}$$

We now observe that the first term on the right hand side of (3.3b) is a Fourier transform of an object which in the x_0 variable possesses singularities that are no worse than discontinuities. Hence as $q_0 \to \infty$, this term should vanish, according to the Riemann-Lesbegue lemma. Therefore, in the limit as $q_0 \to \infty$, (3.3b) reproduces (3.2). This derivation also indicates how anomalous, non-canonical results may be obtained from the BJL definition. It may turn out that a *canonical* computation of $\dot{A}$, by use of the operator equations of motion of the theory, may give an expression which leads to a singular behavior of $\int d^4x\, e^{iqx} < \alpha|T\dot{A}(x)B(0)|\beta >$. Thus, in spite of the formalism, this term may survive in the large q_0 limit, and add a non-vanishing contribution to the *canonical* value of $i\int d^3x\, e^{-i\vec{q}\cdot\vec{x}} < \alpha|[A(0,\vec{x})B(0)]|\beta >$. This results in a non canonical value for $\lim_{q_0 \to \infty} q_0 T(q)$, and is interpreted as a non-canonical value for the (by definition) commutator.

The second derivation is Bjorken's original presentation.[1] Define

$$\rho(q_0,q) = \int d^4x\, e^{iqx} < \alpha|A(x)B(0)|\beta > \tag{3.4a}$$

$$\bar{\rho}(q_0,\vec{q}) = \int d^4x\, e^{iqx} < \alpha|B(0)A(x)|\beta > \quad . \tag{3.4b}$$

By use of the integral representation for the step function,

$$\theta(x_0) = \frac{i}{2\pi}\int d\alpha\, e^{-ix_0\alpha} \frac{1}{\alpha+i\epsilon} \quad , \tag{3.5}$$

the T product (3.1) may be written in the form

$$T(q) = \frac{i}{2\pi}\int_{-\infty}^{\infty} \frac{d\alpha}{\alpha+i\epsilon} [\rho(q_0-\alpha,\vec{q}) + \bar{\rho}(q_0+\alpha,\vec{q}]$$

$$= \frac{i}{2\pi}\int_{-\infty}^{\infty} dq_0{}' \left[\frac{\rho(q_0{}',\vec{q})}{q_0-q_0{}'+i\epsilon} - \frac{\bar{\rho}(q_0{}'\vec{q})}{q_0-q_0{}'-i\epsilon}\right] \quad . \tag{3.6a}$$

This representation for the T product is called the Low representation. Multiplying by q_0 and passing to the $q_0 \to \infty$ limit, leaves

(3.6b) $$\lim_{q_0 \to \infty} q_0 T(q) = \frac{i}{2\pi} \int_{-\infty}^{\infty} dq_0{}' [\rho(q_0{}',\vec{q}) - \bar{\rho}(q_0{}',\vec{q})] \ .$$

According to the definitions (3.4), (3.6b) is just (3.2).

$$\lim_{q_0 \to \infty} q_0 T(q) = i \int d^4x \ e^{-i\vec{q}\cdot\vec{x}} \delta(x_0) \left(< \alpha | A(x) B(0) | \beta > \right.$$

$$\left. - < \alpha | B(0) A(x) | \beta > \right)$$

(3.6c) $$= i \int d^3x \ e^{-i\vec{q}\cdot\vec{x}} < \alpha | [A(0,\vec{x}), B(0)] | \beta >$$

This derivation is useful in that it shows explicitly that the ETC, as given by BJL definition, is physically interesting. The point is that according to (3.6b) the commutator has been expressed in terms of the spectral functions ρ and $\bar{\rho}$. These in turn are related to directly measurable matrix elements of A and B.

$$\rho(q_0,\vec{q}) = \sum_n (2\pi)^4 \ \delta^4 \ (q+p_\alpha - p_n) < \alpha | A(0) | n > < n | B(0) | \beta >$$

(3.7) $$\bar{\rho}(q_0,\vec{q}) = \sum_n (2\pi)^4 \ \delta^4 \ (q+p_n - p_\beta) < \alpha | B(0) | n > < n | A(0) | \alpha >$$

The third derivation is due to Johnson and Low[2]. Consider the position space T product

(3.8) $$t(x) = \int \frac{d^4q}{(2\pi)^4} e^{-iqx} \ T(q)$$

The ETC may be defined as

(3.9) $$C(0,\vec{x}) = \lim_{x_0 \to 0^+} t(x) - \lim_{x_0 \to 0^-} t(x) \ .$$

As $x_0 \to 0^+$, the q_0 integral in (3.8) may be extended into the complex q_0 plane, by closing the contour in the lower half plane; similarly as $x_0 \to 0^-$, the integral can be closed in the upper half plane. In the limit $x_0 \to 0^\pm$, the integrals along the real q_0 axis cancel between the two terms in (3.9), and one is left with

$$C(0,\vec{x}) = \oint \frac{dq_0}{2\pi} \int \frac{d^3q}{(2\pi)^3}\, e^{i\vec{q}\cdot\vec{x}}\, T(q) \quad . \tag{3.10a}$$

Assume now that T(q) has an expansion in inverse powers of q_0. The clockwise contour integral of $(q_0)^{-n}$ is $-2\pi i\, \delta_{n,1}$; so that only the $1/q_0$ part of T(q) contributes. Call that part $T_{-1}(\vec{q})$. We are left with

$$C(0,\vec{x}) = -i\int \frac{d^3q}{(2\pi)^3}\, e^{i\vec{q}\cdot\vec{x}}\, T_{-1}(\vec{q}) \quad . \tag{3.10b}$$

The Fourier transform of this is equivalent to (3.2). Johnson and Low further show that if the leading singularity of T(q) at large q_0 is $(1/q_0)\log q_0$, then the previous derivation remains correct, and the equal time commutator is logarithmically divergent. Similarly if the singularity is of the form $q_0 \log q_0 = \frac{1}{q_0}(q_0^2 \log q_0)$; this should be interpreted as a quadratically divergent commutator.

It should be remarked here that when the BJL theorem was first discovered it was applied in many instances to a *derivation* of the high energy behavior of amplitudes. The idea was to write

$$T(q) \sim \frac{i}{q_0}\int d^3x\, e^{-i\vec{q}\cdot\vec{x}} < \alpha|[A(0,\vec{x}),B(0)]|\beta > \tag{3.11}$$

and to evaluate the right hand side of (3.11) canonically. However, the use of the canonical commutator in this context is unjustified, as Johnson and Low already demonstrated in their paper.[2] The fact that the high energy behavior of amplitudes *is not,* in general, correctly given by this technique has recently been described as "the breakdown of the BJL theorem".

Note that the BJL theorem defines the commutator from the T product, rather than from the covariant T* product. However, in perturbation one calculates only the covariant object. Hence the T product must be separated. This is achieved by remembering that the difference between T and T* is local in position space, hence it is a polynomial of q_0 in momentum space. Therefore, before applying the BJL technique to the expressions calculated in perturbation theory, all polynomials in q_0 must be dropped.

It is clear that the expansion in inverse powers of q_0 can be extended beyond the first. From (3.3a), it is easy to see that if the [A,B] ETC vanishes, then we have

$$\lim_{q_0 \to \infty} q_0^{\,2} T(q) = -\int d^3x \, e^{-i\vec{q}\cdot\vec{x}} < \alpha|[\dot{A}(0,x), B(0)]|\beta > . \tag{3.12}$$

Again, if this limit is divergent, then this matrix element of the $[\dot{A},B]$ ETC is infinite. Eventually commutators with sufficient number of time derivatives probably are infinite, since it is unlikely that the expansion in inverse powers of q_0 can be extended without limit.

REFERENCES

[1] J. D. Bjorken, *Phys. Rev.* **148**, 1467 (1966).

[2] K. Johnson and F. E. Low, *Prog. Theoret. Phys.* (Kyoto), Suppl. 37-38, 74 (1966).

[3] This method was developed in conversations with Prof. I. Gerstein.

IV. *THE $\pi^0 \to 2\gamma$ PROBLEM*

A. Preliminaries

The neutral pion is observed to decay into 2 photons with a width of the order of 10 eV. This experiment measures the matrix element $M(p,q) = \langle \pi,k|\gamma,p;\gamma',q \rangle$; p and q are the 4 momenta of the photons, $k = p+q$ is the 4 momentum of the pion. M(p,q) has the form

$$\epsilon_\mu(p)\epsilon_\nu{}'(q)T^{\mu\nu}(p,q) \quad ,$$

i.e., $T^{\mu\nu}(p,q)$ is the previous matrix element with the photon polarization vectors $\epsilon_\mu(p)\epsilon_\nu{}'(q)$ removed. The tensor $T^{\mu\nu}$ has the following structure.

$$T^{\mu\nu}(p,q) = \epsilon^{\mu\nu\alpha\beta}p_\alpha q_\beta T(k^2) \tag{4.1}$$

This is dictated by Lorentz covariance and parity conservation (the pion is a pseudoscalar, $T^{\mu\nu}$ must be a pseudo-tensor, hence the factor $\epsilon^{\mu\nu\alpha\beta}$). Gauge invariance ($p_\mu T^{\mu\nu}(p,q) = 0 = q_\nu T^{\mu\nu}(p,q)$) and Bose symmetry ($T^{\mu\nu}(p,q) = T^{\nu\mu}(q,p)$) are seen to hold.

We shall keep q^2 and p^2, the photon variables, on their mass shell $q^2 = p^2 = 0$. The pion variable k^2 is, of course, equal to the pion mass squared μ^2, but for our arguments we allow it to vary away from this point. This continuation off the mass shell may be effected by the usual LSZ method.

$$\begin{aligned} T^{\mu\nu}(p,q) &= \epsilon^{\mu\nu\alpha\beta}p_\alpha q_\beta T(k^2) \\ &= (\mu^2-k^2) \langle 0|\phi(0)|\gamma,p;\gamma',q \rangle \end{aligned} \tag{4.2}$$

Here ϕ is an interpolating field for the pion. We, of course, do not assert that it is *the* pion field – such an object may not exist. It merely is some local operator which has a non-vanishing matrix element between the vacuum and the single pion state, normalized to unity $< 0|\phi(0)|\pi > = 1$.

B. Sutherland-Veltman Theorem

Following Sutherland and Veltman,[1] we now prove that if the divergence of the axial current is used as the pion interpolating field, then $T(0) = 0$, as long as the conventional current algebraic ideas are valid. This is a *mathematical* fact, without direct experimental content. However, since μ^2 is small, compared to all other mass parameters relevant to this problem, one may expect that $T(\mu^2) \approx T(0)$. This smoothness hypothesis is based on the supposition that the divergence of the axial current is a "gentle" operator whose matrix elements do not have any dynamically unecessary rapid variation. This is the content of PCAC, which is a very successful notion in other contexts. Unfortunately, in the present application, one cannot understand the *experimental* fact that $T(\mu^2) \neq 0$.

After Sutherland and Veltman pointed out this *experimental* failure of PCAC, the most widely accepted explanation was that $T(k^2)$ *was* rapidly varying, for unknown reasons. This is not impossible, since it has happened before in current algebra-PCAC applications that a source of rapid variation for a particular amplitude was at first overlooked. However, in the present instance, as the years passed by, no reason was forthcoming to explain the putative rapid variation.

The Sutherland-Veltman argument begins by representing the off mass shell pion amplitude (4.2) by

$$T^{\mu\nu}(p,q) = e^2(\mu^2-k^2) \int d^4x d^4y \, e^{-ipx} \, e^{-iqy}$$

$$< 0|T^* J^\mu(x) J^\nu(y) \phi(0)|0 > . \quad (4.3)$$

Here J^μ is the electromagnetic current. The pion field is replaced by the divergence of the neutral axial current $J_5{}^\alpha$.

$$\phi(0) = \frac{\partial_\alpha J_5{}^\alpha(0)}{F\mu^2} \tag{4.4}$$

$F\mu^2$ is the appropriate factor which assures the proper normalization for the pion field, defined by (4.4).

$$< 0|J_5{}^\alpha(0)|\pi > \equiv ip^\alpha F$$

$$< 0|\partial_\alpha J_5{}^\alpha(0)|\pi > = \mu^2 F \tag{4.5}$$

Thus

$$T^{\mu\nu}(p,q) = \frac{e^2(\mu^2-k^2)}{F\mu^2}\int d^4x d^4y e^{-ipx}\, e^{-iqy}$$

$$< 0|T^*J^\mu(x)J^\nu(y)\partial_\alpha J_5{}^\alpha(0)|0 >$$

$$= \frac{e^2(\mu^2-k^2)}{F\mu^2}\int d^4x d^4y e^{-ipx}\, e^{-iqy}$$

$$\partial_\alpha < 0|T^*J^\mu(x)J^\nu(y)J_5{}^\alpha(0)|0 >$$

$$= \frac{(\mu^2-k^2)}{F\mu^2} k_\alpha T^{\alpha\mu\nu}(p,q); \tag{4.6a}$$

where $T^{\alpha\mu\nu}(p,q)$ is defined by

$$T^{\alpha\mu\nu}(p,q) = -ie^2 \int d^4x d^4y e^{-ipx}\, e^{-iqy}$$

$$< 0|T^*J^\mu(x)J^\nu(y)J_5{}^\alpha(0)|0 > \ . \tag{4.6b}$$

The justification for passing from the first to the second term on the right hand side of (4.6a) is the current algebra satisfied by $J_5{}^0$ and J^μ: apart from possible ST the currents commute.

$$[J_5^{\,0}(0,\vec{x}),J^{\mu}(0)] = 0 + \mathrm{ST} \tag{4.7}$$

The ST is handled by one of three ways. One may simply assume that it is absent; since the ETC does not involve two identical currents, there is no proof that a ST must be present. Alternatively a weaker assumption is that the ST is a c number. It is easy to see that since the vacuum expectation value of a current vanishes, a c number ST would not interfere with passing the derivative through the T* product. Finally the weakest assumption that one can make is Feynman's conjecture – without discussing the nature of any possible ST, it is asserted that the naive procedure is the correct one, due to cancellation of ST with divergences of seagulls.

The tensor $T^{\alpha\mu\nu}(p,q)$ must possess odd parity, because J_5^{α} is a pseudo-vector; it must satisfy the Bose symmetry: $T^{\alpha\mu\nu}(p,q) = T^{\alpha\nu\mu}(q,p)$; finally it must be transverse to p_μ and q_ν: $p_\mu\, T^{\alpha\mu\nu}(p,q) = 0$, $q_\nu\, T^{\alpha\mu\nu}(p,q) = 0$. The last condition follows from the conservation of J^{μ} and the current algebra satisfied by J^0 with J^{μ} and J_5^{α}. Again all these commutators vanish apart from possible ST; the latter being ignored in this calculation.

$$[J^0(0,\vec{x}),J^{\mu}(0)] = 0 + \mathrm{ST} \tag{4.8a}$$

$$[J^0(0,\vec{x}),J_5^{\mu}(0)] = 0 + \mathrm{ST} \tag{4.8b}$$

The following form for $T^{\alpha\mu\nu}(p,q)$ is the most general structure, free from kinematical singularities, satisfying the above requirements (remember that $p^2=q^2=0$).

$$\begin{aligned} T^{\alpha\mu\nu}(p,q) &= \epsilon^{\mu\nu\omega\phi}p_\omega q_\phi k^\alpha F_1(k^2) \\ &+ (\epsilon^{\alpha\mu\omega\phi}q^\nu - \epsilon^{\alpha\nu\omega\phi}p^\mu)p_\omega q_\phi F_2(k^2) \\ &+ (\epsilon^{\alpha\mu\omega\phi}p^\nu - \epsilon^{\alpha\nu\omega\phi}q^\mu)p_\omega q_\phi F_3(k^2) \\ &+ \epsilon^{\alpha\mu\nu\omega}(p_\omega - q_\omega)\tfrac{1}{2}k^2 F_3(k^2) \end{aligned} \tag{4.9}$$

It now follows that

$$k_\alpha T^{\alpha\mu\nu}(p,q) = \epsilon^{\mu\nu\omega\phi}\, p_\omega q_\phi k^2\, [F_1(k^2)-F_3(k^2)]\ . \tag{4.10a}$$

Comparison with (4.6a) and (4.1) finally gives

$$T(k^2) = \frac{(\mu^2-k^2)}{F\mu^2}\, k^2\, [F_1(k^2)-F_3(k^2)]\ . \tag{4.10b}$$

As we mentioned, the F_i are free from kinematical singularities; since we are working to lowest order in electromagnetism, they do not possess dynamical singularities at $k^2 = 0$. Hence we find, as promised, $T(0) = 0$. Note that PCAC is not used to obtain the mathematical statement $T(0) = 0$. This hypothesis becomes necessary only when $T(0)$ is related to $T(\mu^2)$. It will now be shown that even the mathematical prediction is invalid in the σ model.

C. Model Calculation

We calculate[2] the off mass shell pion decay constant, $T(k^2)$, in the σ model where all the assumptions of the Sutherland-Veltman theorem seem to be satisfied.[3] The Lagrangian is

$$\begin{aligned}\mathcal{L} &= \bar\psi(i\gamma^\mu\partial_\mu - m)\psi + \tfrac{1}{2}\partial_\mu\phi\partial^\mu\phi - \tfrac{1}{2}\mu^2\phi^2 + \tfrac{1}{2}\partial_\mu\sigma\partial^\mu\sigma - \tfrac{1}{2}(\mu^2+2\lambda F^2)\sigma^2 \\ &\quad + e\bar\psi\gamma^\mu\psi A_\mu + g\bar\psi(\sigma+\phi\gamma_5)\psi - \lambda[(\phi^2+\sigma^2)^2 - 2F\sigma(\sigma^2+\phi^2)]\ .\end{aligned} \tag{4.11}$$

Here ψ, ϕ and σ are fields for the "proton", "pion" and σ particle, each possessing the respective masses m, μ and $(\mu^2+2\lambda F^2)^{1/2}$. The proton interacts with the pion and σ in a chirally symmetric fashion with strength g. The proton also has an electromagnetic interaction; since we work to lowest order in that interaction, it is sufficient to consider the electromagnetic potential A^μ as an external perturbation. There are also meson self-couplings with strength λ, which are necessary for the consistency of the model, but which do not affect the present discussion. The parameter F is equal to $2mg^{-1}$. All isospin effects are ignored, since they are irrelevant to the argument.

The model possesses a neutral axial current J_5^α whose divergence according to the equations of motion of the theory is the pion field.

$$J_5^\alpha = i\bar{\psi}\gamma^\alpha\gamma^5\psi + 2(\sigma\partial^\alpha\phi - \phi\partial^\alpha\sigma) - F\partial^\alpha\phi \tag{4.12a}$$

$$\partial_\alpha J_5^\alpha = \mu^2 F\phi \tag{4.12b}$$

The electromagnetic current $J^\mu = \bar{\psi}\gamma^\mu\psi$ and the axial current satisfy conventional current commutators. Of course, no ST is present canonically, in the time component algebra so we cannot ascertain whether or not Feynman's conjecture is satisfied.

In this theory the pion can decay into two photons by dissociating first into a virtual proton-antiproton pair, which then emits two photons. The lowest order graphs are those of Fig. IV-1. These have the integral representation

$$T^{\mu\nu}(p,q) = \Gamma^{\mu\nu}(p,q) + \Gamma^{\nu\mu}(q,p)$$

$$\Gamma^{\mu\nu}(p,q) = ige^2\int\frac{d^4r}{(2\pi)^4}\,\mathrm{Tr}\,\gamma^5[\gamma_\alpha r^\alpha+\gamma_\alpha p^\alpha-m]^{-1}$$

$$\gamma^\mu[\gamma_\alpha r^\alpha-m]^{-1}\gamma^\nu[\gamma_\alpha r^\alpha-\gamma_\alpha q^\alpha-m]^{-1} \tag{4.13a}$$

The integral appears to diverge linearly; however, after the trace is performed one is left with a finite expression.

$$\Gamma^{\mu\nu}(p,q) = 4mi\,ge^2\epsilon^{\mu\nu\alpha\beta}\,p_\alpha q_\beta\int\frac{d^4r}{(2\pi)^4}$$

$$[(r+p)^2-m^2]^{-1}[r^2-m^2]^{-1}[r-q)^2-m^2]^{-1} \tag{4.13b}$$

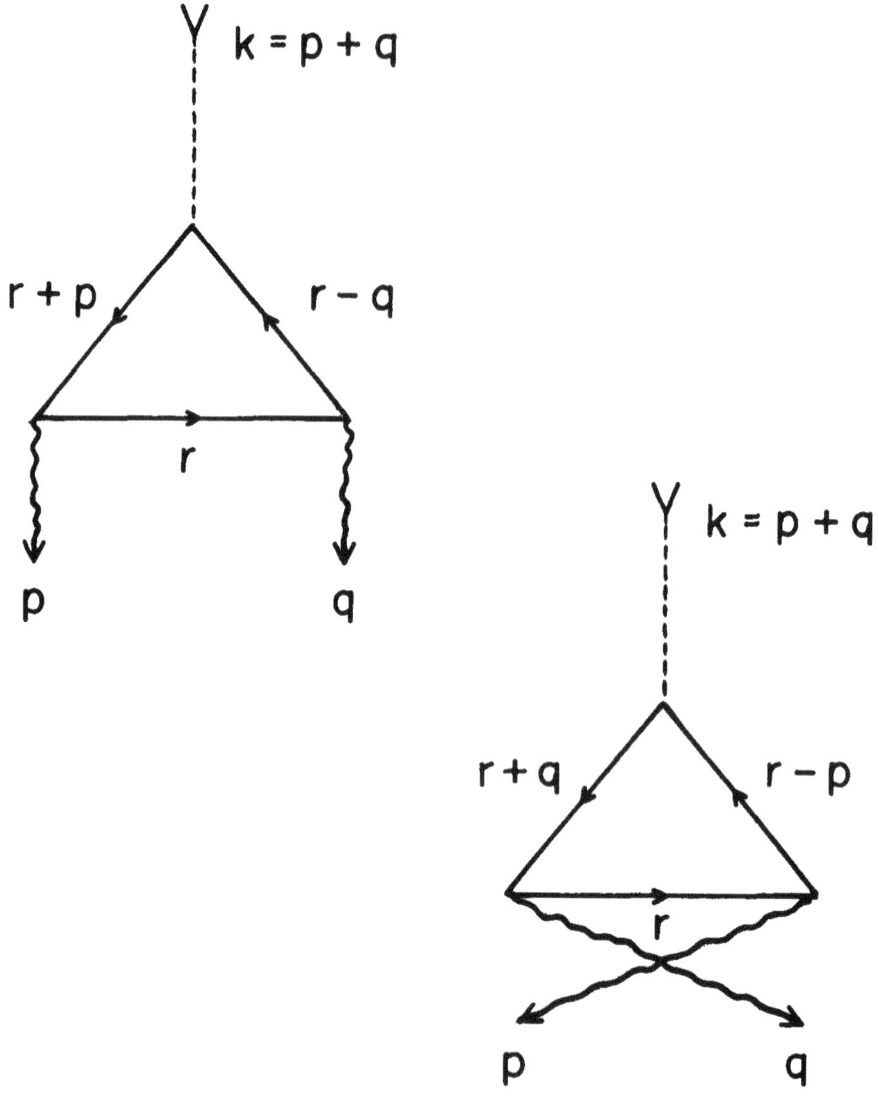

Fig. IV–1

Feynman graphs describing the $\pi^0 \to 2\gamma$ amplitude, in lowest order for the theory given by Eq. (4.11).

The remaining evaluation is elementary.[4] The answer is

$$\Gamma^{\mu\nu}(p,q) = \frac{mg\, e^2}{4\pi^2}\, \epsilon^{\mu\nu\alpha\beta} \int_0^1 dx \int_0^{1-x} dy\, [m^2 - k^2 xy]^{-1} \quad . \tag{4.13c}$$

In the notation (4.1) we find

$$T(k^2) = \frac{mg\, e^2}{2\pi^2} \int_0^1 dx \int_0^{1-x} dy\, [m^2 - k^2 xy]^{-1}$$

$$T(0) = \frac{ge^2}{4\pi^2 m} = \frac{e^2}{2\pi^2 F} \neq 0 \quad . \tag{4.14}$$

For future reference, note that the large m behavior of $T(k^2)$ is $ge^2/4\pi^2 m$.

Our calculation has demonstrated the falseness of the *mathematical* portion of the Sutherland-Veltman theorem. Since $T(k^2)$ is perfectly smooth for small k^2, $T(k^2) \approx T(0)\,(1 + \frac{1}{12}\frac{k^2}{m^2})$, we see that the *experimental* part of that theorem is also incorrect in this model. The reason does not lie in any unexpected rapid variation, but rather in the failure of conventional current algebra.

D. Anomalous Ward Identity

To gain further understanding into the problem, we calculate the function $T^{\alpha\mu\nu}(p,q)$, (4.6b). The relevant Feynman graphs are given in Fig. IV-2. In the first two graphs, Fig. IV-2a, the axial current attaches directly to the Fermion loop. In the last two, Fig. IV-2b, it passes first through the virtual pion with the coupling $2mg^{-1}$, thus acquiring the necessary pion pole. The integral representation is

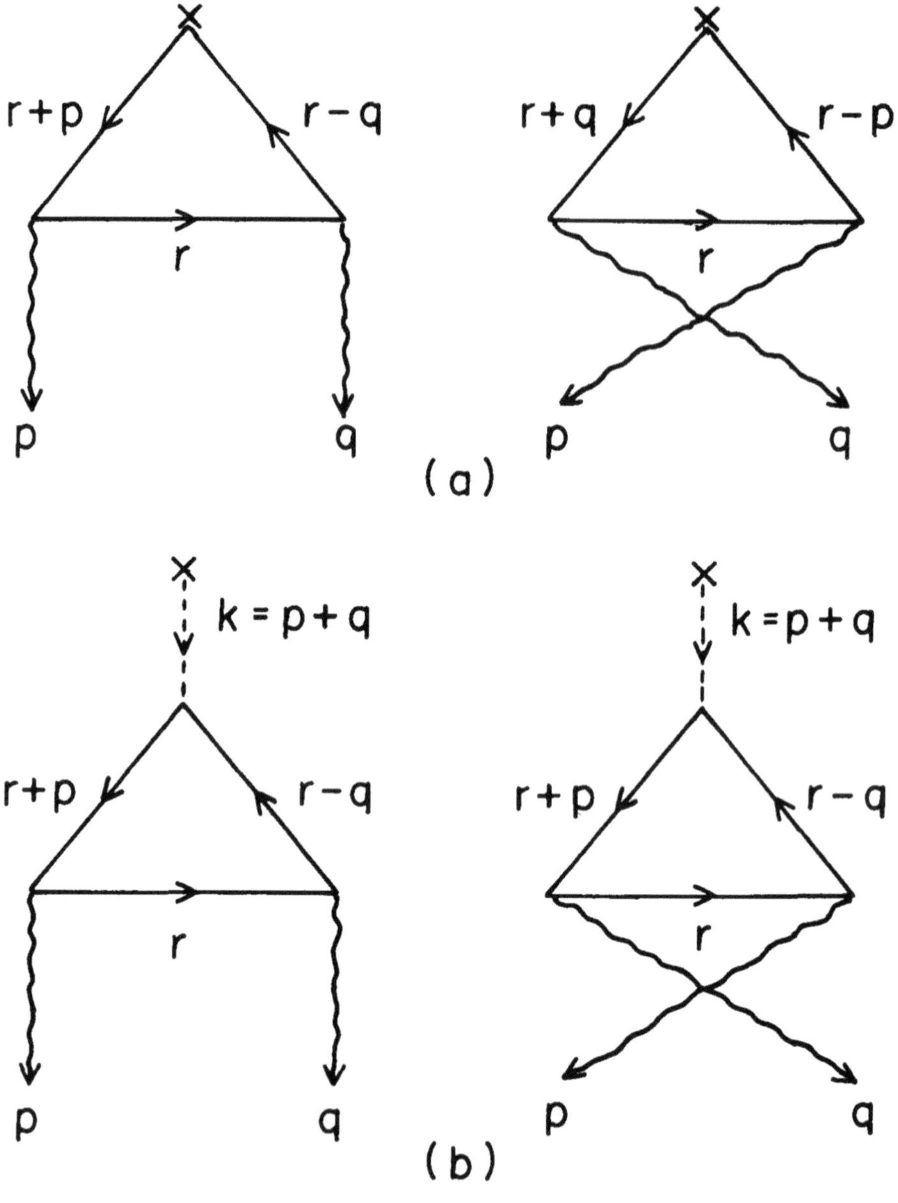

Fig. IV–2

Feynman graphs describing the $J_5^\alpha \to 2\gamma$ amplitude, in lowest order for the theory given by Eq. (4.11).

(4.15a) $$T^{\alpha\mu\nu}(p,q) = T_1{}^{\alpha\mu\nu}(p,q) + T_2{}^{\alpha\mu\nu}(p,q)$$

(4.15b) $$T_1{}^{\alpha\mu\nu}(p,q) = \Gamma^{\alpha\mu\nu}(p,q) + \Gamma^{\alpha\nu\mu}(q,p)$$

(4.15c) $$\Gamma^{\alpha\mu\nu}(p,q) = ie^2 \int \frac{d^4r}{(2\pi)^4} \mathrm{Tr}\gamma^5\gamma^\alpha[\gamma_\beta r^\beta + \gamma_\beta p^\beta - m]^{-1} \gamma^\mu[\gamma_\beta r^\beta - m]^{-1}\gamma^\nu[\gamma_\beta r^\beta - \gamma_\beta q^\beta - m]^{-1}$$

(4.15d) $$T_2{}^{\alpha\mu\nu}(p,q) = -\frac{2mg^{-1}}{k^2-\mu^2} k^\alpha T^{\mu\nu}(p,q) \ .$$

Evidently the verification of the axial Ward identity

(4.16a) $$(F\mu^2)^{-1}(\mu^2-k^2)k_\alpha T^{\alpha\mu\nu}(p,q) = T^{\mu\nu}(p,q),$$

which was used in the derivation of the Sutherland-Veltman theorem, is equivalent to showing

(4.16b) $$k_\alpha T_1{}^{\alpha\mu\nu}(p,q) = 2mg^{-1}T^{\mu\nu}(p,q) \ .$$

The vector Ward identity, i.e., gauge invariance, is also necessary for the theorem. In the present notation it is equivalent to

(4.17) $$p_\mu T_1{}^{\alpha\mu\nu}(p,q) = q_\nu T_1{}^{\alpha\mu\nu}(p,q) = 0 \ .$$

We now demonstrate that both (4.16) and (4.17) cannot be maintained simultaneously for $T_1{}^{\alpha\mu\nu}$ as given by (4.15).

The important property of the graphs in Fig. IV-2a which is responsible for this anomalous behavior is their linear divergence. Unlike in the evaluation of $T^{\mu\nu}$, (4.13), performing the trace does not remove this divergence. In a linearly divergent integral it is illegitimate to shift the variable of integration. This is easily seen on one dimension. Consider

(4.18a) $$\Delta(a) = \int_{-\infty}^{\infty} dx[f(x+a)-f(x)] \quad .$$

If one can shift the integration variable in the first integral $x+a \to x$; one can conclude that $\Delta(a) = 0$. To see that $\Delta(a)$ need not vanish, let us expand the integrand.

(4.18b) $$\Delta(a) = \int_{-\infty}^{\infty} dx[af'(x)+ \frac{a^2}{2} f''(x)+ \ldots]$$

Integrating by parts, we find

(4.18c) $$\Delta(a) = a[f(\infty)-f(-\infty)]+ \frac{a^2}{2}[f'(\infty)-f'(-\infty)]+ \ldots \quad .$$

When the integral $\int_{-\infty}^{\infty} dx\, f(x)$ converges (or at most diverges logarithmically) we have $0 = f(\pm\infty) = f'(\pm\infty) \ldots$, and $\Delta(a)$ vanishes. However, for a linearly divergent integral $0 \neq f(\pm\infty)$, $0 = f'(\pm\infty) \ldots$ and $\Delta(a)$ need not vanish.

(4.18d) $$\Delta(a) = a[f(\infty)-f(-\infty)]$$

Such a contribution is called a "surface term". This state of affairs persists in 4-dimensional, Minkowski space integrals. Consider

(4.19a) $$\Delta^{\mu}(a) = i\int \frac{d^4r}{(2\pi)^4} \left[\frac{r^{\mu}+a^{\mu}}{([r+a]^2-m^2)^2} - \frac{r^{\mu}}{(r^2-m^2)^2} \right] \quad .$$

Here a is an arbitrary 4-vector. The surface term may be evaluated. The result is non vanishing; see Exercise 4.1.

(4.19b) $$\Delta^{\mu}(a) = -\frac{a^{\mu}}{32\pi^2}$$

The consequence of this for our problem is that the integral $\Gamma^{\alpha\mu\nu}$ (4.15c), which contributes to $T_1^{\alpha\mu\nu}$, is not uniquely defined. The point is that in exhibiting (4.15c) we have routed the integration momentum r in a particular fashion: the Fermion leg between the two photons carries r. However any other routing could also be chosen, so that the Fermion leg between the photons carries the 4-momentum r + a, where a is an arbitrary four vector. If the integral were not linearly divergent, then a shift of integration would return this routing to the previous one; but in the present instance such shifts produce surface terms.

In conventional evaluations of divergent Feynman graphs, such ambiguities are usually ignored. Typically cut-cffs are introduced, which eliminate these problems and then the cut-offs are removed by the renormalization procedure. For our purposes we need to keep track of all the possible sources of ambiguity. Thus we replace the expression for $\Gamma^{\alpha\mu\nu}$, (4.15c), by a class of expressions, parametrized by an arbitrary 4-vector a^μ.

$$\Gamma^{\alpha\mu\nu}(p,q|a) = \Gamma^{\alpha\mu\nu}(p,q) + \Delta^{\alpha\mu\nu}(p,q|a) \tag{4.20a}$$

$$\begin{aligned}\Delta^{\alpha\mu\nu}(p,q|a) = ie^2\int\frac{d^4r}{(2\pi)^4}\,&\mathrm{Tr}\gamma^5\gamma^\alpha\{[\gamma_\beta r^\beta+\gamma_\beta a^\beta+\gamma_\beta p^\beta-m]^{-1}\\ &\gamma^\mu[\gamma_\beta r^\beta+\gamma_\beta a^\beta-m]^{-1}\gamma^\nu[\gamma_\beta r^\beta+\gamma_\beta a^\beta-\gamma_\beta q^\beta-m]^{-1}\\ &-[\gamma_\beta r^\beta+\gamma_\beta p^\beta-m]^{-1}\gamma^\mu[\gamma_\beta r^\beta-m]^{-1}\\ &\gamma^\nu[\gamma_\beta r^\beta-\gamma_\beta q^\beta-m]^{-1}\}\end{aligned} \tag{4.20b}$$

The surface term is evaluated, see Exercise 4.2.

$$\Delta^{\alpha\mu\nu}(p,q|a) = -\frac{e^2}{8\pi^2}\epsilon^{\alpha\mu\nu\beta}a_\beta \tag{4.21}$$

The arbitrariness of a_β is limited somewhat by the plausibility requirement that no vectors, other than those already present in the problem at hand, should be introduced. Hence we set $a_\beta = (a+b)p_\beta + bq_\beta$. Corresponding to the class of functions $\Gamma^{\alpha\mu\nu}(p,q|a)$ we have a class of functions $T_1^{\alpha\mu\nu}(p,q|a)$. Accordingly (4.15b) and (4.21)

$$(4.22) \qquad T_1^{\alpha\mu\nu}(p,q|a) = T_1^{\alpha\mu\nu}(p,q) - \frac{e^2}{8\pi^2}\, a\epsilon^{\alpha\mu\nu\beta}(p_\beta - q_\beta) \ .$$

Note that Bose symmetry has been maintained.

Any member of the class of functions $T_1^{\alpha\mu\nu}(p,q|a)$ may be considered "correct". The various functions differ among themselves only by a polynomial in p and q, i.e., by a covariant seagull. We now attempt to determine a by imposing the axial and the vector Ward identities. We are hoping that $T_1^{\alpha\mu\nu}(p,q|a)$, for some definite value of a, will satisfy these identities. It will be seen that no such value for a exists.

It is possible to evaluate $T_1^{\alpha\mu\nu}(p,q)$ as given by (4.15), and therefore to exhibit an explicit formula for $T_1^{\alpha\mu\nu}(p,q|a)$. The evaluation is effected by conventional methods, except it must be always remembered that shifts of integration variables produce non-vanishing, but well defined terms. A remarkable thing that occurs is that the end result is finite; symmetric integration removes the linear divergences as well as the subdominant logarithmic divergence. Thus a finite, unambiguous formula for $T_1^{\alpha\mu\nu}(p,q|a)$ may be arrived at. (The illegitimacy of shifts of integration follows from the *superficial* divergence properties of integral, even if accidentally the result is finite.) The detailed evaluation of $T_1^{\alpha\mu\nu}(p,q)$ has been given in the literature[5]; for our present purposes of verifying the Ward identities we do not need this formula, the integral representation (4.15) will suffice.

Consider first the axial Ward identity. We wish to learn the form of $k_\alpha T_1^{\alpha\mu\nu}(p,q)$. From (4.15c) it follows that

$$k_\alpha \Gamma^{\alpha\mu\nu}(p,q) = (p_\alpha + q_\alpha)\Gamma^{\alpha\mu\nu}$$

$$= ie^2 \int \frac{d^4r}{(2\pi)^4} \operatorname{Tr}\gamma^5(\gamma_\beta p^\beta + \gamma_\beta q^\beta)[\gamma_\beta r^\beta + \gamma_\beta p^\beta - m]^{-1}$$

(4.23a)
$$\gamma^\mu[\gamma_\beta r^\beta - m]^{-1}\gamma^\nu[\gamma_\beta r^\beta - \gamma_\beta q^\beta - m]^{-1} .$$

After rewriting $\gamma_\beta p^\beta + \gamma_\beta q^\beta$ as $2m + (\gamma_\beta p^\beta + \gamma_\beta r^\beta - m) - (\gamma_\beta r^\beta - \gamma_\beta q^\beta + m)$ we have

$$k_\alpha \Gamma^{\alpha\mu\nu}(p,q) = (2mg^{-1}) ige^2 \int \frac{d^4r}{(2\pi)^4} \gamma^5[\gamma_\beta r^\beta + \gamma_\beta p^\beta - m]^{-1}$$

$$\gamma^\mu[\gamma_\beta r^\beta - m]^{-1}\gamma^\nu[\gamma_\beta r^\beta - \gamma_\beta q^\beta - m]^{-1}$$

$$+ ie^2 \int \frac{d^4r}{(2\pi)^4} \operatorname{Tr}\gamma^5\gamma^\mu[\gamma_\beta r^\beta - m]^{-1}\gamma^\nu[\gamma_\beta r^\beta - \gamma_\beta q^\beta - m]^{-1}$$

$$- ie^2 \int \frac{d^4r}{(2\pi)^4} \operatorname{Tr}\gamma^5[\gamma_\beta r^\beta - \gamma_\beta q^\beta + m]\,[\gamma_\beta p^\beta + \gamma_\beta r^\beta - m]^{-1}$$

(4.23b)
$$\gamma^\mu[\gamma_\beta r^\beta - m]^{-1}\gamma^\nu[\gamma_\beta r^\beta - \gamma_\beta q^\beta - m]^{-1} .$$

The first integral is recognized as $2mg^{-1}$ times $\Gamma^{\mu\nu}(p,q)$; see (4.13). In the third integral $\gamma_\beta r^\beta - \gamma_\beta q^\beta + m$ may be taken through the γ^5, thus changing the overall sign and the sign of m. Then the cyclicity of the trace allows one to transpose that term to the end of that expression, thus cancelling the last propagator. We are now left with

$$k_\alpha \Gamma^{\alpha\mu\nu}(p,q) = 2mg^{-1}\Gamma^{\mu\nu}(p,q)$$

$$+ ie^2 \int \frac{d^4r}{(2\pi)^4} \operatorname{Tr}\gamma^5\gamma^\mu[\gamma_\beta r^\beta - m]^{-1}\gamma^\nu[\gamma_\beta r^\beta - \gamma_\beta q^\beta - m]^{-1}$$

(4.23c)
$$+ ie^2 \int \frac{d^4r}{(2\pi)^4} \operatorname{Tr}\gamma^5[\gamma_\beta p^\beta + \gamma_\beta r^\beta - m]^{-1}\gamma^\mu[\gamma_\beta r^\beta - m]^{-1}\gamma^\nu$$

Each of the two integrals must vanish since it is impossible to form a two index pseudotensor which depends on only one vector. We find therefore

$$k_\alpha T_1^{\alpha\mu\nu}(p,q) = 2mg^{-1}T^{\mu\nu}(p,q)$$

$$k_\alpha T_1^{\alpha\mu\nu}(p,q|a) = 2mg^{-1}T^{\mu\nu}(p,q) + \frac{ae^2}{4\pi^2}\epsilon^{\mu\nu\alpha\beta}p_\alpha q_\beta \tag{4.24}$$

The conclusion is that in order to satisfy the axial Ward identity the routing of the integration variable must be as in Fig. 2; i.e., a must be set to zero. Note that this verification required no shifts of integration variable. The vector Ward identity, i.e., gauge invariance cannot be established in the same fashion, as we now demonstrate.

We wish to learn the form of $p_\mu T_1^{\alpha\mu\nu}(p,q)$. According to (4.15)

$$\begin{aligned} p_\mu T_1^{\alpha\mu\nu}(p,q) &= p_\mu \Gamma^{\alpha\mu\nu}(p,q) + p_\mu \Gamma^{\alpha\mu\nu}(q,p) \\ &= ie^2 \int \frac{d^4 r}{(2\pi)^4}\, \mathrm{Tr}\, \gamma^5 \gamma^\alpha [\gamma_\beta r^\beta + \gamma_\beta p^\beta - m]^{-1} \\ &\qquad \gamma_\beta p^\beta\, [\gamma_\beta r^\beta - m]^{-1} \gamma^\nu\, [\gamma_\beta r^\beta - \gamma_\beta q^\beta - m]^{-1} \\ &\quad + ie^2 \int \frac{d^4 r}{(2\pi)^4}\, \mathrm{Tr}\, \gamma^5 \gamma^\alpha [\gamma_\beta r^\beta + \gamma_\beta q^\beta - m]^{-1} \\ &\qquad \gamma^\nu [\gamma_\beta r^\beta - m]^{-1} \gamma_\beta p^\beta [\gamma_\beta r^\beta - \gamma_\beta p^\beta - m]^{-1} . \end{aligned} \tag{4.25a}$$

Use of the identities

$$[\gamma_\beta r^\beta + \gamma_\beta p^\beta - m]^{-1} \gamma_\beta p^\beta [\gamma_\beta r^\beta - m]^{-1} = [\gamma_\beta r^\beta - m]^{-1} - [\gamma_\beta r^\beta + \gamma_\beta p^\beta - m$$

$$[\gamma_\beta r^\beta - m]^{-1} \gamma_\beta p^\beta [\gamma_\beta r^\beta - \gamma_\beta p^\beta - m]^{-1} = [\gamma_\beta r^\beta - \gamma_\beta p^\beta - m]^{-1} - [\gamma_\beta r^\beta - m]^{-1}$$

allows (4.25a) to be written as

$$p_\mu T_1^{\alpha\mu\nu}(p,q) = ie^2\int\frac{d^4r}{(2\pi)^4}\,\mathrm{Tr}\gamma^5\gamma^\alpha[\gamma_\beta r^\beta - m]^{-1}\gamma^\nu[\gamma_\beta r^\beta - \gamma_\beta q^\beta - m]^{-1}$$

$$- ie^2\int\frac{d^4r}{(2\pi)^4}\,\mathrm{Tr}\gamma^5\gamma^\alpha[\gamma_\beta r^\beta + \gamma_\beta p^\beta - m]^{-1}\gamma^\nu[\gamma_\beta r^\beta - \gamma_\beta q^\beta - m]^{-1}$$

$$+ ie^2\int\frac{d^4r}{(2\pi)^4}\,\mathrm{Tr}\gamma^5\gamma^\alpha[\gamma_\beta r^\beta + \gamma_\beta q^\beta - m]^{-1}\gamma^\nu[\gamma_\beta r^\beta - \gamma_\beta p^\beta - m]^{-1}$$

(4.25b)
$$- ie^2\int\frac{d^4r}{(2\pi)^4}\,\mathrm{Tr}\gamma^5\gamma^\alpha[\gamma_\beta r^\beta + \gamma_\beta q^\beta - m]^{-1}\gamma^\nu[\gamma_\beta r^\beta - m]^{-1} .$$

The first and last integrals in (4.25b) vanish because they are two index pseudotensors depending on one vector. The remaining two integrals could be made to cancel against each other if shifts of integration were allowed. Unfortunately such shifts are lead to a finite contribution. The value of the surface term evaluated is; see Exercise 4.3; one finds

(4.25c)
$$p_\mu T_1^{\alpha\mu\nu}(p,q) = \frac{e^2}{4\pi^2}\epsilon^{\alpha\mu\nu\beta}p_\mu q_\beta .$$

Therefore

(4.26a)
$$p_\mu T_1^{\alpha\mu\nu}(p,q|a) = \frac{e^2}{4\pi^2}\epsilon^{\alpha\mu\nu\beta}p_\mu q_\beta[1+\frac{a}{2}] .$$

Bose symmetry, which has been maintained all along, insures a similar Ward identity in the ν index.

(4.26b)
$$q_\nu T_1^{\alpha\mu\nu}(p,q|a) = -\frac{e^2}{4\pi^2}e^{\alpha\mu\nu\beta}p_\nu q_\beta[1+\frac{a}{2}]$$

It is seen that the choice for a which insures the vector Ward identity, $a = -2$, is different from the choice that insures the axial Ward identity, $a = 0$. The conclusion is that there is no way of evaluating $T_1^{\alpha\mu\nu}(p,q)$ so that both Ward identities are satisfied. This remarkable result is even more striking when it is remembered that $\Gamma_1^{\alpha\mu\nu}(p,q)$ is not divergent in the explicit evaluation.

One might inquire whether it is possible to add to $T_1^{\alpha\mu\nu}$ a further seagull, which then would restore both Ward identities. If such a seagull were to exist, one would gladly insert it into the definition of $T_1^{\alpha\mu\nu}$ even though it did not arise "naturally" from the integration. It should be clear that no such further additions are possible. Any seagull one adds must be a three index pseudotensor, and a polynomial in p and q. Bose symmetry limits it to be proportional to $\epsilon^{\alpha\mu\nu\beta}(p_\beta - q_\beta)$. This is precisely the arbitrariness which we have previously allowed for; see (4.22); and it is not sufficient to establish both Ward identities.

Faced with the impossibility of maintaining both Ward identities, we must decide which one we shall accept and which one we shall abandon, i.e., we wish to choose a. It is recognized that the vector Ward identity is a consequence of gauge invariance, while the axial Ward identity is a consequence of an equation of motion, $\partial_\mu J_5^\mu = F\mu^2\phi$. Clearly the former is a much more important principle, and a should be set equal to -2. If there were a physical principle which assured the conservation of the axial current as well, we would be faced with a much more problematical situation. Thus we should be grateful that massless neutral pions do not, in fact, occur in nature. We conclude, therefore, that the reason for the violation of the Sutherland-Veltman theorem, $T(0) = 0$, is the violation of the axial Ward identity. Once a modified Ward identity is used, the Sutherland-Veltman theorem is modified, and the new conclusion agrees with the explicit evaluation. With the choice for a which assures gauge invariance, the Ward identities are

$$p_\mu T^{\alpha\mu\nu}(p,q) = q_\nu T^{\alpha\mu\nu}(p,q) = 0 \tag{4.27a}$$

and

$$k_\alpha T^{\alpha\mu\nu}(p,q) = \frac{F\mu^2}{\mu^2-k^2} T^{\mu\nu}(p,q) - \frac{e^2}{2\pi^2} \epsilon^{\mu\nu\alpha\beta} p_\alpha q_\beta \ . \tag{4.27b}$$

The Sutherland-Veltman derivation is now modified at the crucial step (4.6a). Instead of that equation, we have

$$\epsilon^{\mu\nu\alpha\beta} p_\alpha q_\beta T(k^2) = T^{\mu\nu}(p,q)$$

$$= \frac{\mu^2-k^2}{F\mu^2}\left[k_\alpha T^{\alpha\mu\nu}(p,q) + \frac{e^2}{2\pi^2}\epsilon^{\mu\nu\alpha\beta} p_\alpha q_\beta\right] \tag{4.28}$$

The first term in the brackets is as before; therefore

$$T(k^2) = \frac{\mu^2-k^2}{F\mu^2}\left[k^2[F_1(k^2)-F_3(k^2)] + \frac{e^2}{2\pi^2}\right]$$

$$T(0) = \frac{e^2}{2\pi^2 F} \ . \tag{4.29}$$

This agrees with the explicit calculations, (4.14).

The phenomenon of the violation of a Ward identity in perturbation theory should be familiar from quantum electrodynamics. For example, the vacuum polarization tensor and the photon-photon scattering amplitude, as calculated perturbatively in spinor electrodynamics, are not transverse to the photon momenta as they should be. The conventional way of restoring gauge invariance is by the Pauli-Villars regulator technique. It is instructive to demonstrate the workings of that technique in the present context.

Recall that according to the Pauli-Villars regulator method, an amplitude involving a loop integration is considered to be a function of the mass of the particles circulating in the loop. A "regulated" amplitude is defined as the difference between the given amplitude and the same amplitude with the mass evaluated at a "regulator" mass. Finally the physical amplitude is regained by letting the regulator mass pass to infinity. Thus for the pion decay amplitude we have

$$T^{\mu\nu}_{\text{Reg}}(p,q) = T^{\mu\nu}(p,q|m) - T^{\mu\nu}(p,q|M) \tag{4.30a}$$

$$T^{\mu\nu}_{\text{Physical}}(p,q) = \lim_{M\to\infty} T^{\mu\nu}_{\text{Reg}}(p,q) \ . \tag{4.30b}$$

According to (4.14), $T^{\mu\nu}(p,q|M)$ vanishes for large M as $\epsilon^{\mu\nu\alpha\beta} p_\alpha q_\beta \dfrac{ge^2}{4\pi^2 M}$, hence

$$T^{\mu\nu}_{\text{Physical}}(p,q) = T^{\mu\nu}(p,q|m) \ . \tag{4.30c}$$

This is as it should be, since $T^{\mu\nu}(p,q)$ was evaluated unambiguously from a finite integral. For the axial current amplitude on the other hand, we have

$$T^{\alpha\mu\nu}_{1,\text{Reg}}(p,q|a) = T_1^{\alpha\mu\nu}(p,q|a|m) - T_1^{\alpha\mu\nu}(p,q|a|M) \tag{4.31a}$$

$$T^{\alpha\mu\nu}_{1,\text{Physical}}(p,q) = \lim_{M\to\infty} T^{\alpha\mu\nu}_{1,\text{Reg}}(p,q|a) \ . \tag{4.31b}$$

Consider now the vector Ward identity. According to (4.26a)

$$\begin{aligned} p_\mu T^{\alpha\mu\nu}_{1,\text{Reg}}(p,q|a) &= \frac{e^2}{4\pi^2} \epsilon^{\alpha\mu\nu\beta} p_\mu q_\beta [1+\tfrac{a}{2}] \\ &- \frac{e^2}{4\pi^2} \epsilon^{\alpha\mu\nu\beta} p_\mu q_\beta [1+\tfrac{a}{2}] = 0 \end{aligned} \tag{4.32a}$$

(4.32b) $$p_\mu T^{\alpha\mu\nu}_{1,\text{Physical}}(p,q) = 0 \ .$$

For the axial Ward identity, we have according to (4.24) and (4.30c)

$$k_\alpha T^{\alpha\mu\nu}_{1,\text{Reg}}(p,q|a) = 2mg^{-1}T^{\mu\nu}(p,q|m) + \frac{ae^2}{4\pi^2}\epsilon^{\mu\nu\alpha\beta} p_\alpha q_\beta$$

$$- 2Mg^{-1}T^{\mu\nu}(p,q|M) - \frac{ae^2}{4\pi^2}\epsilon^{\mu\nu\alpha\beta} p_\alpha q_\beta$$

(4.33a) $$= 2mg^{-1}T^{\mu\nu}_{\text{Physical}}(p,q) - 2Mg^{-1}T^{\mu\nu}(p,q|M)$$

(4.33b) $$k_\alpha T^{\alpha\mu\nu}_{1,\text{Physical}}(p,q) = 2mg^{-1}T^{\mu\nu}_{\text{Physical}}(p,q) - \lim_{M\to\infty} 2Mg^{-1}T^{\mu\nu}(p,q|M) \ .$$

Since $\lim_{M\to\infty} 2Mg^{-1}T^{\mu\nu}(p,q|M) = \frac{e^2}{2\pi^2}\epsilon^{\mu\nu\alpha\beta} p_\alpha q_\beta$, we are left with

(4.33c) $$k_\alpha T^{\alpha\mu\nu}_{1,\text{Physical}}(p,q) = 2mg^{-1}T^{\mu\nu}_{\text{Physical}}(p,q) - \frac{e^2}{2\pi^2}\epsilon^{\mu\nu\alpha\beta} p_\alpha q_\beta \ .$$

It is seen that the Pauli-Villars technique automatically evaluates the gauge invariant expression for the amplitude. It selects $a = -2$, which, as we have seen, leads to a violation of the axial Ward identity.

In conclusion we remark that the troubles we found with the matrix element of the axial current are not restricted to the σ model. It is clear that it is the triangle graph which leads to difficulties. Such a graph occurs in quantum electrodynamics, in a quark model, indeed in any model which possesses an axial current which is bilinear in Fermion fields. This observation will permit us to generalize the present results beyond the specific σ model.[6]

E. Anomalous Commutators

We have seen that the evaluation of the vector, vector, axial vector triangle graph, Fig. IV-2a, results in a formula for $T_1^{\alpha\mu\nu}(p,q)$ which does not satisfy the Ward identities one would naively expect. Our next task is to understand the breakdown of the Ward identities in terms of the anomalous commutators, which must be responsible for this state of affairs.[7]

According to the BJL theorem, the ETC between the various currents may be evaluated from the high energy behavior of the triangle graph. Evidently we now must go off the photon mass shell $p^2 = q^2 = 0$, so that the time component of the 4 momentum can be sent to infinity independently, as is required by the BJL technique. It turns out, for our purposes, to be sufficient to go off mass shell for one photon only. Thus we are led to consider

$$\bar{T}_1^{\alpha\mu}(p,q) = -ie\int d^4x\ e^{-ipx} < 0|TJ^\mu(x)\bar{J}_5^\alpha(0)|\gamma q >$$

(4.34a)
$$= \bar{T}_1^{\alpha\mu\nu}(p,q)\epsilon_\nu{}'(q)$$

$$\bar{T}_1^{\alpha\mu\nu}(p,q) = -ie\int d^4x d^4y\ e^{-ipx}\ e^{-iqy}$$

(4.34b)
$$< 0|TJ^\mu(x)J^\nu(y)\bar{J}_5^\alpha(0)|0 > \Big|_{\substack{q^2=0 \\ p^2\neq 0}} .$$

Here the bar over $\bar{T}_1^{\alpha\mu}$ and $\bar{T}_1^{\alpha\mu\nu}$ serves to remind that one photon is on the mass shell. The bar over $\bar{J}_5^\alpha$ indicates that we are not considering the full axial current of the σ model, (4.12a), but only the part bilinear in the nucleon fields; thus the lowest order matrix element involves only the problematical triangle graph, Fig. IV-2a. Note also that we are interested in the T product, *not* the covariant T* product. It is the former object that determines the ETC by the BJL definition.

According to the discussion of Chapter III, the following formula for the ETC will enable us to calculate it.

$$\lim_{p_0 \to \infty} \quad p_0 \bar{T}_1^{\alpha\mu}(p,q) =$$

(4.35) $$-e \int d^3x \, e^{i\vec{p}\cdot\vec{x}} < 0|[J^\mu(0,\vec{x}),\bar{J}_5^{\alpha}(0)]|\, \gamma q >$$

Our program, therefore, is the following. We evaluate the triangle graph as before, except that the photon with 4-momentum p is now off mass shell. From the explicit formula for that amplitude, which is a covariant T* product, as it must be since it arises from covariant Feynman rules, we extract the non-covariant T product by dropping all seagulls — all polynomials in p_0. This provides us with an explicit formula for $\bar{T}_1^{\alpha\mu}(p,q) = \bar{T}_1^{\alpha\mu\nu}(p,q)\epsilon_\nu'(q)$. Note that the present evaluation does not suffer from the ambiguities which beset the calculation of $T_1^{\alpha\mu\nu}(p,q)$ in the previous subsection. The reason is that all the previously encountered ambiguities are seagulls, which we are neglecting.

The evaluation of the relevant triangle graph appears in the literature.[8] The integral yields a finite result as before, and the limit indicated in (4.35) is performed. The resulting commutators are summarized by the following formulas.

(4.36a) $$[J^0(t,\vec{x}),J_5^0(t,\vec{y})] = 2c\tilde{F}^{0j}(y)\partial_j\delta(\vec{x}-\vec{y})$$

(4.36b) $$[J^0(t,\vec{x}),J_5^i(t,\vec{y})] = c\tilde{F}^{ij}(y)\partial_j\delta(\vec{x},\vec{y})$$

(4.36c) $$[J^i(t,\vec{x}),J_5^0(t,\vec{y})] = -c\partial_j\tilde{F}^{ij}(x)\delta(\vec{x},\vec{y})+c\tilde{F}^{ij}(y)\partial_j\delta(\vec{x},\vec{y})$$

Here $c = ie/8\pi^2$ and $\tilde{F}^{\mu\nu}$ is the antisymmetric, conserved electromagnetic dual tensor: $\tilde{F}^{\mu\nu} = \epsilon^{\mu\nu\alpha\beta}(\partial_\alpha A_\beta - \partial_\beta A_\alpha)$. In offering (4.36), we do not imply that we have derived the ETC by any operator technique. All that is meant is that the limit (4.35) is non-vanishing and the non-zero expression may be regained by evaluating the appropriate matrix element of (4.36). For example, explicit computation shows that

$$\lim_{p_0 \to \infty} p_0 \bar{T}_1{}^{00}(p,q) = \frac{-e^2}{2\pi^2} \epsilon^{0\mu\nu\alpha} p_\mu q_\nu \epsilon_\alpha{}'(q) \quad . \tag{4.37}$$

The right hand side of (4.37) is also obtained when (4.36a) is inserted into the right hand side of (4.35). Therefore, properly speaking, all we have shown is that the ETC (4.36) has non-canonical contributions whose vacuum-one photon matrix element is equal to the same matrix element of the operators appearing in the right hand side of (4.36).

We see that the ETC between time components has acquired a non-canonical ST, Eq. (4.36a). Therefore according to the discussion of Chapter II, the Feynman conjecture may not be satisfied for both Ward identities. To see that indeed the conjecture is violated,[9] the ETC (4.36) is expressed in the formalism of Chapter II.

$$[J^\mu(x), J_5{}^\nu(y)]\delta([x-y]\cdot n) = C^{\mu\nu}(x;n)\delta^4(x-y) + S^{\mu\nu|\alpha}(y;n)$$

$$P_{\alpha\beta}\partial^\beta \delta^4(x-y) \tag{4.38a}$$

The ST, according to (4.36), is

$$S^{\mu\nu|\alpha}(y;n) = c\tilde{F}^{\mu\gamma}(y)[g_\gamma{}^\alpha n^\nu + g^{\alpha\gamma} n_\gamma] + c\tilde{F}^{\nu\gamma}(y)[g_\gamma{}^\alpha n^\mu + g^{\alpha\mu} n_\gamma] \quad . \tag{4.38b}$$

Equation (4.38b) may be expressed as a total divergence.

$$S^{\mu\nu|\alpha}(y;n) = c\tilde{F}^{\mu\gamma}(y) \frac{\delta}{\delta n_\alpha}[n_\gamma n^\mu] + c\tilde{F}^{\nu\gamma}(y) \frac{\delta}{\delta n_\alpha}[n_\gamma n^\mu] \tag{4.38c}$$

Hence the seagull which covariantizes the T product of J^μ and $J_5{}^\alpha$ is given by

$$\tau^{\mu\nu}(x,y;n) = \int^{n} S^{\mu\nu|\alpha}(y;n')dn'_{\alpha}\delta^4(x-y) + \tau_0^{\mu\nu}(x,y)$$

$$= \tau^{\mu\nu}(y;n)\delta^4(x-y) + \tau_0^{\mu\nu}(x,y)$$

$$\tau^{\mu\nu}(y;n) = c\tilde{F}^{\mu\gamma}(y)n_\gamma n^\nu + c\tilde{F}^{\nu\gamma}(y)n_\gamma n^\mu \quad . \tag{4.39}$$

In the above, as before, $\tau_0^{\mu\nu}(x,y)$ is a covariant seagull, as yet undermined. The covariant quantities $I_1^{\mu\nu}$ and $I_2^{\mu\nu}$ defined from $\tau^{\mu\nu}(y;n)$ by (2.47) may now be evaluated.

$$n_\mu\tau^{\mu\nu}(y;n) = n_\mu I_1^{\mu\nu}(y) = c\tilde{F}^{\nu\gamma}(y)n_\gamma$$

$$n_\nu\tau^{\mu\nu}(y;n) = n_\nu I_2^{\mu\nu}(y) = c\tilde{F}^{\mu\gamma}(y)n_\gamma \tag{4.40a}$$

Evidently we have

$$I_2^{\mu\nu}(y) = -I_1^{\mu\nu}(y) = c\tilde{F}^{\mu\nu}(y) \quad . \tag{4.40b}$$

Finally to evaluate the covariant seagull we make use of the relations (2.48). These require that the following combinations be free of gradients of δ function

$$\frac{\partial}{\partial x^\mu}\tau_0^{\mu\nu}(x,y) - c\tilde{F}^{\mu\nu}(y)\partial_\mu\delta^4(x-y) \quad , \tag{4.41a}$$

$$\frac{\partial}{\partial y^\nu}\tau_0^{\mu\nu}(x,y) - c\tilde{F}^{\mu\nu}(y)\partial_\nu\delta^4(x-y) \quad . \tag{4.41b}$$

The first of the above equations assures the validity of Feynman's conjecture in the μ, vector Ward identity; while the second effects this state of affairs in the ν, axial Ward identity.

It may be verified that a solution to both conditions (4.41) is

(4.42) $$\tau_0^{\mu\nu}(x,y) = c\tilde{F}^{\mu\nu}(y)\delta^4(x-y) + 4c\epsilon^{\mu\nu\alpha\beta}A_\alpha(y)\partial_\beta\delta^4(x-y) \ .$$

Hence it appears that Feynman's conjecture can be satisfied in both indices. However, the seagull (4.42) is unacceptable for the following reason. The explicit dependence of the seagull on the vector potential A_α indicates that gauge invariance has been lost. Recall that all the operators, which we are here considering, are to be sandwiched between the vacuum and one photon state. If one of these operators is the vector potential, then this matrix element will not be gauge invariant. Therefore we must reject this seagull, and content ourselves with one which allows one or the other of the two Ward identities to satisfy Feynman's conjecture. Such a seagull may easily be shown to be

(4.43) $$\tau_0^{\mu\nu}(x,y) = -(1+a)c\tilde{F}^{\mu\nu}(y)\delta^4(x-y) \ .$$

Here a is an arbitrary parameter, which is determined only when it is decided which Ward identity is to be satisfied. The choice $a = -2$ effects cancellation of Schwinger terms and seagulls in the μ, vector identity; while $a = 0$ performs this service in ν, axial identity. If the former choice is made, one then finds

(4.44) $$\frac{\partial}{\partial y^\nu} T^* J^\mu(x)J_5^{\nu}(y) = T^* J^\mu(x)\partial_\nu J_5^{\nu}(y) + 2c\tilde{F}^{\mu\nu}(y)\partial_\nu\delta^4(x-y) \ .$$

The second term on the right hand side of (4.44) is the anomaly.

Thus we have understood why the Ward identities are not satisfied; the ETC between J^μ and J_5^α departs from its canonical value, acquiring non-canonical contributions. These non-canonical terms are consequences of the intrinsic singularities of local field theory. They have the property that naive current algebraic manipulations become invalid.

F. Anomalous Divergence of Axial Current

The anomalies of the triangle graph, which we have understood in terms of non-canonical commutators and modified Ward identities, may also be shown to lead to a modified divergence equation of the neutral, gauge invariant axial current.[10]

$$\partial_\mu J_5^\mu = J_5 + \frac{e^2}{16\pi^2} F^{\mu\nu} \tilde{F}_{\mu\nu} . \tag{4.45}$$

Here J_5 is the naive value of the divergence, derived by application of the equations of motion of whatever model we have under consideration.

We consider the Fermion part of the axial current.

$$J_5^\mu(x) = i\bar{\psi}(x)\gamma^5\gamma^\mu\psi(x) \tag{4.46}$$

Since it is known that the equal time anti-commutator of ψ with $\bar{\psi}$ involves a three dimensional δ function, we must expect that $\lim_{x \to y} \bar{\psi}(x)\psi(y)$ is singular. Hence the definition (4.46) for $J_5^\mu(x)$ is necessarily singular. To regulate this singularity, a small separation is introduced in a preliminary definition for J_5^μ.

$$J_5^\mu(x|\epsilon) = i\bar{\psi}(x+\epsilon/2)\gamma^5\gamma^\mu\psi(x-\epsilon/2) \tag{4.47a}$$

In the presence of electromagnetism, which we shall always consider to be described by an *external* field (i.e., we work to lowest order in electromagnetism), the definition (4.47a) is not gauge invariant. If the electromagnetic potential A^μ is replaced by $A^\mu + \partial^\mu\Lambda$, where Λ is arbitrary, and the Fermion fields are allowed to change correspondingly, $\psi(x) \to e^{ie\Lambda(x)}\psi(x)$, then no changes should occur in quantities of physical interest. The formula (4.47a) does not have this property. A modified expression can be constructed which is gauge invariant.

$$J_5^{\mu}(x|\epsilon|a) = i\bar{\psi}(x+\epsilon/2)\gamma^5\gamma^{\mu}\psi(x-\epsilon/2)$$

$$\exp iea \int_{x-\epsilon/2}^{x+\epsilon/2} A^{\alpha}(y)dy_{\alpha} \tag{4.47b}$$

In (4.47b) a should be set equal to 1 for gauge invariance. However, we prefer to leave this constant unspecified for the time being. The local physical current is obtained by choosing ϵ to be small, averaging over the directions of ϵ and letting $\epsilon^2 = \epsilon_{\mu}\epsilon^{\mu} \to 0$. The method of defining singular products of operators by introducing a small separation is called the "point splitting technique".

We now wish to calculate the divergence of (4.47b). To do so, we need the equation of motion for ψ. We shall here assume that the only interaction is with the external electromagnetic field. More general, interactions have been discussed in the literature.[10]

$$i\gamma_{\mu}\partial^{\mu}4 = m\psi - e\gamma_{\mu}A^{\mu}\psi \tag{4.48}$$

By virtue of (4.48), the divergence of $J_5^{\mu}(x|\epsilon|a)$ is

$$\partial_{\mu}J_5^{\mu}(x|\epsilon|a) = J_5(x|\epsilon|a) - ieJ_5^{\mu}(x|\epsilon|a)$$

$$[A_{\mu}(x+\epsilon/2) - A_{\mu}(x-\epsilon/2) - a\partial_{\mu} \int_{x-\epsilon/2}^{x+\epsilon/2} A_{\nu}(y)dy^{\nu}]$$

$$= J_5(x|\epsilon|a) - ieJ_5^{\mu}(x|\epsilon|a)\epsilon^{\alpha}[\partial_{\alpha}A_{\mu}(x) - a\partial_{\mu}A_{\alpha}(x) + O(\tag{4.49}$$

Here $J_5(x|\epsilon|a)$ is the regulated, split point formula for the naive divergence in this model $2m\bar{\psi}\gamma^5\psi$. The usual naive result $\partial_{\mu}J_5^{\mu} = J_5$ is is regained from (4.49) if ϵ is set to zero, uncritically. Then the last term in (4.49) appears to vanish. This is legitimate when $J_5^\mu(x|\epsilon|a)$ is well behaved as $\epsilon \to 0$. On the other hand, if a matrix element of $J_5^\mu(x|\epsilon|a)$ diverges as $\epsilon \to 0$, a finite result may remain. Since the dimension of J_5^μ is $(\text{length})^{-3}$, one may expect a cubic divergence. However, the pseudovector character of J_5^μ reduces the divergence by two powers, leaving a possible linear divergence. We now show that such a divergence is indeed present, and modifies the naive formula for $\partial_\mu J_5^\mu(x|\epsilon|a)$.

Consider the vacuum expectation value of $\partial_\mu J_5^\mu(x|\epsilon|a)$.

$$< 0|\partial_\mu J_5^\mu(x|\epsilon|a)|0 > \; = \; < 0|J_5(x|\epsilon|a)|0 >$$

$$- \, ie\, \epsilon^\alpha < 0|J_5^\mu(x|\epsilon|a)|0 > [\partial_\alpha A_\mu(x) - a\partial_\mu A_\alpha(x) + 0(\epsilon)] \tag{4.50}$$

The vacuum element of $J_5^\mu(x|\epsilon|a)$ is non-vanishing because it is computed in the presence of an external electromagnetic field. We have for the last term in (4.50)

$$- \, ie\epsilon^\alpha < 0|J_5^\mu(x|\epsilon|a)|0 > \; = \epsilon^\alpha < 0|\bar\psi(x+\epsilon/2)\gamma^5\gamma^\mu\psi(x-\epsilon/2)|0 > \exp iea \int_{x-\epsilon/2}^{x+\epsilon/2} A_\alpha(y)dy^\alpha$$

$$= -\mathrm{Tr}\gamma^5\gamma^\mu\epsilon^\alpha < 0|T\psi(x-\epsilon/2)\bar\psi(x+\epsilon/2)|0 > \exp iea \int_{x-\epsilon/2}^{x+\epsilon/2} A^\alpha(y)dy_\alpha$$

$$= -\mathrm{Tr}\gamma^5\, \gamma^\mu\epsilon^\alpha\, G(x-\epsilon/2, x+\epsilon/2) \exp iea \int_{x-\epsilon/2}^{x+\epsilon/2} A^\alpha(y)dy_\alpha \; . \tag{4.51}$$

The Fermion propagator function, $G(x,y)$, in the external field A^μ, has been introduced. In offering (4.51), ϵ^0 is taken to be positive.

$G(x,y)$ possesses an expansion in powers of A^μ which may be summarized graphically as in Fig. IV-3. The double line represents G; the single line is the free Fermion propagator $S(x)$; while the $\times$ represents an interaction with the external field. $S(x)$ behaves as $1/x^3$ as $x \to 0$. Therefore the successive terms in the series for $G(x,y)$ behave, when $x \to y$, as $(x-y)^{-3}$, $(x-y)^{-2}$, $(x-y)^{-1}$, $\log(x-y)$, etc. For our calculation of $G(x-\frac{1}{2}\epsilon,\, x+\frac{1}{2}\epsilon)$ we need terms which do not vanish, for small ϵ, when multiplied by ϵ. Therefore we set

$$G(x-\tfrac{1}{2}\epsilon,\, x+\tfrac{1}{2}\epsilon) = S(-\epsilon) + ie \int d^4y\; S(x-\tfrac{1}{2}\epsilon-y)\gamma^\alpha S(y-x-\tfrac{1}{2}\epsilon)A_\alpha(y)$$

$$- e^2 \int d^4y\, d^4z\; S(x-\tfrac{1}{2}\epsilon-y)\,\gamma^\alpha\; S(y-z)\gamma^\beta$$

$$S(z-x-\tfrac{1}{2}\epsilon)A_\alpha(y)A_\beta(z) + O(\log\epsilon) \quad . \tag{4.52a}$$

$$S(x) = i\int \frac{d^4p}{(2\pi)^4}\, e^{-ipx}\, [\gamma_\alpha p^\alpha - m]^{-1} \quad . \tag{4.52b}$$

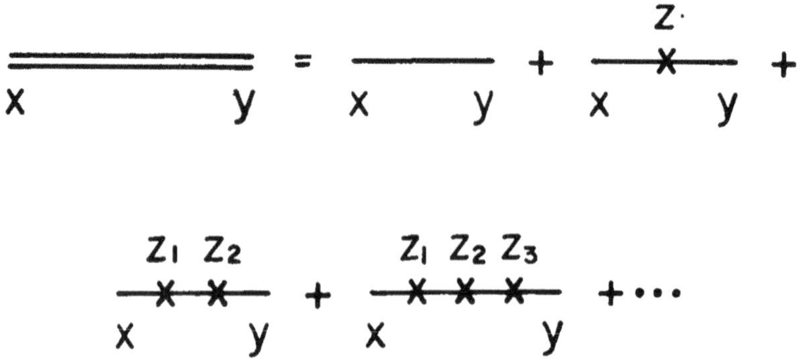

Fig. IV-3

Fermion propagator $G(x,y)$ in an external field.

By C invariance, only the contribution to G which is linear in A^μ is of interest. That term in (4.52a) has the following momentum representation.

$$ie\int \frac{d^4p\,d^4q}{(2\pi)^8}\, e^{i\epsilon p}\, e^{-ixq}\, S(p+\tfrac{1}{2}q)\gamma^a S(p-\tfrac{1}{2}q)A_a(q) \tag{4.53}$$

Therefore (4.51) becomes

$$-\mathrm{Tr}[\epsilon^a\gamma^5\gamma^\mu G(x-\tfrac{1}{2}\epsilon,x+\tfrac{1}{2}\epsilon)]$$

$$= -ie\,\mathrm{Tr}\gamma^5\gamma^\mu\int \frac{d^4p\,d^4q}{(2\pi)^8}\,\epsilon^a\epsilon^{i\epsilon p}\,e^{-ixq}$$

$$S(p+\tfrac{1}{2}q)\gamma^\nu S(p-\tfrac{1}{2}q)A_\nu(q) + 0(\epsilon\log\epsilon)$$

$$= e\,\mathrm{Tr}\gamma^5\gamma^\mu\int \frac{d^4p\,d^4q}{(2\pi)^8}\, e^{i\epsilon p}e^{-ixq}\, A_\nu(q)\frac{\partial}{\partial p_a}$$

$$S(p+\tfrac{1}{2}q)\gamma^\nu S(p-\tfrac{1}{2}q) + 0(\epsilon\log\epsilon) \quad . \tag{4.54a}$$

The last equality follows by integration by parts. We now set ϵ to zero. The p integral is just a surface term; it is easily evaluated by the symmetric methods exemplified in the exercises. The remaining q integral inverts the Fourier transform of $A_\nu(q)$. The final result for (4.54a) is

$$-\mathrm{Tr}[\epsilon^a\gamma^5\gamma^\mu G(x-\tfrac{1}{2}\epsilon,x+\tfrac{1}{2}\epsilon)]_{\epsilon\to 0} = -\frac{e}{16\pi^2}\tilde{F}^{\mu a}(x) \quad . \tag{4.54b}$$

Therefore returning to (4.51) and letting ϵ go to zero, we have

$$< 0|\partial^\mu J_\mu{}^5(x|a)|0 > \; = \; < 0|J_5(x|a)|0 > +$$

$$+ \frac{e^2}{16\pi^2}\frac{(1+a)}{2}\tilde{F}^{\mu\nu}(x)F_{\mu\nu}(x) \quad . \tag{4.55}$$

It is seen that when the gauge invariant definition is selected, $a = 1$, then the divergence of the axial current contains an anomalous term. The naive divergence equation is regained at the expense of gauge invariance if $a = -1$. One may give a simple, heuristic argument which illuminates the origin of this anomalous divergence term. Consider the naive axial current in the model discussed in this subsection. In order to assure gauge invariance, a Pauli-Villars regulator field Ψ is introduced, and correspondingly a regulated axial current is defined.

$$J_5^\mu\Big|_{\text{Reg.}} = J_5^\mu - \mathcal{J}_5^\mu \tag{4.56a}$$

$\mathcal{J}_5^\mu$ is constructed from the regulator fields, in the same fashion as J_5^μ is constructed from the usual fields. The physical axial current is regained by letting the mass M of the regulator field pass to infinity. The divergence of (4.56a) is

$$\partial_\mu J_5^\mu\Big|_{\text{Reg.}} = 2m\bar{\psi}\gamma^5\psi - 2m\bar{\Psi}\gamma^5\Psi \ . \tag{4.56b}$$

Now when $M \to \infty$, the regulator field contribution to (4.56b) may leave a non-vanishing remainder if the matrix elements of $\bar{\Psi}\gamma^5\Psi$ behave as M^{-1} for large M. Detailed calculation shows that this indeed occurs.

Note that the anomalous divergence does not directly affect our previous derivation of the Sutherland-Veltman theorem for $\pi^0 \to 2\gamma$. The amplitude $T^{\mu\nu}$, which is considered in (4.3), is already $O(e^2)$, the two photons having been contracted out of the state. Hence to order e^2 we need not inquire into any modification of $\partial_\mu J_5^\mu$. The anomaly in that argument came from the commutators and seagull, as was explicitly demonstrated. Nevertheless it is possible to use an anomalous divergence equation to give an alternate derivation of the true Sutherland-Veltman theorem.[11] The photons are not contracted out of their state, and we consider (4.2) in conjunction with (4.45).

$$T^{\mu\nu}(p,q) = \epsilon^{\mu\nu\alpha\beta}\, p_\alpha q_\beta T(k^2) = (\mu^2-k^2) < 0|\phi(0)|\gamma,p;\gamma',q >$$

$$= \frac{e^2}{16\pi^2 F\mu^2}(k^2-\mu^2) < 0|\tilde{F}^{\mu\nu}(0)F_{\mu\nu}(0)|\gamma,p;\gamma',q >$$

$$+ \frac{1}{F\mu^2}(\mu^2-k^2)\partial_\alpha < 0|J_5{}^\alpha(0)|\gamma,p;\gamma',q > \tag{4.57a}$$

The last term in (4.57a) is the divergence of a gauge invariant, three index pseudotensor, hence the original Sutherland-Veltman argument applies. We conclude that it will not contribute to T(0). Note that in this derivation we do not pull the divergence through any T* product, so we need not concern ourselves with commutators. The matrix element of the anomaly may be evaluated to lowest order in electromagnetism. Its value is

$$\frac{e^2}{16\pi^2 F\mu^2}(k^2-\mu^2) < 0|\tilde{F}^{\mu\nu}(0)F_{\mu\nu}(0)|\gamma,p;\gamma'q > =$$

$$= \frac{e^2}{2\pi^2 F}\frac{(\mu^2-k^2)}{\mu^2}\epsilon^{\mu\nu\alpha\beta}\, p_\alpha q_\beta \quad . \tag{4.57b}$$

Therefore T(0), as before, is $\frac{e^2}{2\pi^2 F}$.

This exercise shows that the anomalies in commutators, which are encountered in the original derivation, and the anomalous divergence are two sides of the same coin. One must be present when the other is.

G. Discussion

We conclude this treatment of the anomalies of the neutral axial-vector current with a discussion of various disconnected, but important topics.

1) Consider massless spinor electrodynamics.[12] The present arguments indicate that it is impossible to define a conserved gauge invariant

axial current, in spite of the fact that chirality is a symmetry of the theory. Nevertheless there does exist a conserved, gauge invariant axial charge.[11] This charge is constructed as follows. Define

$$\tilde{J}_5^{\mu} = J_5^{\mu} - \frac{e^2}{8\pi^2} \tilde{F}^{\mu\nu} A_\nu \quad . \tag{5.58a}$$

J_5^{μ} is gauge invariant, but its conservation is broken by the anomaly. On the other hand $\tilde{J}_5^{\mu}$ is conserved, but not gauge invariant. The charge Q_5 constructed from $\tilde{J}_5^{\mu}$ is time independent.

$$Q_5 = \int d^3x \tilde{J}_5^{0}(x) \tag{5.58b}$$

Performing a gauge transformation on Q_5: $\delta A_\nu = \partial_\nu \Lambda$, we see that Q_5 is gauge invariant, even though $\tilde{J}_5^{\mu}$ is not.

$$\delta Q_5 = \int d^3x \left(- \frac{e^2}{8\pi^2} \tilde{F}^{0\nu}(x) \right) \partial_\nu \Lambda(x)$$

$$= \int d^3x \left(- \frac{e^2}{8\pi^2} \tilde{F}^{0i}(x) \right) \partial_i \Lambda(x)$$

$$= \int d^3x \frac{e^2}{8\pi^2} \Lambda(x) \partial_i \tilde{F}^{0i}(x)$$

$$= 0 \tag{4.58c}$$

The conservation and antisymmetry of $\tilde{F}^{\mu\nu}$ has been used.

Therefore, in spite of the trouble with the *local* axial current, globally axial symmetry can be implemented in the model. This has the consequence that any property of the model, based on axial symmetry, will be maintained in perturbation theory. For example, the anomaly in the divergence cannot be used to generate a mass for the election.[13]

In massless electrodynamics, one may describe the anomaly as a clash between two symmetry principles: gauge invariance and chirality. In perturbation theory it is impossible to maintain both, though either one can be satisfied. Such a class between the conservation of two symmetry currents has been encountered before in the model field theory of spinor electrodynamics in two dimensions.[14]

2) An important question is whether or not higher order effects modify the anomaly. An argument may be given to the end that in spinor electrodynamics and in the σ model, they do not.[11] The argument is as follows: for definiteness we consider the former theory. To 4th order in e, the axial-vector, vector, vector triangle graph has the insertions represented in Fig. IV-4. If the photon integration is carried out *after* the Fermion loop integration, then the Fermion loop integral is completely convergent. Therefore all shifts of integration, which are required to verify the Ward identities may be performed with impunity. Thus, it is argued, that no anomalies will be present in this or higher orders.

This argument may be criticized because it is somewhat formal.[15] The rules of renormalized perturbation theory require one to perform the photon integrals first, and then the renormalized vertex and propagator corrections are to be inserted into the triangle skeleton. In order to resolve this question, explicit calculations for the graphs in Fig. IV-4 have been performed, and they support the formal argument.[16] It should be noticed, however, that the 4th order contribution is characterized by the fact that all indicated corrections are either self-energy or vertex *insertions*. This feature is not true in higher orders, and perhaps something new will be found there.[17]

3) The models in which the axial anomaly has been exposed: spinor electrodynamics and σ model, do not have much dynamical significance for hadron physics. However, as we have stated before, *any* theory with Fermion fields out of which the axial vector current is constructed, will possess the anomaly, as long as electromagnetixm is coupled minimally.

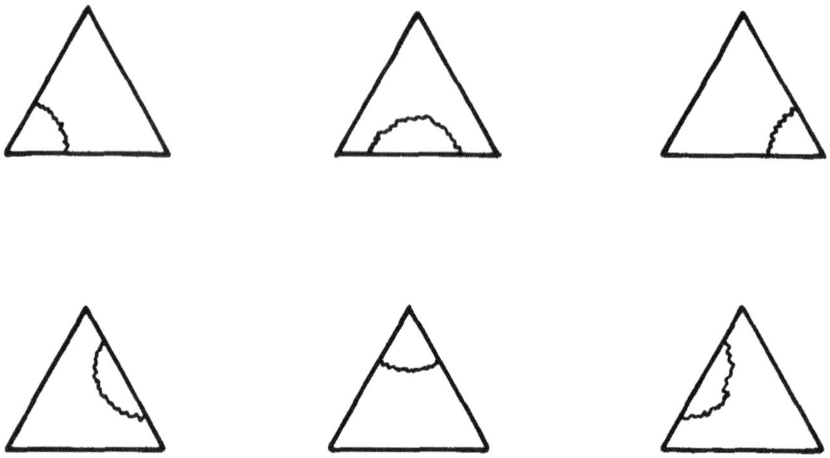

Fig. IV–4

Radiative corrections to the axial-vector, vector vector triangle graph.

Therefore, it is to be expected that in general PCAC should be modified in the presence of electrodynamics. Thus when we consider the neutral member of the octet of axial currents, $\mathcal{F}_3^{\mu}(x)$, (this current is ½ of the previously defined J_5^{μ}) PCAC should be modified by

$$\partial_\mu \mathcal{F}_3^{\mu} = F\mu^2\phi_3 + c\,\frac{e^2}{16\pi^2}\,F^{\mu\nu}\tilde{F}_{\mu\nu} \tag{4.59}$$

Here c is constant which, of course, we cannot derive theoretically in a general fashion. The contribution to c from the triangle graph is determined by the coupling of Fermion fields to the axial and vector currents. In the σ model this contribution is ½. In a general quark triplet model where the charges of the quarks are Q, Q-1 and Q-1, the triangle graph contributes Q-½ to c.[11]

By use of the PCAC hypothesis, and the corrected Sutherland-Veltman theorem, c may be determined experimentally. The currently published $\pi^0 \to 2\gamma$ width of 7.37 ± 1.5 eV sets $|c|$ at 0.44. Further experimental analysis indicates that most likely the sign of c is positive. Thus the theoretical value for c as given by the σ model, triangle graph, $c = \frac{1}{2}$, is in good agreement with the data. If one assumes that in quark models the entire value of c is determined by the triangle graph – a bold hypothesis since one does not know the nature of quark dynamics – $Q = 1$ is preferred, and the coventional quarks with $Q = 2/3$ are excluded.

Perhaps at the present stage of understanding of hadron physics, one should not expect to be able to calculate c theoretically. Just as the coefficient of the usual term in $\partial_\mu \mathcal{F}_3^\mu$ is taken from experiment – F_μ^2 has not been calculated – so also we should content ourselves with an experimental determination of the anomaly.

When c is fitted to the pion data, and a model for SU(3) x SU(3) symmetry breaking is adopted, modified Sutherland-Veltman theorems for $\eta \to 2\gamma$ and $X \to 2\gamma$ may be derived. Such calculations have been performed in the context of the $(3,\bar{3}) \oplus (\bar{3},3)$ symmetry breaking scheme.[18] The results for the η width are consistent with experiment (~ 1 keV), while the X width comes out remarkably enhanced beyond 80 keV. Present experimental data (< 360 keV) provides no check. Such a check would be very interesting, since this large value is very difficult to understand from any different point of view.

4) One may wonder why we speak of a modification of PCAC; why one cannot continue using the divergence of the axial current as the pion interpolating field. The answer lies in part in our model calculations where we found that the term $\tilde{F}^{\mu\nu}F_{\mu\nu}$ is manifestly not smooth when its matrix elements vary off the pion mass shell. It is this property which we have abstracted, and which we assume holds in nature, as well as in models. Our assumption is supported by the observation that the dimension of the anomaly is 4, and there is no reason to believe that such an

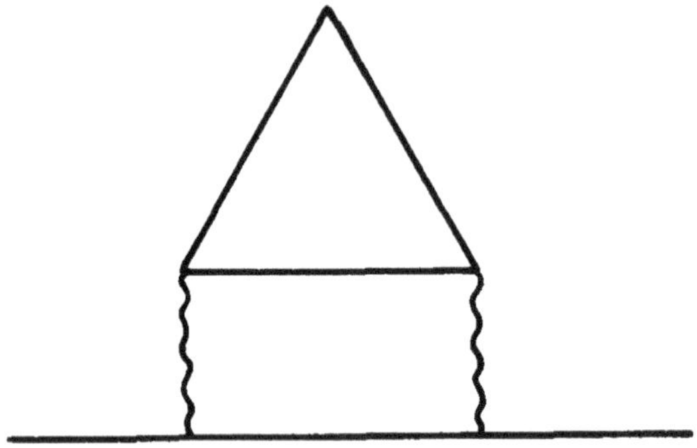

Fig. IV–5

Divergent contribution to axial vector vertex function.

operator is smooth. Finally, by adopting the present philosophy, the experimentally unsatisfactory prediction of Sutherland and Veltman is avoided.

5) The discovery of anomalous Ward identities in the present context engendered a systematic study of all useful Ward identities in SU(3) $\otimes$ SU(3) models.[19] Although other anomalous Ward identities have been found, they seem to be without interest. The only other triangle graph anomaly is in the triple axial vector vertex, which has not, as yet, been used in physical predictions.

6) In spinor electrodynamics, the proper electromagnetic vertex function is renormalized by the same infinite constant which effects electron wave function renormalization. This desirable state of affairs is a consequence of the Ward identity satisfied by that function. Before the discovery of anomalous PCAC it was thought that the proper vertex function of the axial vector current possessed this property as well, as consequence of the axial Ward identity.[20] The anomaly has destroyed this result; the axial vector vertex function remains infinite after wave function

renormalization. The offending graph is the one of Fig. IV-5. One consequence of this is that radiative corrections to neutrino-lepton elastic scattering are infinite.[11]

REFERENCES

[1] D. G. Sutherland, *Nucl. Phys.* B2, 433 (1967); M. Veltman, *Proc. Roy. Soc.* A301, 107 (1967).

[2] The argument here follows J. S. Bell and R. Jackiw, *Nuovo Cimento* 60A, 47 (1969).

[3] M. Gell-Mann and M. Lévy, *Nuovo Cimento* 16, 705 (1960).

[4] The evaluation of this graph was first performed by J. Steinberger, *Phys. Rev.* 76, 1180 (1949). He considered pion decay in the old PS-PS model of π-nucleon interaction. In that theory the pion decays into two photons; the lowest order graphs being given by Fig. 1. It presumably is only an accident that this completely implausible calculation gives a result in excellent agreement with experiment.

[5] This careful and unambiguous evaluation is given in Ref. 2.

[6] A historical note is here in order. The first person to calculate the $\pi^0 \to 2\gamma$ process in field theory was J. Steinberger, Ref. 4. In addition to the PS-PS calculation, where the π-N vertex is γ^5, he also calculated the same amplitude in the PV-PS model, where the π-N vertex is $ik_\alpha \gamma^5 \gamma^\alpha$. The second calculation is identical to our Pauli-Villars regulator method evaluation of $T^{\alpha\mu\nu}$. Steinberger then attempted to verify the equivalence theorem between PS-PS and PV-PS theory, which is based on the formal Ward identity (4.16b); and of course failed to do so. He noted this puzzle, and then ceased being a theoretical physicist. Two years later, J. Schwinger, *Phys. Rev.* 82, 664 (1951), gave an analysis and resolution of the problem. This work was essentially forgotten, and its significance for modern ideas of current

algebra and PCAC was not appreciated. The problem was rediscovered in the σ model by J. S. Bell and R. Jackiw, Ref. 2; and independently and simultaneously in spinor electrodynamics by S. L. Adler, *Phys. Rev.* 177, 2426 (1969). Schwinger's analysis is similar to the one we shall present in subsection F.

[7] This presentation follows R. Jackiw and K. Johnson, *Phys. Rev.* 182, 1459 (1969). The same anomalous commutators have also been given by S. L. Adler and D. G. Boulware, *Phys. Rev.* 184, 1740 (1969)

[8] L. Rosenberg, *Phys. Rev.* 129, 2786 (1963). Rosenberg's formula appears also in Adler's paper, Ref. 6.

[9] That Feynman's conjecture is violated in $\pi^0 \to 2\gamma$ was first pointed out by R. Jackiw and K. Johnson, Ref. 7. The present analysis follows the paper of D. J. Gross and R. Jackiw, *Nucl. Phys.* B14, 269 (1969).

[10] Equation (4.45) was first derived by J. Schwinger, Ref. 3. The same form was given by S. L. Adler, Ref. 6, on the basis of an investigation in spinor electrodynamics. The presentation in this section is analogous to Schwinger's and has been given in the contemporary literature by R. Jackiw and K. Johnson, Ref. 7; C. R. Hagen, *Phys. Rev.* 177, 2622 (1969) and B. Zumino, *Proceedings of Topical Conference on Weak Interactions*, CERN, Geneva, p. 361 (1969).

[11] This analysis is due to S. L. Adler, Ref. 6.

[12] It is possible that such a theory has physical significance. It has been suggested by K. Johnson, R. Willey and M. Baker, *Phys. Rev.* 163, 1699 (1967), that the bare mass of the electron is zero and that the physical mass is entirely of dynamical origin.

[13] For example, it might be thought the anomaly allows a mass to be generated in a massless theory without the intervention of Goldstone Bosons. That this is not the case in theories with Abelian vector mesons has been demonstrated by H. Pagels (private communication).

[14] K. Johnson, *Phys. Letters* 5, 253 (1963).

[15] R. Jackiw and K. Johnson, Ref. 7.

[16] S. L. Adler and W. Bardeen, *Phys. Rev.* 182, 1517 (1969).

[17] Ambiguities in Ward identities containing two loop integrations have been found by E. Abers, D. Dicus and V. Teplitz, *Phys. Rev.* D3, 485 (1971).

[18] S. L. Glashow, R. Jackiw and S. Shei, *Phys. Rev.* 187, 1916 (1969). An alternate treatment of SU(3) × SU(3) symmetry breaking together with axial current anomalies has been considered by G. Gounaris, *Phys. Rev.* D1, 1426 (1970) and *Phys. Rev.* D2, 2734 (1970). Gounaris' theory makes use of a detailed phenomenological Lagrangian. Correspondingly his results are more precise than those of the above reference, though, of course, they are model dependent. Gounaris gives predictions for off mass shell photon processes, as well as for the processes considered by Glashow, Jackiw and Shei. Agreement is obtained for $\pi^0 \to 2\gamma$ and $\eta \to 2\gamma$; but the prediction for $X^0 \to 2\gamma$ is decreased by a factor of 10.

[19] K. Wilson, *Phys. Rev.* 181, 1909 (1969); I. Gerstein and R. Jackiw, *Phys. Rev.* 181, 1955 (1969); W. Bardeen, *Phys. Rev.* 184, 1849 (1969); R. W. Brown, C. C. Shih and B. L. Young, *Phys. Rev.* 186, 1491 (1969); D. Amati, G. Bouchiat and J. L. Gervais, *Nuovo Cimento* 65A, 55 (1970).

[20] G. Preparata and W. Weisberger, *Phys. Rev.* 175, 1965 (1968).

V. ELECTROPRODUCTION SUM RULES

A. Preliminaries

In the electroproduction experiments, an electron is scattered off a nucleon target, typically a proton. The hadronic final states that are observed, are thought to arise, in the context of lowest order electromagnetism, from the inelastic interaction between an off mass shell photon and the proton. The process is depicted in Fig. V-1. By measuring total inelastic cross sections, that is by summing over all final states, one obtains a determination of the one proton connected matrix element of the commutator function of the electromagnetic current.[1]

$$C^{\mu\nu}(p,q) = \int d^4x \; e^{iqx} < p|[J^\mu(x), J^\nu(0)]|p > \tag{5.1}$$

Here $|p>$ is the target state of 4 momentum p^μ, $p^2 = m^2$. We shall assume that an averaging has been performed over the spin states of the proton. The momentum transferred by the lepton pair to the photon is q which is spacelike, $q^2 < 0$, $q_0 > 0$. Since the proton is the lowest state, we have $q^2 \geq -2qp$. We use the symbol ν for $q \cdot p$ and ω for $-q^2/2q \cdot p$. Because the photon is off its mass shell, it has transverse and longitudinal components. Correspondingly one speaks of two total cross sections – the transverse σ_T and the longitudinal σ_L.

Since J^μ is conserved $C^{\mu\nu}(q,p)$ must be transverse in q. A gauge invariant, Lorentz covariant decompositions of $C^{\mu\nu}(q,p)$ is

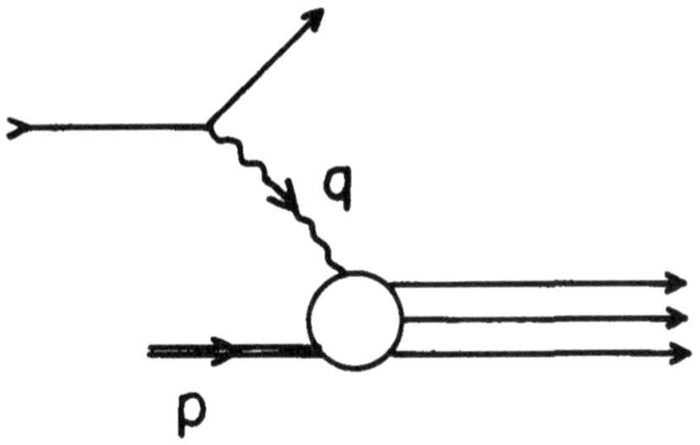

Fig. V-1

Inelastic electron-proton scattering.

$$C^{\mu\nu}(q,p) = -(g^{\mu\nu} - \frac{q^\mu q^\nu}{q^2})\,\tilde{F}_1$$

$$(5.2) \qquad + \frac{1}{p\cdot q}\,(p^\mu - q^\mu \frac{q\cdot p}{q^2})(p^\nu - q^\nu \frac{q\cdot p}{q^2})\,\tilde{F}_2$$

The notation here is slightly unconventional; in the literature $\tilde{F}_1$ is usually called W_1 and $\tilde{F}_2$ corresponds to νW_2. The Lorentz invariant functions $\tilde{F}_i$ depend on q^2 and ν; frequently we shall choose ω and ν or ω and q^2 as the independent variables. The combination $\tilde{F}_2 - 2\omega\tilde{F}_1$ is called $\tilde{F}_L$. The total transverse and longitudinal cross sections are directly expressible in terms of the $\tilde{F}_i$. It is easy to show that the functions $\tilde{F}_i$ are dimensjonless. $\tilde{F}_2$ and $\tilde{F}_L$ are even in ν, $\tilde{F}_1$ is odd.

A remarkable fact can be deduced from the experimental data. At large values of ν and q^2, the $\tilde{F}_i$ cease to depend on ν and q^2 separately; they become functions of only the ratio q^2/ν. (This is even more remarkable when it is observed that the energies at which this phenomenon occurs are not particularly large on the relevant mass scale.) This phenomenon is called the "scaling behavior" and leads to the following hypothesis. It is assumed that the experimental data indicates that as ν gets large, at fixed ω, the limit of $\tilde{F}_i(\omega,\nu)$ exists.[2] We shall here accept this scaling hypothesis

$$\lim_{\nu \to \infty} \tilde{F}_i(\omega,\nu) = F_i(\omega) < \infty \quad . \tag{5.3}$$

One can show that $F_2(\omega) \geq 2\omega F_1(\omega) \geq 0$ since the F_i are proportional to limiting terms of non negative cross sections. The region of large ν and q^2, at fixed ratio has been described as "deep inelastic".

The scaling hypothesis about the existence of scaling functions $F_i(\omega)$ has led to some very interesting applications of current algebra. It has been possible to derive sum rules, which relate properties of the $F_i(\omega)$ to ETC of electromagnetic currents. Our program is to indicate the derivation of the sum rules. Then we shall show that these applications of current algebraic reasoning are not verified by explicit calculations. Thus we shall demonstrate the existence of commutator anomalies in the present context.

B. Derivation of Sum Rules, Naive Method

We shall now derive the two sum rules which will be the objects of the present discussion. Consider the 0i component of (5.2) in the frame where the three vectors $\vec{q}$ and $\vec{p}$ are parallel, $\vec{p} = |\vec{p}|\hat{q}$. Since the starting point and end results are covariant, no loss of generality occurs by working in a particular frame. Now pass into the deep inelastic region by letting $p_0 \to \infty$, subject to the constraints $p^2 = m^2$, i.e., $p_0 \simeq |\vec{p}|$; and ω constant, i.e.,

$$\omega = -\frac{q_0{}^2 - \vec{q}^{\,2}}{2(|\vec{p}|q_0 - |\vec{p}||\vec{q}|)} = -\frac{1}{2|\vec{p}|}(q_0 + |\vec{q}|) \; ,$$

$$q_0 = -2p_0\omega - |\vec{q}| \; .$$

In this limit ν gets large as $-2p_0{}^2\omega$.

From (5.2) we have

(5.4a) $$C^{0i}(q,p) = -\frac{q^0 q^i}{2\omega\nu}\tilde{F}_1 + \frac{q^i}{\nu}\left(p^0 + \frac{q^0}{2\omega}\right)\left(\frac{|\vec{p}|}{|\vec{q}|} + \frac{1}{2\omega}\right)\tilde{F}_2 \,.$$

In (5.4a) the vector $\vec{p}$ has been replaced by the vector $|\vec{p}|\hat{q}$. In the second term in the second parentheses $|\vec{p}|/|\vec{q}|$ dominates $1/2\omega$, since $|\vec{p}|$ is getting large and $|\vec{q}|$ is fixed. Also

$$p^0 + \frac{q^0}{2\omega} \sim -\frac{|\vec{q}|}{2\omega} \,.$$

Thus the following expression for C^{0i} may be used in the deep inelastic limit.

(5.4b) $$C^{0i}(q,p) \to -\frac{q^0 q^i}{2\omega\nu}\tilde{F}_1 - \frac{q^i|\vec{p}|}{2\omega\nu}\tilde{F}_2$$

Next replace $\frac{q^0}{\nu}$ by $1/p_0$ and $|\vec{p}|/\nu$ by $-1/2p_0\omega$. We therefore find

(5.4c) $$2p_0 C^{0i}(q,p) \to \tfrac{1}{2} q^i \frac{F_L(\omega)}{\omega^2} \,.$$

An alternate formula for $C^{0i}(q,p)$, in the deep inelastic region, may be given from (5.1).

(5.5) $$2p_0 C^{0i}(q,p) \to 2p_0 \int d^4x \, e^{-2ip_0\omega x_0} \, e^{-i|\vec{q}|x_0} \, e^{-i\vec{q}\cdot\vec{x}} \langle p|[J^0(x),J^i(0)]|p\rangle$$

Combining (5.5) with (5.4c), and integrating over ω gives

$$\lim_{p_0 \to \infty} \int d\omega \; 2p_0 C^{0i}(q,p) = \tfrac{1}{2} q^i \int_{-\infty}^{\infty} d\omega \, \frac{F_L(\omega)}{\omega^2}$$

$$= \lim_{p_0 \to \infty} 2\pi \int d^4x \; 2p_0 \delta(2p_0 x_0) e^{-i|\vec{q}|x_0}$$

$$e^{-i\vec{q}\cdot\vec{x}} < p|[J^0(x), J^i(0)]|p >$$

(5.6) $$= 2\pi \int d^3x \; e^{-i\vec{q}\cdot\vec{x}} \lim_{p_0 \to \infty} < p|[J^0(0,\vec{x}), J^i(0)]|p > .$$

This is the desired ST sum rule,[3] to which we referred in Chapter II. It shows that, for consistency of the result, the ST matrix element must be a Lorentz scalar; otherwise the $p_0 \to \infty$ limit diverges.

(5.7a) $$< p|[J^0(0,\vec{x}), J^i(0)]|p > = i < p|S|p > \partial^i \delta(\vec{x})$$

(5.7b) $$< p|S|p > = \frac{1}{4\pi} \int_{-\infty}^{\infty} d\omega \, \frac{F_L(\omega)}{\omega^2}$$

The integration extends over negative ω which are unphysical for electroproduction. However, if we assume that the symmetry of $\tilde{F}_L(\omega,\nu) = \tilde{F}_L(-\omega,-\nu)$ is maintained at infinite ν, as well as finite ν, we conclude that $F_L(\omega)$ is even in ω. Also the integration cuts off at $\omega = 1$, since $q^2 + 2\nu > 0$.

(5.8) $$< p|S|p > = \frac{1}{2\pi} \int_{0}^{1} d\omega \, \frac{F_L(\omega)}{\omega^2}$$

If valid, this sum rule measures the q number ST. (Only the q number part of the ST contributes, since the connected matrix element is under consideration here – the vacuum expectation value of S has been subtracted out.) At the present time F_L is consistent with zero, a strong indication that the ST is a c number. Alternatively, if it is known that the ST is a c number, (5.8) predicts that F_L vanishes since F_L is non-negative. However, if F(0) does not vanish sufficiently rapidly, so that the sum rule diverges it is not certain whether one should interpret that as a consistent statement about the infinite nature of this matrix element of the ST, or whether the sum rule is invalid. We shall return to the question of validity below.

The second sum rule which we consider is the Callan-Gross relation.[4] One begins with the i,j components of $C^{\mu\nu}$ but $\vec{q}$ is no longer taken parallel to p; indeed $\vec{q}$ is set to zero.

$$C^{ij}(q,p) = \delta^{ij}\,\tilde{F}_1 - \frac{p^i p^j}{2{p_0}^2\omega}\,\tilde{F}_2 \tag{5.9a}$$

p_0 is sent to infinity in the deep inelastic region.

$$2\omega C^{ij}(q,p) \to \delta^{ij} 2\,\omega F_1(\omega) - \hat{p}^i\hat{p}^j F_2(\omega)$$

$$\to -\hat{p}^i\hat{p}^j F_L(\omega) + (\delta^{ij} - \hat{p}^i\hat{p}^j) 2\omega F_1(\omega) \tag{5.9b}$$

The alternate expression for C^{ij} follows from (5.1).

$$2\omega C^{ij}(q,p) \to 2\int d^4x\,\omega\, e^{-i2\omega x_0 p_0} < p|[J^i(x), J^j(0)]|p > \tag{5.10}$$

Combining (5.9b) and (5.10) and integrating over ω yields

$$\lim_{p_0\to\infty} \int_{-\infty}^{\infty} d\omega\, 2\omega C^{ij}(q,p) = (\delta^{ij} - \hat{p}^i\hat{p}^j) \int_{-\infty}^{\infty} d\omega\, 2\omega F_1(\omega)$$

$$-\hat{p}^i\hat{p}^j \int_{-\infty}^{\infty} d\omega\, F_L(\omega) =$$

$$= i\pi \lim_{p_0 \to \infty} \frac{1}{p_0^2} \int d^4x \, \delta'(x_0) \; < p|[J^i(x), J^j(0)]|p >$$

$$(5.11a) \qquad = -i\pi \lim_{p_0 \to \infty} \frac{1}{p_0^2} \int d^3x < p|[\dot{J}^i(0,\vec{x}), J^j(0)]|p > \; .$$

The final sum rule is

$$(\delta^{ij} - \hat{p}^i\hat{p}^j) \int_0^1 d\omega \, 2\omega F_1(\omega) - \hat{p}^i\hat{p}^j \int_0^1 d\omega F_L(\omega)$$

$$(5.11b) \qquad = -i\frac{\pi}{2} \lim_{p_0 \to \infty} \frac{1}{p_0^2} \int d^3x \; < p|[\dot{J}^i(0,\vec{x}), J^j(0)]|p > \; .$$

The assumed properties of the F_i have been used to write the ω integral over physical ω.

Note that the $\vec{x}$ integration picks out only the coefficient of the δ function in the ETC; all ST disappear upon integration. Also the limiting procedure $p_0 \to \infty$ selects only those parts of the ETC which are components of a second rank Lorentz tensor. The matrix element of such objects can be bilinear in $\vec{p}$, hence will survive in the limit. Matrix elements of Lorentz scalars must be independent of $\vec{p}$, hence they will vanish in the $p_0 \to \infty$ limit, when multiplied by p_0^{-2}. For consistency of (5.11b) it must be assumed that the $< p|[\dot{J}^i, J^j]|p >$ ETC does not possess any tensors of rank 4 or higher in the coefficient of the δ function. Such tensors behave as $|\vec{p}|^4$, and (5.11b) would diverge.

This sum rule may also be used in two ways. It converts experimental data into information about the commutator. Alternatively if the commutator is known, it makes predictions about the results of experiment. It is the latter route which was chosen by Callan and Gross. By computing the $[\dot{J}^i, J^j]$ in a wide variety of quark models, they found that; see Exercise 5.1

(5.12) $\int d^3x < p|[\dot{J}^i(0,\vec{x}'(J^j(0)]|p > = A(\delta^{ij}\vec{p}^2 - p^ip^j) + B\,\delta^{ij}$.

It therefore follows from (5.11b) that $\int_0^1 d\omega F_L(\omega)$ and therefore $F_L(\omega)$ should identically vanish in these quark models. (The same conclusion is, of course, obtained from the Schwinger term sum rule, if the $[J^0,J^i]$ ETC is evaluated canonically.)

Clearly the Callan-Gross sum rule is more convergent at $\omega = 0$ than the ST sum rule. Unfortunately its predictive power is weaker, since knowledge is required of a commutator whose form can be deduced only after an application of canonical equations of motion for fields as well as canonical commutation relations. Further sum rules may be derived which relate integrals over $F_i(\omega)$, weighted by positive powers of ω, to ETC of currents involving many time derivatives. As the relevant power of ω increases, the convergence in ω is improved. The price paid is that the relevant commutator can be determined only after many applications of the equations of motion.

The present derivations, although very simple and direct, are heuristic in the sense that they require an interchange of limit and integral. Over this mathematical manipulation we have no control, and we cannot assess its validity. Also the technique of passing into the deep inelastic region by requiring p_0 and q_0 to become large at fixed $\vec{q}$, forces q^2 to be time-like, which is outside the physical region for this process. We now give an alternate derivation which exposes some of the mathematical underpinnings.

C. Derivation of Sum Rules, Dispersive Method

To effect the second derivation,[5] we consider the forward Compton amplitude, whose absorptive part is the commutator function (5.1).

$$T^{*\mu\nu}(q,p) = i\int d^4x\, e^{iqx} < p|T^*J^\mu(x)J^\nu(0)|p >$$

(5.13) $$= -\left(g^{\mu\nu} - \frac{q^\mu q^\nu}{q^2}\right) T_1 + \left(p^\mu - q^\mu \frac{q\cdot p}{q^2}\right)\left(p^\nu - q^\nu \frac{q\cdot p}{q^2}\right) T_2$$

The first assumption which is required for the derivation of the sum rules is that the matrix element of the ST (if any) is a Lorentz scalar. It then follows that the seagull (if any) in the T* product is present only in the i,j components; see Exercise 2.7. It is to be recalled that this condition was met in our previous derivation.

The T_i satisfy fixed q^2 dispersion relations in ν. Such dispersion relations may be converted, by a change of variable, $\nu = -q^2/2\omega$, into dispersion relations in ω, for fixed q^2. Therefore T_i is taken to be a function of q^2 and ω. In order to derive the two sum rules it is necessary to assume that T_2 is unsubtracted.

(5.14) $$T_2(\omega,q^2) = \frac{2\omega^2}{\pi q^2} \int_0^1 d\omega' \frac{\tilde{F}_2(\omega',q^2)}{\omega'^2-\omega^2}$$

To derive the ST sum rule, it is necessary to assume that the combination $T_L = T_1 + \frac{\nu^2}{q^2} T_2$ is unsubtracted.

(5.15a) $$T_L(\omega,q^2) = \frac{\omega^2}{2\pi} \int_0^1 \frac{d\omega'}{\omega'^2} \frac{\tilde{F}_L(\omega',q^2)}{\omega'^2-\omega^2}$$

The Callan-Gross sum rule may be derived from a weaker hypothesis. It suffices to assume that T_1 requires at most one subtraction, as long as the subtraction does not grow with q^2.

$$T_1(\omega,q^2) = T(q^2) + \frac{1}{\pi} \int_0^1 d\omega' \frac{\omega'\tilde{F}_1(\omega',q^2)}{\omega^2-\omega'^2}$$

(5.15b) $$\lim_{q^2\to\infty} T(q^2) < \infty$$

$T(q^2)$ is the value of T_1 at $\nu = 0$: $T(q^2) = T_1(\infty, q^2)$. Equation (5.15b) is certainly true if (5.14) and (5.15a) hold; it can be valid even when (5.15a) needs subtractions. The dispersive representations (5.14) and (5.15b) are consistent with conventional Regge lore which predicts that both $\bar{F}_2$ and $\omega\bar{F}_1$ possess a finite limit as $\omega \to 0$. Evidently (5.15a) is valid only if $\bar{F}_L$ vanishes sufficiently rapidly for small ω (large ν).

We now apply the BJL limit to selected components of $T^{*\mu\nu}$. First consider T^{*0i}. Since there is no seagull in this object, the T^* product coincides with the T product. Therefore from (5.13) we have

$$\lim_{q_0 \to \infty} q_0 T^{0i}(q,p) = -\int d^3x\, e^{-i\vec{q}\cdot\vec{x}} \langle p|[J^0(0,\vec{x}), J^i(0)]|p\rangle$$

$$= \lim_{q_0 \to \infty} \left[\frac{q_0{}^2 q^i}{q^2} T_1 + q_0(p_0 - q_0\frac{\nu}{q^2})(p^i - q^i\frac{\nu}{q^2})T_2\right]$$

$$= q^i \lim_{q_0 \to \infty} T_1 + p^i\, \vec{p}\cdot\vec{q} \lim_{q_0 \to \infty} T_2 \ . \tag{5.16a}$$

Since the matrix element of the ST, by hypothesis, has no $\vec{p}$ dependence, $\lim_{q_0 \to \infty} T_2$ must be zero. The same conclusion can be drawn from (5.14). (In the BJL limit $\omega^2 \to \infty$, $\tilde{F}_2(\omega', q^2) \to F_2(\omega')$.) Equation (5.16a) may be now rewritten as

$$-\int d^3x\, e^{-i\vec{q}\cdot\vec{x}} \langle p|[J^0(0,\vec{x}), J^i(0)]|p\rangle$$

$$= q^i \lim_{q_0 \to \infty} (T_1 + \frac{\nu^2}{q^2} T_2) = q^i \lim_{q_0 \to \infty} T_L \ . \tag{5.16b}$$

The BJL limit of T_L is evaluated from the dispersion formula (5.15a), and the derivation of the ST sum rule is complete.[6]

(5.16c) $$\lim_{q_0 \to \infty} T_L = -\frac{1}{2\pi} \int_0^1 \frac{d\omega'}{\omega'^2} F_L(\omega')$$

To derive the Callan-Gross sum rule, one begins with T^{*ij}. The first task is to extract the seagull and arrive at the T product so that the BJL theorem may be applied. It is sufficient for our purposes to examine T^{ij} in the rest frame of q.[7] Since the BJL limit of T_2 is zero, we have, from (5.13)

(5.17a) $$\lim_{q_0 \to \infty} T^{*ij}\Big|_{\vec{q}=0} = -g^{ij} \lim_{q_0 \to \infty} T_1\Big|_{\vec{q}=0} .$$

Therefore the seagull, which is that portion of a T* product which does not vanish in the BJL limit, is

(5.17b) $$-g^{ij} \lim_{q_0 \to \infty} T_1\Big|_{\vec{q}=0} .$$

According to the dispersive representation for T_1, the BJL limit of that object is $\lim_{q_0 \to \infty} T(q^2)$. Defining an asymptotic expansion at large q^2 for the subtraction term

(5.17c) $$T(q^2) \approx a + b/q^2 ,$$

we learn from (5.17a) that the T product is

(5.17d) $$T^{ij}\Big|_{\vec{q}=0} = T^{*ij}\Big|_{\vec{q}=0} + g^{ij}a = -g^{ij}T_1\Big|_{\vec{q}=0} + p^i p^j T_2\Big|_{\vec{q}=0} + g^{ij}a .$$

We now apply the BJL limit to $q_0{}^2 T^{ij}$.

$$\lim_{q_0 \to \infty} q_0{}^2 T^{ij}\Big|_{\vec{q}=0} = -g^{ij} \lim_{q_0 \to \infty} q_0{}^2 (T_1 - a)\Big|_{\vec{q}=0}$$

$$+ p^i p^j \lim_{q_0 \to \infty} q_0{}^2 T_2\Big|_{\vec{q}=0}$$

(5.18a) $$= -i \int d^3x < p|[\dot{J}^i(0,\vec{x}), J^j(0)]|p >$$

The BJL limits occurring in (5.18a) are evaluated from the dispersive representations of the respective function. We find

$$\lim_{q_0 \to \infty} q_0{}^2 T_2\Big|_{\vec{q}=0} = -\frac{2}{\pi} \int_0^1 d\omega' F_2(\omega')$$

(5.18b) $$\lim_{q_0 \to \infty} q_0{}^2 (T_1 - a)\Big|_{\vec{q}=0} = b + \frac{4}{\pi} p_0{}^2 \int_0^1 d\omega' \omega' F_1(\omega') \;.$$

A division by $p_0{}^2$, which is then sent to ∞, completes the derivation of the Callan-Gross sum rule.

The present derivation is especially instructive, since it demonstrates explicitly that it is the BJL definition of the commutator that is relevant to a high energy sum rule. If there are subtractions beyond the ones assumed, the present method is incapable of yielding a sum rule; see Exercise 5.2.

D. Model Calculation

In a quark-vector gluon model, the above two sum rules, when coupled with canonical evaluations of the relevant commutators, predict that F_L vanishes. (Of course the ST cannot be evaluated canonically; however, as we shall discuss below, it is known that the non-canonical ST is a c number in this model, at least to low orders of perturbation theory.) As an explicit check of canonical reasoning, and of the sum rules, we evaluate F_L to lowest order perturbation theory – a non-vanishing result is found.

The model is described by the Lagrangian

$$\mathcal{L} = i\bar{\psi}\gamma_\mu\partial^\mu\psi - m\bar{\psi}\psi - \frac{1}{4}B^{\mu\nu}B_{\mu\nu} + \frac{1}{2}\mu^2 B^\mu B_\mu + g\bar{\psi}\gamma^\mu\psi B_\mu \tag{5.19}$$

A vector meson B^μ of mass μ interacts with a Fermion ψ of mass m through a Yukawa type coupling of strength g; $B^{\mu\nu} = \partial^\mu B^\nu - \partial^\nu B^\mu$. The current to which the meson couples is conserved, therefore the theory is renormalizable. All considerations of internal symmetry are being ignored. Since we are interested in high-energy limits, be they deep inelastic or BJL, it is possible to ignore the Fermion mass. Therefore, to simplify all the calculations, we henceforth set m to zero.

To check the formal results, we need the cross section for the production of a meson by an off mass shell photon. The relevant one-boson amplitude, to lowest order perturbation theory, is given by the diagrammatic representation of Fig. V–2. The wavy line is the photon with 4 momentum q. The solid line is the Fermion, with incoming momentum p. The Boson, represented by the dotted line, carries off 4 momentum r. By a standard and unambiguous application of the Feynman rule method for calculating amplitudes, the spin averaged transverse and longitudinal cross section for the relevant process can be obtained. From these results, the $\tilde{F}_i$ can be deduced. Passing to the deep inelastic limit, one finds

$$F_L(\omega) \propto g^2\theta(1-\omega^2)\omega^2 \ . \tag{5.20}$$

Thus the conclusion of formal reasoning is not maintained: it is found that $F_L(\omega)$ is non vanishing. This is an unambiguous consequence of the dynamics. Note that neither sum rule diverges; $F_L(\omega)$ is sufficiently regular at the origin.

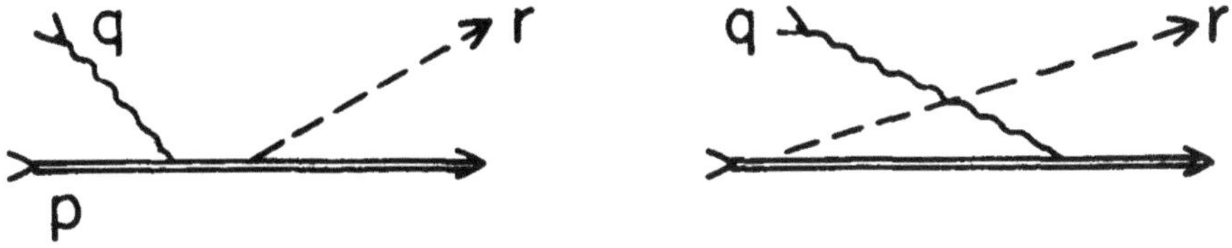

Fig. V-2

Lowest order Feynman graphs for electroproduction in the model given by Eq. (5.19).

The other independent scaling function does not exist in this model. It is found that

(5.21) $$\lim_{\substack{\nu \to \infty \\ \omega \text{ fixed}}} \tilde{F}_2(\omega,\nu) \propto g^2 \log \nu/\mu^2 + \text{finite terms.}$$

It might be objected that since scaling is not exhibited in this model, we have no business comparing sum rules which are derived with the help of the scaling hypothesis. This objection does not survive close scrutiny. To this order, one can unambiguously separate off the effects of the breakdown of the scaling hypothesis, and demonstrate that the failure of the formal reasoning is due to other causes. This is possible, since the longitudinal amplitude separately satisfies scaling, and no reference need be made to the transverse amplitude which violates scaling.

The failure of the ST sum rule is easy to understand. Examination of the explicit function $\tilde{F}_L(\omega,q^2)$, as given by the perturbative calculation, indicates that the no subtraction hypothesis is wrong in the example. That is, even though the ST is indeed a c number, when calculated by the BJL limit, see below, the formula relating it to an integral over F_L was improperly derived.[9,10] Therefore this failure is not due to anomalous commutators. Nevertheless, it is a particularly interesting example: there is no evidence in the deep inelastic limit of the need for a subtraction – the integral over F_L converges. We shall not discuss the ST sum rule any further since it is outside the scope of our main interest. The Callan-Gross sum rule is violated because the $[\dot{J}^i, J^j]$ ETC, when calculated by the BJL limit, as it must be in this application, does not have the form given by canonical reasoning (5.12). We now turn to a demonstration of this.[11]

E. Anomalous Commutators

To compute the $[\dot{J}^i, J^j]$ ETC, to lowest order in the perturbation, we need to know the Compton amplitude $T^{*\mu\nu}$ to this order. The relevant unrenormalized graphs are those of Fig. V–3. The physical amplitude is obtained by multiplying by Z^{-1}, the inverse of the wave-function renormalization constant. Although for the Callan-Gross application, only the forward Compton amplitude is relevant, we compute here the non-forward amplitude as well. Evidently we must also calculate the vertex function and the propagator to second order. In addition we shall need, in subsequent applications, the axial vertex function. The computation is performed by use of the Landau form for the Boson propagator $D^{\mu\nu}(p) = -i(g^{\mu\nu} - p^\mu p^\nu/p^2)/(p^2-\mu^2)$. We continue to keep the Fermions massless. A convenient feature of this model is *finite* wave-function renormalization. Hence it is possible to speak of well defined unrenormalized Fermion fields.

The following Green's functions are computed to lowest non-trivial order of perturbation theory.

Unrenormalized Fermion propagator

$$G(p) = \int d^4x \, e^{ipx} < 0|T^*\psi(x)\bar{\psi}(0)|0> . \tag{5.22}$$

Unrenormalized, improper vector vertex function

$$F^\mu(p,q) = \int d^4x d^4y \, e^{ipx} \, e^{-iqy} < 0|T^*\psi(x)\bar{\psi}(y)J^\mu(0)|0> . \tag{5.23}$$

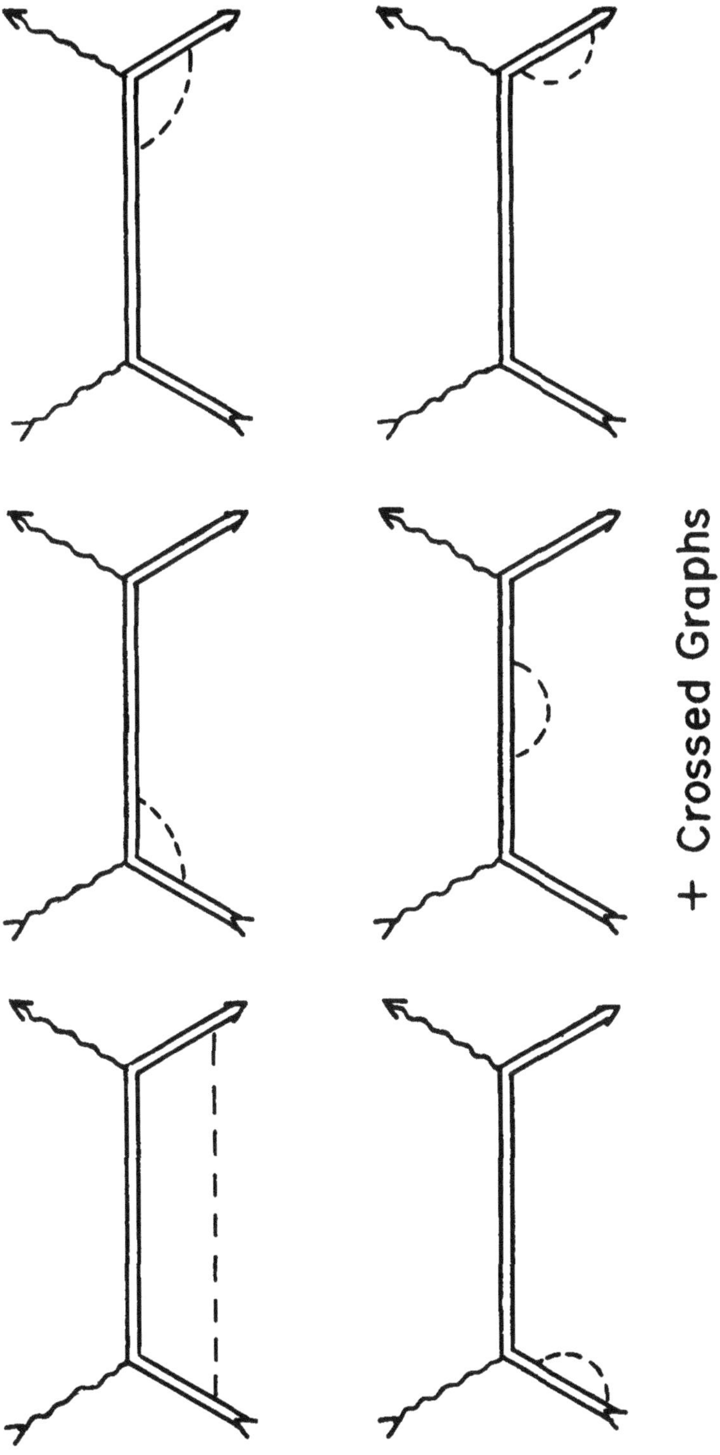

Fig. V–3

Higher order corrections to the Compton amplitude.

Unrenormalized improper axial vector vertex function

$$F_5^{\mu}(p,q) = \int d^4x d^4y \, e^{ipx} e^{-iqy} < 0|T^*\psi(x)\bar{\psi}(y)J_5^{\mu}(0)|0 > \quad . \tag{5.24}$$

Compton amplitude

$$T^{*\mu\nu}(q) = i\int d^4x \, e^{iqx} < p|T^*J^{\mu}(x)J^{\nu}(0)|p'> \quad . \tag{5.25}$$

Once these functions have been determined explicitly, several commutators can be computed by the BJL limit. They are not all relevant to the Callan-Gross sum rule, but they are sufficiently interesting to deserve mention. It turns out that none of the T* products possess seagulls; thus they are equivalent to T products, and the BJL theorem may be applied directly. The commutators which we determine are the following.[12]

$$-i \lim_{p_0 \to \infty} p_0 G(p) = \int d^3x \, e^{-i\vec{p}\cdot\vec{x}} < 0|[\psi(0,\vec{x}),\bar{\psi}(0)]_+|0 > \tag{5.26}$$

$$-i \lim_{p_0 \to \infty} p_0 F^{\mu}(p,q)\Big|_{q \text{ fixed}} = \int d^3x \, d^4y \, e^{-i\vec{p}\cdot\vec{x}} e^{-iqy}$$

$$< 0|T[\psi(0,\vec{x}),J^{\mu}(0)]\bar{\psi}(y)|0 > \tag{5.27}$$

$$-i \lim_{q_0 \to \infty} q_0 T^{\mu\nu}(q) = i\int d^3x \, e^{-i\vec{q}\cdot\vec{x}}$$

$$< p|[J^{\mu}(0,\vec{x}),J^{\nu}(0)]|p' > \tag{5.28}$$

$$- \lim_{q_0 \to \infty} q_0^2 T^{ij}(q)\Big|_{\substack{\vec{q}=0 \\ p=p' \\ \text{spin averaged}}} =$$

$$= i\int d^3x < p|[\dot{J}^i(0,\vec{x}),J^j(0)]|p >\Big|_{\text{spin averaged}} \tag{5.29}$$

The last commutator is the Callan-Gross commutator. No $1/q_0$ term is present in the forward amplitude, since the $[J^i,J^j]$ ETC between diagonal, spin averaged states vanishes. The matrix elements of the axial current, calculated in (5.24), are required in an evaluation of the right hand side of (5.28) for the ij component, which involves $\epsilon^{ijk}J_k{}^5$. When the limits are explicitly performed the following results are obtained.[13]

$$[\psi(0,\vec{x}),\bar{\psi}(0)]_+ = (1-\frac{3g^2}{32\pi^2})\,\gamma^0\delta(\vec{x}) \tag{5.30}$$

$$[\psi(0,\vec{x}),J^0(0)] = \psi(0,\vec{x})\delta(\vec{x}) \tag{5.31a}$$

$$[\psi(0,\vec{x}),J^i(0)] = (1-\frac{g^2}{8\pi^2})\,\gamma^0\gamma^i\psi(0,\vec{x})\delta(\vec{x}) \tag{5.31b}$$

$$< p|[J^0(0,\vec{x}),J^0(0)]|p'> = 0 \tag{5.32a}$$

$$< p|[J^0(0,\vec{x}),J^i(0)]|p'> = 0 \tag{5.32b}$$

$$< p|[J^i(0,\vec{x}),J^j(0)]|p'> = i\epsilon^{ijk}(1-\frac{3g^2}{16\pi^2}) < p|J_5{}^k(0,\vec{x})|p'> \delta(\vec{x}) \tag{5.32c}$$

$$\int d^3x < p|[\dot{J}^i(0,\vec{x}),J^j(0)]|p>\Big|_{\text{spin averaged}} = A(\delta^{ij}\vec{p}^2-p^ip^j)+B\delta^{ij}+B'\vec{p}^2\delta^{ij} \tag{5.33}$$

It is seen that most commutators are non-canonical; they have corrections which are interaction dependent. It is remarkable that all the non-canonical results are finite. Equation (5.30) indicates that the basic canonical (anti) commutator of the field variable with its conjugate momentum is not preserved. This is not particularly surprising; remember that the fields ψ are unrenormalized, and it is well known that, in general, the commutator of these fields ceases to exist due to infinite wave function renormalization Z. It is only in our particular model, where Z is finite, that a well defined result is obtained; although it is noteworthy that the answer even when finite, is not the naive one. Equation (5.31b) shows that the field with the spatial components of the current density has a non-canonical commutator. Nevertheless the charge density with the field has the canonical commutator (5.31a). The $[J^0,J^0]$ and $[J^0,J^i]$ ETC are canonical. The latter result shows that here the ST has no Fermion matrix element. This substantiates our previously made assertion that the ST term sum rule (5.8), if valid, predicts the vanishing of F_L in this model. Equation (5.32c) is the remarkable statement that the quark algebra of spatial components of currents is not maintained in perturbation theory. This has the consequence that certain radiative corrections to β decay, which were thought to be finite on the basis of the quark algebra, are, in fact, infinite.[1] Finally the term $B'\vec{p}^2\delta^{ij}$ in (5.33) is non-canonical. The $\vec{p}^2$ factor in that term, allows it to contribute to the Callan-Gross sum rule. This spoils the canonical prediction that $F_L = 0$.

It is instructive to scrutinize the canonical derivation of the commutators given above, to pinpoint the precise step which is not validated by explicit calculation.[15] For definiteness we consider the $[\psi,J^i]$ ETC. One way to compute this commutator canonically is to exhibit the canonical expression for $J^i = \bar{\psi}\gamma^i\psi$ and then to use the canonical (anti) commutator relations for ψ with ψ and $\bar{\psi}$. Since we wish to minimize the use of the unreliable canonical formalism we prefer an alternate derivation which uses only reliable commutators, and which exposes the reason for the occurrence of an interaction dependent modification. For present purposes it is sufficient to consider only the term proportional to the δ

function in the $[\psi, J^i]$ ETC. (Our above explicit calculation does not give evidence for the existence of gradient terms.) Thus we examine

$$\int d^3y\, [\psi(t,\vec{x}), J^i(t,\vec{y})] = \int d^3y\, y^i[\partial_j J^j(t,\vec{y}), \psi(t,\vec{x})] =$$

$$\int d^3y\, y^i[\psi(t,\vec{x}), \dot{J}^0(t,\vec{y})] = \int d^3y\, y^i \frac{\partial}{\partial t} [\psi(t,x), J^0(t,\vec{y})]$$

(5.34a)
$$- \int d^3y\, y^i[\dot{\psi}(t,\vec{x}), J^0(t,\vec{y})] \ .$$

An integration by parts and the conservation of J^μ were used to pass from the first to the last expression in (5.34a). We now may use the canonical result for the $[J^0,\psi]$ ETC to evaluate the first of the two terms appearing in the last equality of (5.34a). Here the use of the canonical value is legitimate; explicit calculation did not cast any doubts on it. To determine the last occurring ETC in (5.34a), an expression for $\dot{\psi}$ is required. The equation of motion for ψ is by definition $i\gamma_\mu \partial^\mu \psi = m\psi + g\eta$; $\dot{\psi} = -\gamma^0\gamma^i\partial_i\psi - im\gamma^0\psi - ig\gamma^0\eta$. Here η is the source of the Fermion field; in our model it is equal to $B_\mu\gamma^\mu\psi$, but we prefer to leave it unspecified for the moment. Equation (5.34a) is now seen to be equivalent to

$$\int d^3y\, [\psi(t,\vec{x}), J^i(t,\vec{y})] =$$

$$\int d^3y\, y^i \Big[\dot{\psi}(t,\vec{x})\delta(\vec{x}-\vec{y}) + \gamma^0\gamma^k\partial_k\psi(x)\delta(\vec{x}-\vec{y})$$

$$+ \gamma^0\gamma^k\psi(x)\partial_k\delta(\vec{x}-\vec{y}) + im\gamma^0\psi(x)\delta(\vec{x}-\vec{y})$$

(5.34b)
$$+ ig\gamma^0[\eta(t,\vec{x}), J^0(t,\vec{y})] \Big] \ .$$

Performing the indicated time differentiation and using the equations of motion again, converts (5.34b) to

$$\int d^3y[\psi(t,\vec{x}),J^i(t,\vec{y})] = \gamma^0\gamma^i\psi(x)$$

(5.34c) $$+ \; ig\,\gamma^0 \int d^3y \; y^i \left[[\eta(t,\vec{x}),J^0(t,\vec{y})]-\eta(x)\delta(\vec{x}-\vec{y})\right] .$$

Thus it is seen that the canonical result is obtained only if the second term on the right hand side of (5.34c) vanishes. This object *does* vanish if the canonical formula for $\eta(x)$ is taken: $\eta(x) = B_\mu(x)\gamma^\mu\psi(x)$; $[B_\mu(0,\vec{x})\gamma^\mu\psi(t,\vec{x}),J^0(t,\vec{y})] = B_\mu(x)\gamma^\mu\psi(x)\delta(\vec{x}-\vec{y}) = \eta(x)\delta(\vec{x}-\vec{y})$. However the expression $B_\mu(x)\psi(x)$ involves the product of two quantum mechanical operators at the same space time point and is, in general, undefined. Evidently the "correct" form for $\eta(x)$ differs from the canonical one in such a way that the $[\eta,J^0]$ ETC is modified from its canonical form. It is easy to see that the following form for the $[\eta,J^0]$ ETC will reproduce the value given in (5.31b) for the $[\psi,J^i]$ ETC.

(5.35) $$[\eta(t,\vec{x}),J^0(t,\vec{y})] = \eta(x)\delta(\vec{x}-\vec{y}) + \frac{i}{8\pi^2}\, g\gamma^i\psi(x)\partial_i\delta(\vec{x}-\vec{y})$$

F. Discussion

We have demonstrated the occurrence of anomalous commutators in the quark model. The use of a vector gluon interaction is not important; similar results are obtained with scalar or pseudoscalar gluons. It is important to note that the local current algebra involving at least one time component has not been put into question. It therefore follows that vector Ward identities for Green's functions involving J^μ are not modified. This, of course, is a consequence of the fact that the symmetry implied by the conservation of J^μ can be maintained in perturbation theory. However the space component algebra, as well as all higher commutators possess interaction dependent modifications. Such commutators are not used in derivations of Ward identities and low energy theorems. Rather they give rise, as we have seen, to high energy sum rules. It is these sum rules that are now found to be untrustworthy.

REFERENCES

[1] For an excellent discussion of electroproduction kinematics see L. S. Brown, lectures delivered at the Summer Institute for Theoretical Physics, University of Colorado, Boulder (1969) to be published in *Lectures in Theoretical Physics*, ed. W. E. Brittin, B. W. Downs and J. Downs, Interscience (New York).

[2] This hypothesis was enunciated by J. D. Bjorken, *Phys. Rev.* 179, 1547 (1969).

[3] This was derived by R. Jackiw, B. Van Royen and G. B. West, *Phys. Rev.* D2, 2473 (1970). See also L. S. Brown, Ref. 1 and J. M. Cornwall, D. Corrigan and R. E. Norton, *Phys. Rev. Letters* 24, 1141 (1970).

[4] C. G. Callan, Jr. and D. J. Gross, *Phys. Rev. Letters* 22, 156 (1969).

[5] The dispersive derivation for the ST sum rule is presented by R. Jackiw, R. Van Royen and G. B. West, Ref. 3. The Callan-Gross sum rule was derived in Ref. 4 by essentially the present method; the only difference being that the possibility of q number ST was ignored.

[6] A technical comment: In order to avoid manipulations with the $\tilde{F}_i$ in the unphysical, timelike domain of q^2, q_0 should be sent to $i\infty$ rather than to ∞. It can be shown that the BJL definition of the commutator holds in this limit as well.

[7] This does not take us out of the physical region of the $\tilde{F}_i$, since q_0 can be imaginary; see Ref. 6.

[8] The treatment here follows R. Jackiw and G. Preparata, *Phys. Rev. Letters* 22, 975 (1969), (E) 22, 1162 (1969) and *Phys. Rev.* 185, 1748 (1969). Identical results were found by S. L. Adler and Wu-Ki Tung, *Phys. Rev. Letters* 22, 978 (1969) and *Phys. Rev.* D1, 2846 (1970).

[9] The "naive" derivation of the ST sum rule presumably fails due to improper interchange of limit with integral.

[10] J. M. Cornwall, D. Corrigan and R. E. Norton, Ref. 3, were able to derive a weaker ST sum rule, under hypotheses which are less stringent than the no subtraction hypothesis employed here. However, in the case that $F_L(\omega)/\omega^2$ is regular at $\omega = 0$, their result is the same as the ST sum rule given in the text. Therefore the present model offers a counterexample to their argument as well. An analysis of this question has been given by Anthony Zee, *Phys. Rev.* D3, 2432 (1971).

[11] It is striking that the same model provides the same function, $F_L(\omega)$, as a counterexample to two common "regularity" assumptions: no subtraction in dispersion relations, validity of canonical reasoning. This suggests that there may be a relation between these two ideas. It would be most interesting to explore this relation.

[12] The presentation here is that of R. Jackiw and G. Preparata, *Phys. Rev.* 185, 1929 (1969).

[13] Equations (5.30), (5.31) and (5.33) were obtained by R. Jackiw and G. Preparata, Refs. 8 and 12. Simultaneously S. L. Adler and Wu-Ki Tung, Ref. 8, obtained (5.32) and (5.33). The current commutators (5.32) were previously studied in a different model by A. I. Vainshtein and B. L. Ioffe, *Zh. Eksperim. i Toer. Fiz. Pisma v Redaktjiyu* 6, 917 (1967) [English translation: *Soviet Phys. JETP Letters* 6, 341 (1967)].

[14] S. L. Adler and Wu-Ki Tung, Ref. 8.

[15] The development here is due to R. Jackiw and G. Preparata, Ref. 12.

VI. DISCUSSION OF ANOMALIES IN CURRENT ALGEBRA

A. Miscellaneous Anomalies

In addition to the results which were presented in the last two chapters, various other commutators have been calculated in perturbation theory with the help of the BJL definition. We shall not discuss these calculations in detail, but merely quote three especially interesting conclusions.

(i) In the vector gluon model, which was introduced in the previous chapter, the connected matrix element of the ST between single Boson states vanishes.[1]

$$\text{(6.1)} \qquad < B|[J^0(t,\vec{x}),J^i(t,\vec{y})]|B > \; = \; 0$$

This, together with the previously quoted result that also the single Fermion matrix element of the ST is zero, see (5.32b), strongly indicates that the ST is a c number. The relevant graph, whose high energy behavior yields (6.1), is the off-mass shell Boson-Boson scattering amplitude, Fig. VI–1. The result (6.1) is interesting because earlier calculations of this object, by methods other than the BJL definition, yielded a q number ST, bilinear in the Boson field.[2] This shows, as was asserted previously, that alternate calculations of the ETC may give different results. Only the BJL definition has so far proven itself to be of physical interest.

(ii) The vacuum expectation value of the ST in the same model has been shown to be a quadratically divergent term, proportional to the first derivative of the δ function. In addition, a finite term proportional to the triple derivative of the δ function has been uncovered.[3]

$$[J^0(t,\vec{x}),J^i(t,\vec{y})] = iS_1\partial^i\delta(\vec{x}-\vec{y})+iS_3\partial^i\nabla^2\delta(\vec{x}-\vec{y}) \tag{6.2}$$

S_1 is quadratically infinite; S_3 is well defined. The relevant graph which yields (6.2) is the vacuum polarization, Fig. VI–2. The quadratic divergence of S_1 is correlated with the well known quadratic divergences of the relevant Feynman integral. The existence of the finite term involving S_3, is also a consequence of the presence of a dominant, quadratically divergent contribution. The triple derivative term violates both the operator proof of the known existence of such objects, given in Chapter II, (2.24); as well as the commonly accepted formula, derived in Exercise 3.2, for the vacuum expectation value of the ST. (That formula exhibits only one derivative of the δ function.) It is clear that the origin of the contradiction is the quadratic divergence of the $[J^0,J^i]$ ETC. The above two arguments proceed as if the commutator were well defined. When it is divergent the arguments may be circumvented; see Exercice 6.1.

Since the vector-gluon model is formally identical to spinor electrodynamics when the Boson mass is zero, and since none of the current commutators discussed here depends on the Boson mass, we may also conclude that the ST in spinor electrodynamics is a quadratically divergent c number proportional to a derivative of the δ function, and a finite c number proportional to three derivatives of the δ function. It should be recalled that all the calculations which we are here discussing are performed to lowest non-trivial order in perturbation theory.[4]

(iii) The third discovery which we mention is the remarkable conclusion that the Jacobi identity for three spatial components of current densities is in general invalid in the quark model.[5] Explicit calculation shows that

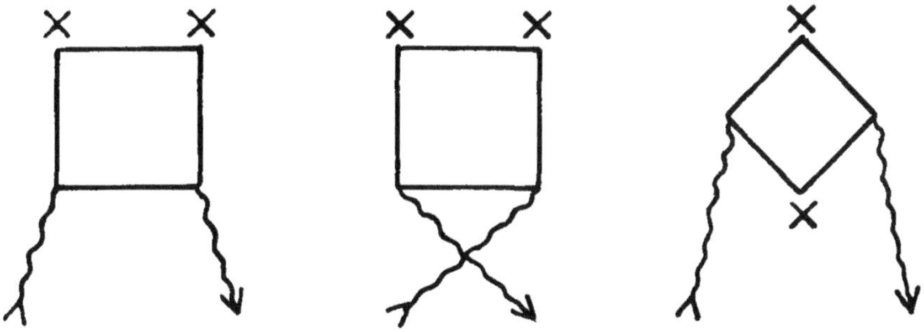

Fig. VI-1

Graphs from which the one-Boson matrix element of the ST is determined.

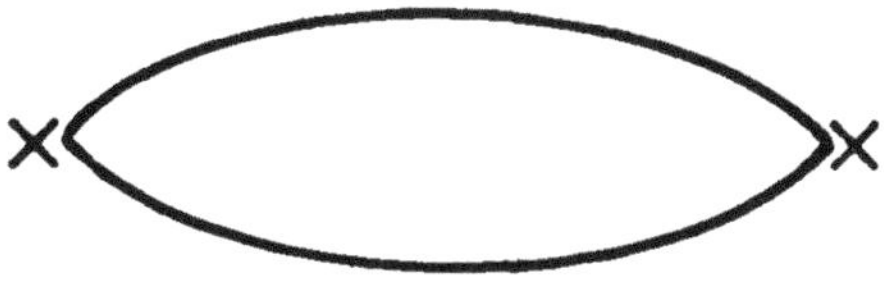

Fig. VI-2

Graph from which the vacuum expectation value of the ST is determined.

$$[J_a^{\mu}(t,\vec{x}),[J_b^{\nu}(t,\vec{y}),J_c^{\omega}(t,\vec{z})]] \neq$$

$$[[J_a^{\mu}(t,\vec{x}),J_b^{\nu}(t,\vec{y})],J_c^{\omega}(t,\vec{z})] + [J_b^{\nu}(t,\vec{x}),[J_a^{\mu}(t,\vec{x}),J_c^{\mu}(t,\vec{z})]]$$

$$\mu,\nu,\omega = \text{spatial indices} \quad . \tag{6.3}$$

(A double commutator may also be defined by the BJL technique; see Exercise 6.2). This anomaly, interesting in its own right, is relevant to the fact that it is possible to prove, with the help of the Jacobi identity, that the ST in the quark model is a q number.[6] Since we previously concluded that explicit calculation shows the ST to be a c number, we must now accept the violation of the Jacobi identity.

B. Non-Perturbative Arguments for Anomalies

Almost all the calculations which we have performed to substantiate the existence of commutator anomalies are based on perturbation theory. The important question arises whether these anomalies are peculiar to perturbation theory, or are they an essential feature of the complete theory? Although our knowledge of the structure of a complete local field theory is very limited, we may nevertheless assert that all available evidence points to the existence of anomalies even outside the perturbative framework.

In the first place, the necessary existence of a ST was established without reference to perturbation theory. Hence in a Fermion theory, where a ST is not given by canonical reasoning, at least the ST anomaly must be present. Furthermore the results quoted in subsection A(iii) above, indicate that this anomaly must be sufficiently singular so that either the ST is a q number, or the Jacobi identity is violated.

For other anomalies no one has constructed arguments which are as general as those for the ST. However it is clear that some of the anomalies must be present to *all* orders in perturbation theory. Consider for example the off mass shell pion decay amplitude $T(k^2)$. In renormalized perturbation theory this is a perfectly well defined and unambiguous quantity. Therefore if $T(0)$ does not vanish in lowest order, higher orders cannot cancel the first nonvanishing result, when the coupling constant is arbitrary. Even if the perturbative series does not converge, it is hard to see how the series can be interpreted to sum to zero for all values of the coupling. This argument, although valid to arbitrary order of perturbation theory, can be circumvented nevertheless, if it is asserted that there exists another solution to the theory which has no counterpart in perturbation theory. Such a point of view cannot be dismissed, however this speculation must be characterized as a dynamical assumption of the highest order. Therefore it remains true that the discovery of anomalies has shown current algebra to be dependent on unexpected dynamical assumptions.

For the anomalies associated with high-energy sum rules, like the Callan-Gross, lowest order calculations are less reliable indicators. For example, it is not impossible that $\tilde{F}_L(\omega,\nu)$, which should be vanishing with $\nu \to \infty$ in the quark-gluon model, behaves for large ν as $g^2\omega^2\nu^{-ag^2}$, $a > 0$. In lowest order, this form is non-vanishing; however as $\nu \to \infty$, the complete function vanishes. However, higher order calculations of $\tilde{F}_L(\omega,\nu)$ have been performed, and no evidence of such damping is found.[7] Of course since all orders have not been calculated, one is free to assume that this damping does occur, but again this must be characterized as a dynamical assumption.

Another class of evidence for anomalies derives from physically unrealistic, but solvable, models. It has been shown that the complete solution of the Lee model[8] and of the Thirring model[9] possesses commutator and divergence anomalies. It should be clear that the fundamental reason for the occurrence of anomalies in the renormalized theory is the fact that the underlying unrenormalized theory is divergent. Noone knows with certainty whether these divergences are illusory, though all tentative evidence is that they are not. Indeed axiomatists who have examined and proven the existence of solutions to relativistic field theories, albeit in unrealistic two dimensional space-time, have found that the kind of divergences which they rigorously show to be present, are accurately described by perturbation theory.[10]

Some anomalies can perhaps disappear for a *particular* value of the coupling constant. It has been shown, for example, that in spinor electrodynamics, the canonical commutator of the unrenormalized electromagnetic potentials,

$$i[A^i(t,\vec{x})-\partial^i A^0(t,\vec{x}),A^j(t,\vec{y})] = -g^{ij}\delta(\vec{x}-\vec{y}) \quad , \tag{6.4}$$

can be maintained in the complete theory if the unrenormalized charge satisfies an eigenvalue condition.[11] The condition is expressed by saying that a certain numerical function of the unrenormalized coupling constant must vanish; i.e., the coupling constant must be chosen to be a zero of this function. Unfortunately it is not known whether the relevant function possesses a zero – up to sixth order it does not. Moreover it is not likely that *all* anomalies would be removed even if the eigenvalue condition is met.

One might entertain the hope that the constraints of non-linear unitarity relations, which are not conveniently realized in perturbation theory, enforce sufficient damping on the high-energy behavior of various Green's function, so that the various untoward results disappear in the complete solution. A preliminary investigation of this possibility has produced negative results: no mechanism for damping has been found.[12]

The occurrence of anomalies in selected Ward identities may be understood in the following way. The "naive" Ward identities are a consequence of the relevant symmetries of the dynamics. In field theory, where the dynamics are described by a Lagrangian, these symmetries must be invariances of the Lagrangian. However, even when the Langrange function, expressed in terms of unrenormalized fields, possesses the appropriate invariance, the solution to the theory may not respect this symmetry. The reason is that the unrenormalized Lagrangian is undefined due to the infinities of the unrenormalized theory. A well defined theory emerges only after regularization. Thus the symmetry and the Ward identities are preserved only if the regulator method respects the invariance. Typically regulator procedures introduce large regulator masses into the problem. Therefore we should anticipate the possibility of the violation of symmetries which depend on the absence of mass terms, such as chirality, and scale invariance. (Scale invariance is discussed in the next chapter.)

C. Models without Anomalies

Can one somehow overcome these anomalies; that is can a theory be set up where all results are naive? Only very limited investigations have been devoted to this question. The analysis, presented in the previous chapter in connection with the $[\psi, J^i]$ ETC, offers a hint how one may modify the theory so that this commutator retains its naive value. Evidently one must modify the dynamics so that the source of the Fermion field, when commuted with J^0, no longer produces the gradient term as in (5.35), but coincides with the naive result.

Another example of such a modification can be given in reference to the failure of the Sutherland-Veltman theorem. By adding a direct interaction between the pion and the electromagnetic field of the form $\phi \tilde{F}^{\mu\nu} F_{\mu\nu}$, the non-zero value for the off mass shell decay amplitude for massless pions may be cancelled away, yielding the zero value as is required by the formal argument. Alternatively one may prescribe non-standard regulator techniques, which eliminate the non-zero coupling constant. All these modifications however probably lead to non-renormalizable perturbation series.[13]

D. Discussion

It seems extremely probable, therefore, that if current algebra is derived from a local field theory, one cannot assume that canonical reasoning gives the correct results for current commutators and equations of motion. The anomalies are of two kinds: they may be sufficiently violent so that not only the canonical form of the commutators, is destroyed, but also the Ward identities must be modified; other anomalies merely modify the commutators. The modified Ward identities lead to modified low energy theorems. It is likely that all instances of this have been discovered; they always involve the axial current and a catalog can be given.[14] The gentler anomalies which do not destroy naive Ward identities seem to occur in all commutators with the exception of the time component algebra. Thus those results of current algebra which follow only from the $[J_a^0, J_b^0]$ ETC appear secure. The sum rules based on the space component algebra, on the other hand, have been circumvented.

Ultimately the question whether our anomalies are present in nature, rather than in formal system, must be answered experimentally. The experimental predictions that have been obtained are rather limited. The Ward identity anomalies use PCAC and the possibility always remains that it is the latter idea, rather than the commutators that are at fault. Perhaps the experimental failure of the Sutherland-Veltman theorem is due to a rapid variator of $T(k^2)$ after all![15] Nevertheless some, model dependent, tests have been proposed, and it will be very interesting to see how they turn out; see Chapter IV, Section G.

The matrix elements of commutators can be measured experimentally, as the ST and Callan-Gross sum rules indicate. Unfortunately, with the excepti of the Ward identity violations, the anomalies affect only those commutators which are already model dependent on the canonical level. Thus if F_L comes out to be non-vanishing experimentally, we still shall not know whether this is due to a non-canonical value for the $[\dot{J}^i, J^i]$ ETC, or whether the quark model, which canonically gives $F_L = 0$, is not the correct expression for the underlying dynamics.

The most positive consequence of the anomalies remains the fact that the Sutherland-Veltman theorem for π^0 decay can be circumvented without introducing mysteriously rapid variation. On the other hand, the anomalies do not seem relevant to other problematic predictions of current algebra and PCAC: $\eta \to 3\pi$ and $K\ell_3$ decays.[16]

Should it develop eventually that commutator anomalies must be rejected, we believe that this can be done only if current commutators are divorced from Lagrangian field theory. At the present time schemes are being developed to provide a non-Lagrangian basis for current algebra.[17]

REFERENCES

[1] D. G. Boulware and R. Jackiw, *Phys. Rev.* 186, 1442 (1969).

[2] R. A. Brandt, *Phys. Rev.* 166, 1795 (1968). Similarly Brandt calculates the anomalous commutators relevant to the Sutherland-Veltman theorem, see (4.36); and he obtains results different from those given by the BJL theorem; *Phys. Rev.* 180, 1490 (1969). Brandt's technique is that of split points. The current is defined from the fields by introducing a small separation ϵ: $J_\epsilon^\mu(x) = \bar{\psi}(x+\epsilon/2)\gamma^\mu \psi(x-\epsilon/2)$. The commutator is evaluated first for finite ϵ, and then ϵ is set to zero (ϵ is a 4 vector).

$$[J^\mu(t,\vec{x}),J^\nu(t,\vec{y})] = \lim_{\substack{\epsilon \to 0 \\ \epsilon' \to 0}} [J_\epsilon^\mu(t,\vec{x}),J_\epsilon'^\nu(t,\vec{y})]$$

The reason for the discrepancy between this calculation and that of the BJL method has recently been explained by M. Chanowitz, *Phys. Rev.* D2, 3016 (1970). Chanowitz observes that, in the usual split point calculations, the limit of the *time* components of ϵ and ϵ' is taken to zero first. It is then possible to use canonical commutators to evaluate the current commutators, where the currents are still defined with *spatially* separated points. Finally the spatial separation is set to zero. On the other hand, if the commutator is calculated first for unequal times, $\epsilon_0 = \epsilon_0' \neq 0$; and then ϵ_0 and ϵ_0' are sent to zero *after* the spatial separation has been eliminated, then the BJL result is obtained. Therefore, according to Chanowitz

$$[J^\mu(t,\vec{x}),J^\nu(t,\vec{y})]_{BJL} = \lim_{\substack{\epsilon_0\to 0\\ \epsilon_0'\to 0}} \; \lim_{\substack{\vec{\epsilon}\to 0\\ \vec{\epsilon}\to 0}} [J_\epsilon^{\mu}(t,\vec{x}),J_{\epsilon'}^{\nu}(t,\vec{y})]$$

$$\neq \lim_{\substack{\vec{\epsilon}\to 0\\ \vec{\epsilon}'\to 0}} \; \lim_{\substack{\epsilon_0\to 0\\ \epsilon_0'\to 0}} [J_\epsilon^{\mu}(t,\vec{x}),J_{\epsilon'}^{\nu}(t,\vec{y})]$$

Note that our use of split point techniques, (4.47), was for purposes of calculating the *divergence* of the axial current, and not commutators. Such applications have never been cast into doubt. The split point technique has also been criticized by J. S. Bell and R. Jackiw, *Nuovo Cimento* **60**, 47 (1969), and D. G. Boulware and R. Jackiw, Ref. 1.

[3] D. G. Boulware and R. Jackiw, Ref. 1; R. A. Brandt, Ref. 2, also obtained a third derivative term by use of the split-point method.

[4] The non-occurrence of q number ST in spinor electrodynamics has also been noted by T. Nagylaki, *Phys. Rev.* **158**, 1534 (1967) and D. G. Boulware and J. Herbert, *Phys. Rev.* **D2**, 1055 (1970).

[5] K. Johnson and F. E. Low, *Progr. Theoret. Phys. (Kyoto) Suppl.* **37–38**, 74 (1966). When all space time indices select the time component: $\mu = \nu = \omega = 0$, no violation of the Jacobi identity has been found so far.

[6] F. Bucella, G. Veneziano, R. Gatto and S. Okubo, *Phys. Rev.* **149**, 1268 (1966).

[7] R. Jackiw and G. Preparata, *Phys. Rev.* **185**, 1748 (1969); S. L. Adler and Wu-Ki Tung, *Phys. Rev.* **D1**, 2846 (1970).

[8] J. S. Bell, *Nuovo Cimento* **47A**, 616 (1967).

[9] K. Wilson, *Phys. Rev.* D2, 1473 (1970); H. Georgi and J. Rawls, *Phys. Rev.* D3, 874 (1971); B. Schroer and J. Lowenstein, *Phys. Rev.* D3, 1981 (1971).

[10] For a summary, see, for example, K. Hepp in *Proceedings of the 8th Symposium*, N. Svartholm, ed. Interscience (John Wiley and Sons), New York (1969).

[11] M. Gell-Mann and F. E. Low, *Phys. Rev.* 95, 1300 (1954); M. Baker and K. Johnson, *Phys. Rev.* 183, 1292 (1969).

[12] K. Bitar and N. Khuri, *Phys. Rev.* D3, 462 (1971).

[13] Non standard regulator techniques are discussed by J. S. Bell and R. Jackiw, Ref. 2. The observation that these methods may lead to infinities is due to S. L. Adler, *Phys. Rev.* 177, 2426 (1969). Non minimal coupling is introduced by R. Jackiw and K. Johnson, *Phys. Rev.* 182, 1459 (1969).

[14] K. Wilson, *Phys. Rev.* 181, 1909 (1969); I. Gerstein and R. Jackiw, *Phys. Rev.* 181, 1955 (1969); W. Bardeen, *Phys. Rev.* 184, 1849 (1969); R. W. Brown, C.-C. Shih and B.-L. Young, *Phys. Rev.* 186, 1491 (1969); D. Amati, G. Bouchiat and J. L. Gervais, *Nuovo Cimento* 65A, 55 (1970).

[15] R. A. Brandt and G. Preparata, *Ann. Phys. (N.Y.)* 61, 119 (1970) proposed a non-conventional interpretation of PCAC, which allows for such rapid variation.

[16] For both these processes, current algebra and PCAC predict a suppression of an appropriate quantity. Experimentally no such suppression occurs. For $\eta \to 3\pi$ see D. G. Sutherland, *Nucl. Phys.* B2, 433 (1967); for $K\ell_3$ decays see C. G. Callan and S. B. Treiman, *Phys. Rev. Letters* 16, 153 (1966). Recently K. Wilson has provided a tentative explanation for the breakdown of PCAC in the η decay problem; see *Phys. Rev.* 179, 1499 (1969). R. A. Brandt and G. Preparata, Ref. 15, and *Nuovo Cimento Letters* 4, 80 (1970) have

proposed a non-conventional model of PCAC and SU(3) × SU(3) breaking, which escapes these problematical predictions. Their model has been criticized by M. Weinstein, *Phys. Rev.* D3, 481 (1971).

[17] K. Wilson, Ref. 16.

VII. APPROXIMATE SCALE SYMMETRY

A. Introduction

The experimental discovery of scaling in the electroproduction data has revived the long cherished notion, among symmetry minded physicists, that at high energies masses can be ignored, and a new symmetry sets in: the symmetry of scale or dilatation invariance. Let us examine how the experimental observations may lead one to consider the topic of scale transformations.

As was stated in Chapter V, the electroproduction measurements determine the commutator function.

$$C^{\mu\nu}(q,p) = \int \frac{d^4x}{(2\pi)^4} e^{iqx} < p|[J^\mu(x),J^\nu(0)]|p >$$

$$= -(g^{\mu\nu} - \frac{q^\mu q^\nu}{q^2}) \tilde{F}_1 + (p^\mu - q^\mu \frac{p\cdot q}{q^2})(p^\nu - q^\nu \frac{p\cdot q}{q^2}) \frac{\tilde{F}_2}{p\cdot q} \tag{7.1}$$

It can be verified that the $\tilde{F}_i$ are dimensionless; see Exercice 7.1. (Recall that dimensions are measured in units of mass; $\hbar = c = 1$.) The $\tilde{F}_i$ depend on the kinematical variables q^2 and ν. They also may be considered to be functions of whatever *dimensional* parameters determine the fundamental dynamics of the process. Examples of such additional parameters are masses of the "fundamental" particles that may be relevant. In addition it may be that the fundamental physical theory which governs natural phenomena is characterized by dimensional coupling constants. In that case these objects should also be included in the list of dimensional parameters on which the $\tilde{F}_i$ can depend. However, the hypothesis which enables one to proceed from the experimental fact of scaling of the $\tilde{F}_i$, to the theoretical considerations of scale invariance, is that there are *no* fundamental dimensional coupling constants. In that case the $\tilde{F}_i$ depend only on q^2, ν and the masses; and since the functions are dimensionless, they must depend on these variables through dimensionless ratios: $\tilde{F}_i = \tilde{F}_i = (q^2/m^2, \nu/m^2, \ldots)$. Here m^2 is some mass in the problem, and the dots indicate possible other mass ratios. One says that m "sets the scale" for the problem.

When the kinematical variables q^2 and ν are large, it is not implausible to suppose that the mass dependence of $\tilde{F}_i$ is unimportant. Hence the deep inelastic limit (q^2 and ν both large) is equivalent to the zero mass limit. In that case $\tilde{F}_i$ becomes a function only of the dimensionless ratio q^2/ν, and this is what is observed experimentally. When there are no dimensional parameters, other than the kinematical ones q^2 and ν, there is no scale against which q^2 and ν can be measured. Clearly a simultaneous rescaling of q and ν leaves q^2/ν unchanged. In position space this corresponds to rescaling the position variable x – in the absence of dimensional parameters such as rescaling leaves the theory invariant. Thus we are led to study the effects of the scale or dilatation transformation $x^\mu \rightarrow e^{-\rho} x^\mu$.

In the next subsection we shall derive the canonical theory of these transformations, as well as of the related conformal transformations. The currents that are conserved in the case of symmetry will be calculated. We shall then show how the canonical theory leads to Ward identities which may be used to derive theorems about the high energy behavior of Green's functions. Finally it will be demonstrated that in perturbation theory these results are false – there are anomalies which destroy them.[1]

B. Canonical Theory of Scale and Conformal Transformations

A scale transformation of the coordinates x^μ takes x^μ into $x^{\mu'} = e^{-\rho}x^\mu$. This is to *coordinate* transformation, in the same class of transformations as a translation: $x^\mu \to x^{\mu\,\prime} = x^\mu + a^\mu$, or a Lorentz transformation $x^\mu \to x^{\mu\,\prime} = \Lambda^{\mu\nu}x_\nu$. It is also useful, for reasons which will presently be obvious, to consider conformal coordinate transformations $x^\mu \to x^{\mu\,\prime} = (x^\mu - c^\mu x^2)\;(1-2cx + c^2x^2)^{-1}$. Here ρ, a^μ, $\Lambda^{\mu\nu}$ and c^μ are parameters which specify the dilatation, translation, Lorentz transformation and conformal transformation, respectively. Infinitesimally these transformations are

(7.2a) translation $$\delta_T{}^\alpha x^\mu = g^{\alpha\mu}$$

(7.2b) Lorentz transformation $$\delta_L{}^{\alpha\beta}x^\mu = g^{\alpha\mu}x^\beta - g^{\beta\mu}x^\alpha$$

(7.2c) dilatation $$\delta_D x^\mu = -x^\mu$$

(7.2d) conformal transformation $$\delta_C{}^\alpha x^\mu = 2x^\alpha x^\mu - g^{\alpha\mu}x^2 \;.$$

In exhibiting the formulas (7.2) we have not included the infinitesimal parameters specifying the transformation; this practice will be followed throughout.

The set of 15 transformations given in (7.2) forms a 15 parameter Lie group, called the conformal group. This is a generalization of the 10 parameter Poincaré group, formed from the 10 transformations (7.2a) and (7.2b). By considering the combined action of various infinitesimal transformations in different orders, the Lie algebra of the group may be abstracted. Upon defining P^α, $M^{\alpha\beta}$, D and K^α to be respectively the generators of translations, Lorentz transformations, dilatations and conformal transformations, we find

(7.3a) $$i[P^\alpha, P^\beta] = 0$$

(7.3b) $$i[M^{\alpha\beta},P^{\gamma}] = g^{\alpha\gamma}P^{\beta} - g^{\beta\gamma}P^{\alpha}$$

(7.3c) $$i[M^{\alpha\beta},M^{\mu\nu}] = g^{\alpha\mu}M^{\beta\nu} - g^{\beta\mu}M^{\alpha\nu} + g^{\alpha\nu}M^{\mu\beta} - g^{\beta\nu}M^{\mu\alpha}$$

(7.3d) $$i[D,P^{\alpha}] = P^{\alpha}$$

(7.3e) $$i[D,K^{\alpha}] = -K^{\alpha}$$

(7.3f) $$i[M^{\alpha\beta},K^{\gamma}] = g^{\alpha\gamma}K^{\beta} - g^{\beta\gamma}K^{\alpha}$$

(7.3g) $$i[P^{\alpha},K^{\beta}] = -2g^{\alpha\beta}D + 2M^{\alpha\beta}$$

(7.3h) $$i[D,D] = i[D,M^{\alpha\beta}] = i[K^{\alpha},K^{\beta}] = 0 \ .$$

The first three commutators define the Lie algebra of the Poincare group. In offering (7.3d) to (7.3h) we make no claim that these commutators are realized in nature. They merely reflect the combination rules for the transformations (7.2). Indeed the commutator (7.3d) violates physical considerations. For if (7.3d) holds in nature, then it is also true that $e^{i\alpha D}P^2\, e^{i\alpha D} = e^{2\alpha}P^2$. This implies that the mass spectrum is either continuous or all the masses vanish; neither eventuality is acceptable.[3]

Before establishing the conditions under which a Lagrangian $\mathcal{L}$, which depends on the fields ϕ, is scale invariant or conformally invariant, we must decide how the fields ϕ transform under the dilatation and conformal transformations. For translations and Lorentz transformations the rules are the standard ones.

(7.4a) $$\delta_T{}^{\alpha}\phi(x) = i[P^{\alpha},\phi(x)] = \partial^{\alpha}\phi(x)$$

(7.4b) $$\delta_L{}^{\alpha\beta}\phi(x) = i[M^{\alpha\beta},\phi(x)] = (x^{\alpha}\partial^{\beta} - x^{\beta}\partial^{\alpha} + \Sigma^{\alpha\beta})\phi(x)$$

For the remaining operations, the following choice is consistent with (7.3), and we adopt it.

(7.4c) $$\delta_D \phi(x) = (d+x\cdot\partial)\phi(x)$$

(7.4d) $$\delta_C{}^{\alpha}\phi(x) = (2x^{\alpha}x^{\nu}-g^{\alpha\nu}x^2)\,\partial_{\nu}\phi(x) + 2x_{\nu}(g^{\nu\alpha}d-\Sigma^{\nu\alpha})\phi(x)$$

Here d is a constant; it is called the scale dimension of the field ϕ.

It is now easy, following the general discussion of Chapter II, to establish the variation of $\mathcal{L}$.

(7.5a) $$\delta_D\mathcal{L} = \partial_{\nu}[x^{\nu}\mathcal{L}]-4\mathcal{L} + \pi_{\mu}(d+1)\phi^{\mu} + \frac{\delta\mathcal{L}}{\delta\phi}\, d\phi$$

(7.5b) $$\delta_C{}^{\alpha}\mathcal{L} = \partial_{\nu}[(2x^{\alpha}x^{\nu}-g^{\alpha\nu}x^2)\mathcal{L}]+2x^{\alpha}[\frac{\delta\mathcal{L}}{\delta\phi}\, d\phi+\pi_{\mu}(d+1)\phi^{\mu}-4\mathcal{L}]+V^{\mu}$$

V^{μ} is called the field virial and is given by

(7.6) $$V^{\mu} = \pi_{\alpha}[g^{\alpha\mu}d-\Sigma^{\alpha\mu}]\phi \quad .$$

In deriving (7.5), Lorentz and translation invariance of $\mathcal{L}$ are used.

Examining (7.5a) we see that dilatation invariance requires that

(7.7) $$-4\mathcal{L} + \pi_{\mu}(d+1)\phi^{\mu} + \frac{\delta\mathcal{L}}{\delta\phi}\, d\phi = 0 \quad .$$

If d is chosen to be 3/2 for Fermion fields, and 1 for Boson fields, then the kinetic energy term of the Lagrangian satisfies (7.7). These values for d correspond to the natural dimensions of fields, in units of mass. It is easy to see that (7.7) requires that the scale dimension of $\mathcal{L}$ be 4; i.e., that there be no dimensional parameters in $\mathcal{L}$. Clearly mass terms violate (7.7). Examples of interactions that satisfy (7.7) are ϕ^4 or $\bar{\psi}\psi\phi$.

For conformal invariance, (7.5b) shows that *two* conditions must be met. Firstly, scale invariance must obtain; i.e., (7.7) must be true. This is already seen from (7.3g): if P^α, K^β and $M^{\alpha\beta}$ are symmetry generators, so also must be D. Secondly, the field virial must be a total divergence

(7.8) $$V^\mu = \partial_\alpha \sigma^{\alpha\mu}$$

It is remarkable that for all renormalizable theories (7.8) is true, even though scale invariance of course is broken. Equation (7.8) is also true in all theories involving spins ≤ 1 without derivative coupling. This fact indicates that once a model for scale symmetry breaking has been adopted, one automatically is provided with a model for conformal symmetry breaking.

The currents associated with these transformations are now determined by the general techniques of Chapter II. The *canonical* dilatation current D_c^μ, and the *canonical* conformal current $K_c^{\alpha\mu}$ are

(7.9a) $$D_c^\mu = x_\alpha \theta_c^{\mu\alpha} + \pi^\mu d\phi$$

(7.9b) $$K_c^{\alpha\mu} = (2x^\alpha x_\nu - g_\nu{}^\alpha x^2)\theta_c^{\mu\nu} + 2x_\nu \pi^\mu (g^{\nu\alpha} d - \Sigma^{\nu\alpha})\phi - 2\sigma^{\alpha\mu} .$$

Note that the currents are expressed in terms of the *canonical* energy momentum tensor $\theta_c^{\mu\alpha}$. We have argued in Chapter II that the Belinfante tensor $\theta_B^{\mu\alpha}$ has a greater significance than $\theta_c^{\mu\alpha}$, therefore (7.9) should be expressed in terms of it. Inserting the formula for $\theta_B^{\mu\alpha}$ in terms of $\theta_c^{\mu\alpha}$ and the appropriate superpotential, (2.17), one finds after some tedious steps

(7.10a) $$D_c^\mu = x_\alpha \theta_B^{\mu\alpha} + V^\mu + \partial_\beta X^{\beta\mu}$$

(7.10b) $$K_c^{\alpha\mu} = [2x^\alpha x_\nu - g_\nu{}^\alpha x^2]\theta_B^{\nu\mu} + 2x^\alpha \partial_\nu \sigma_+^{\mu\nu} - 2\sigma_+^{\alpha\mu} + \partial_\beta X^{\beta\mu\alpha}$$

Here $\sigma_+^{\mu\nu}$ is the symmetric part of $\sigma^{\mu\nu}$; $X^{\beta\mu}$ and $X^{\beta\mu\alpha}$ are objects which can be explicitly given – the only property that need concern us here is that both are anti-symmetric in β and μ. Hence they are superpotentials which may be dropped without loss of physical content. Thus we are left with the Belinfante dilatation and conformal current

(7.11a) $$D_B^{\mu} = x_\alpha \theta_B^{\mu\alpha} + V^{\mu}$$

(7.11b) $$K_B^{\alpha\mu} = [2x^\alpha x_\nu - g_\nu{}^\alpha x^2]\theta_B{}^{\nu\mu} + 2x^\alpha \partial_\nu \sigma_+^{\mu\nu} - 2\sigma_+^{\alpha\mu} .$$

These currents may be further simplified by the following consideration. Instead of the Belinfante tensor, one may use, for discussions of translations and Lorentz transformations, the "new, improved energy momentum tensor" $\theta^{\mu\alpha}$.[2] This object is defined by adding to $\theta_B^{\mu\alpha}$ the following superpotential

(7.12a) $$\theta^{\mu\alpha} = \theta_B^{\mu\alpha} + \frac{1}{2}\partial_\lambda\partial_\rho X^{\lambda\rho\mu\nu}$$

$$X^{\lambda\rho\mu\nu} = g^{\lambda\rho}\sigma_+^{\mu\nu} - g^{\lambda\mu}\sigma_+^{\rho\nu} - g^{\lambda\nu}\sigma_+^{\mu\rho}$$

(7.12b) $$+ g^{\mu\nu}\sigma_+^{\lambda\rho} - \frac{1}{3}g^{\lambda\rho}g^{\mu\nu}\sigma_+{}^\alpha{}_\alpha + \frac{1}{3}g^{\lambda\mu}g^{\rho\nu}\sigma_+{}^\alpha{}_\alpha .$$

The additional superpotential does not destroy the conservation or symmetry properties of $\theta^{\mu\alpha}$, nor does it contribute to the translation or Lorentz generators. Hence $\theta^{\mu\alpha}$ may be used instead of $\theta_B^{\mu\alpha}$ as the Poincaré "current". For purposes of expressing the dilatation and conformal currents, $\theta^{\mu\alpha}$ is more convenient than $\theta_B^{\mu\alpha}$. In terms of $\theta^{\mu\alpha}$, the Belinfante dilatation and conformal currents are

(7.13a) $$D_B^{\mu} = x_\alpha \theta^{\mu\alpha} + \partial_\beta Y^{\beta\mu}$$

(7.13b) $$K_B^{\alpha\mu} = [2x^\alpha x_\nu - g_\nu{}^\alpha x^2]\theta^{\nu\mu} + \partial_\beta Y^{\beta\mu\alpha} .$$

The total divergence terms in (7.13) are again superpotentials; they again may be dropped and we are left with the final expressions for the currents.

(7.14a) $$D^\mu = x_\alpha \theta^{\mu\alpha}$$

(7.14b) $$K^{\mu\alpha} = [2x^\alpha x_\nu - g_\nu{}^\alpha x^2]\theta^{\nu\mu}$$

In addition to providing a compact form for D^μ and $K^{\mu\alpha}$, the new improved tensor has another advantage. It can be shown that in renormalized perturbation theory, its matrix elements are less singular than those of $\theta_B{}^{\mu\nu}$.[2]

The divergence of these currents can be expressed in terms of the trace of $\theta^{\mu\nu}$.

(7.15a) $$\partial_\mu D^\mu = \theta_\mu{}^\mu$$

(7.15b) $$\partial_\mu K^{\mu\alpha} = 2x^\alpha \theta_\mu{}^\mu = 2x^\alpha \partial_\mu D^\mu$$

Thus we see that both scale and conformal invariance is broken by $\theta_\mu{}^\mu$ (in theories where $V^\mu = \partial_\alpha \sigma^{\alpha\mu}$). In the subsequent we shall always assume that $\theta_\mu{}^\mu$ is non-zero by virtue of the presence of mass terms in $\mathcal{L}$. Thus scale and conformal symmetries are broken, as they must be in order to avoid a physically absurd mass spectrum.[3]

Explicit computation shows that V^μ and $\sigma^{\alpha\mu}$ are identically zero for spin ½ and spin 1 fields. For spin zero fields, we have

$$V^\mu = \pi^\mu \phi = \phi^\mu \phi = \frac{1}{2}\, \partial^\mu \phi^2$$

$$\sigma^{\alpha\mu} = \frac{1}{2}\, g^{\alpha\mu}\, \phi^2 .$$

An explicit formula for $\theta^{\mu\nu}$ may now be given.

$$\theta^{\mu\nu} = \theta_B^{\mu\nu} - \frac{1}{6} \sum_{\substack{\text{spin}\\ \text{zero}\\ \text{fields}}} (\partial^\mu \partial^\nu - g^{\mu\nu} \Box)\, \phi^2 \tag{7.17}$$

(The unique role of spin zero has not as yet been understood.)

Although for purposes of transformation theory one may use *any* energy-momentum tensor: $\theta_C^{\mu\alpha}$, $\theta_B^{\mu\alpha}$ or $\theta^{\mu\alpha}$; our use of $\theta^{\mu\alpha}$, implies that we attach a unique physical significance to this object. As discussed in Chapter II, this significance emerges when gravitational interactions are discussed. In the conventional gravity theory of Einstein general relativity, $\theta_B^{\mu\alpha}$ is the source of gravitons. If we wish to work with $\theta^{\mu\alpha}$, the theory must be changed so that $\theta^{\mu\alpha}$ is the source. This modification has been given. The modified theory is consistent with all the present tests of theory of relativity.[4]

The charges $D(t)$ and $K^\alpha(t)$

$$D(t) = \int d^3x \; x_\mu \theta^{\mu 0}(x) \tag{7.18a}$$

$$K^\alpha(t) = \int d^3x \; [2x^\alpha x_\nu - g_\nu{}^\alpha x^2] \theta^{\nu 0}(x) \quad , \tag{7.18b}$$

are not Lorentz covariant in the absence of symmetry. Also they do not satisfy the algebra (7.3); see Exercise 7.2. Nevertheless it is true that $D(t)$ and $K^\alpha(t)$ effect the proper transformation on the fields even in the absence of symmetry; see Exercise 7.3.

$$i[D(t), \phi(x)] = \delta_D \phi(x) \tag{7.19a}$$

$$i[K^\alpha(t), \phi(x)] = \delta_C{}^\alpha \phi(x) \tag{7.19b}$$

C. Ward Identities and Trace Identities

We have assumed $\theta_\mu{}^\mu$ to be given in a Lagrangian model by mass terms; for example when Fermions of mass m and Bosons of mass μ are present, $\theta_\mu{}^\mu = m\bar\psi\psi + \mu^2\phi^2$. Evidently the divergence of the dilatation current is soft; the dimension of the operators occurring in $\theta_\mu{}^\mu$ is at most 3. At the present time there appears to be two interesting possibilities for applications of these ideas to physics. One may attempt to repeat the success of PCAC: dominate $\theta_\mu{}^\mu$ by scalar mesons and unravel the low energy dynamics of these mesons.[5] We shall not consider this point of view, but rather examine the second possible application: determination of high-energy behavior of Green's functions. For definiteness we study the two-point function, the renormalized propagator.

Extracting consequences of our hypotheses about scale and conformal symmetry breaking is most easily accomplished from Ward identities satisfied by matrix elements of D^μ and $K^{\mu\alpha}$. Since these currents are simply related to the energy momentum tensor, it is useful to consider Ward identities satisfied by matrix elements of $\theta^{\mu\nu}$. We now present a derivation of these.

In order to proceed we need to know the commutator of $\theta^{\mu\nu}$ with renormalized field ϕ of scale dimensionality d. Under very general hypothesis, one can show that

(7.20a) $\qquad i[\theta^{00}(0,\vec{x}),\phi(0)] = \partial^0\phi(0)\delta(\vec{x}) + \Sigma^{0i}\phi(0)\partial_i\delta(\vec{x})$

(7.20b) $\quad i[\theta^{0i}(0,\vec{x}),\phi(0)[= \partial^i\phi(0)\delta(\vec{x}) - \frac{d}{3}\phi(0)\partial^i\delta(\vec{x}) + \frac{1}{2}\Sigma^{ij}\phi(0)\partial_j\delta(\vec{x})$.

There are the formal, canonical commutators. No statement is being made concerning their validity in perturbation theory. Since the commutator of $\theta^{\mu\nu}$ with ϕ necessarily contains gradient terms, the T product of $\theta^{\mu\nu}$ with ϕ is not covariant. In order to arrive at the covariant T* product, a covariantizing seagull $\tau^{\mu\nu}$ must be added. Hence we are led to consider

$$F_{ij}^{\mu\nu}(p,q) = \int d^4x\, d^4y\, e^{iqx}e^{ipy} < 0|T^*\theta^{\mu\nu}(x)\phi_i(y)\phi_j(0)|0>$$

(7.21)
$$= \int d^4x\, d^4y\, e^{iqx}e^{ipy} < 0|T^{\mu\nu}(x)\phi_i(y)\phi_j(0)|0> + \tau_{ij}^{\mu\nu}(p,q)$$

(7.22)
$$F_{ij}(p,q) = \int d^4x\, d^4y\, e^{iqx}e^{ipy} < 0|T\theta_\mu{}^\mu(x)\phi_i(y)\phi_j(0)|0> .$$

In the above i,j label the fields: the labels may be space time or internal indices. It is assumed that matrix elements of $\theta_\mu{}^\mu$ require no seagull. The covariantizing seagull $\tau_{ij}^{\mu\nu}$ may be explicitly constructed from the known commutators (7.20) by the method explained in Chapter II. Once the seagull is determined, the Ward identity may be derived. We do not present the details here, but merely record the resulting Ward identity.[6]

(7.23)
$$q_\mu F_{ij}^{\mu\nu}(p,q) = ip^\nu G(p) - i(p+q)^\nu G(p+q) + \frac{i}{2}\, q_\mu \Sigma_{ii'}{}^{\mu\nu}\, G_{i'j}(p+q) + \frac{i}{2}\, q_\mu \Sigma_{jj'}{}^{\mu\nu} G_{ij'}(p)$$

Also a trace identity is obtained.

(7.24)
$$g_{\mu\nu} F_{ij}^{\mu\nu}(p,q) = F_{ij}(p,q) - idG_{ij}(p+q) - idG_{ij}(p)$$

The terms in (7.24) additional to F_{ij} arise from the trace of $\tau_{ij}^{\mu\nu}$. In Eq. (7.23) and (7.24) G_{ij} is the renormalized propagator.

(7.25)
$$G_{ij}(p) = \int d^4x\, e^{ipx} < 0|T\phi_i(x)\phi_j(0)|0>$$

The formulae (7.23) and (7.24) contain all the restrictions that the various space time transformations (Lorentz, scale, and conformal) impose on the propagator. (Had we wished to study the n particle Green's function, we would consider the matrix element of $\theta^{\mu\nu}$ with n fields.) Once a model for scale symmetry breaking, e.g., mass terms is adopted, then one may deduce theorems about G(p). We now show explicitly how these restrictions are contained in Eq. (7.23) and (7.24).

1. *Lorentz transformations* Differentiated (7.23) with respect to q_α and set q to zero. This gives

$$F_{ij}^{\alpha\nu}(p,0) = -ig^{\alpha\nu}\, G_{ij}(p) - ip^\nu \frac{\partial}{\partial p_\alpha} G_{ij}(p)$$

$$+ \frac{i}{2}\Sigma_{ii'}{}^{\alpha\nu}\, G_{i'j}(p) + \frac{i}{2}\Sigma_{ij'}{}^{\alpha\nu} G_{ij'}(p) \quad . \tag{7.26}$$

Since $F_{ij}^{\alpha\nu}$ is symmetric in α and ν, we learn from (7.26) that

$$[p^\nu \frac{\partial}{\partial p_\alpha} - p^\alpha \frac{\partial}{\partial p_\nu}]G_{ij}(p) = \Sigma_{ii'}{}^{\alpha\nu} G_{i'j}(p) + \Sigma_{jj'}{}^{\alpha\nu} G_{ij'}(p) \quad . \tag{7.27}$$

This is the trivial and well known constraint of Lorentz covariance.

2. *Scale transformations* Form the trace of (7.26). We have (suppressing indices)

$$g_{\mu\nu} F^{\mu\nu}(p,0) = -4i\; G(p) - ip^\alpha \frac{\partial}{\partial p^\alpha} G(p) \quad . \tag{7.28a}$$

Combining (7.28a) with (7.24) at $q = 0$ leaves

$$F(p,0) = i(2d-4)G(p) - ip^\alpha \frac{\partial}{\partial p^\alpha} G(p) \quad . \tag{7.28b}$$

This provides a constraint on G(p), once F(p,0) is known, i.e., once we have a model for scale symmetry breaking.

3. *Conformal transformations* Differentiate (7.23) by

$$2\frac{\partial}{\partial q^{\alpha}}\frac{\partial}{\partial q^{\nu}} - g_{\alpha\nu}\frac{\partial}{\partial q^{\beta}}\frac{\partial}{\partial q_{\beta}}$$

and set $q = 0$. This gives

$$2\frac{\partial}{\partial q_{\alpha}} g_{\mu\nu} F_{ij}{}^{\mu\nu}(p,q)\Big|_{q=0} = -8i\frac{\partial}{\partial p_{\alpha}} G_{ij}(p) - 2i\, p_{\beta}\frac{\partial}{\partial p_{\beta}}\frac{\partial}{\partial p_{\alpha}} G_{ij}(p)$$

(7.29a)
$$= ip^{\alpha}\frac{\partial}{\partial p^{\beta}}\frac{\partial}{\partial p_{\beta}} G_{ij}(p) + 2i\,\Sigma_{ii'}^{\alpha\beta}\frac{\partial}{\partial p^{\beta}} G_{i'j}(p).$$

The left hand side of (7.29a) may be evaluated from (7.24). After a rearrangement of terms, we are left with

$$2\frac{\partial}{\partial q_{\alpha}} F_{ij}(p,q)\Big|_{q=0} = i(2d-8)\frac{\partial}{\partial p_{\alpha}} G_{ij}(p) - 2ip_{\beta}\frac{\partial}{\partial p_{\beta}}\frac{\partial}{\partial p_{\alpha}} G_{ij}(p)$$

(7.29b)
$$+ ip^{\alpha}\frac{\partial}{\partial p^{\beta}}\frac{\partial}{\partial p_{\beta}} G_{ij}(p) + 2i\,\Sigma_{ii'}^{\alpha\beta}\frac{\partial}{\partial p^{\beta}} G_{i'j}(p) .$$

This equation may be simplified by using (7.27) and (7.28b). First d is eliminated between (7.29b) and (7.28b). This gives

$$2\frac{\partial}{\partial q_{\alpha}} F_{ij}(p,q)\Big|_{q=0} - \frac{\partial}{\partial p_{\alpha}} F_{ij}(p,0) =$$

(7.29c)
$$i\frac{\partial}{\partial p^{\beta}}\left[p^{\alpha}\frac{\partial}{\partial p_{\beta}} G_{ij}(p) - p^{\beta}\frac{\partial}{\partial p_{\alpha}} G_{ij}(p) + 2\,\Sigma_{ii'}^{\alpha\beta} G_{i'j}(p)\right] .$$

Next we use (7.27)

$$2\frac{\partial}{\partial p_\alpha}F_{ij}(p,q)\Big|_{q=0} - \frac{\partial}{\partial p_\alpha}F_{ij}(p,0) =$$

$$i\frac{\partial}{\partial p_\beta}\left[\Sigma_{ii'}^{\alpha\beta}G_{i'j}(p)-\Sigma_{jj'}^{\alpha\beta}G_{ij'}(p)\right] . \tag{7.29d}$$

Equation (7.29d) determines the constraint on G which follows from a model for conformal symmetry breaking.

D. False Theorems

The constraint equations (7.28b) and (7.29d) are, of course, without content as long as F, the symmetry breaking, is not specified. We now show that if a model for F is taken from the canonical result that $\theta_\mu^{\ \mu}$ is given only by mass terms; and furthermore, if the canonical value for d is accepted, theorems about G can be deduced, which however are contradicted by explicit calculation. We shall only discuss broken scale invariance; no further reference to conformal transformation will be made.

Consider for definiteness the propagator for a theory of spin zero fields with mass μ and a quartic self-interaction, $\lambda\phi^4$, which is scale invariant. The propagator may be written in the form

$$G(p) = \frac{i}{p^2}g(p^2/\mu^2) . \tag{7.30}$$

We find from (7.28b) that g satisfies

$$\frac{p^2}{2}F(p,0) = \frac{p^2}{\mu^2}g'(p^2/\mu^2) + (1-d)g(p^2/\mu^2) . \tag{7.31}$$

As $\mu^2 \to 0$, one might expect that the left hand side above vanishes, since F is the matrix element of θ^μ_μ which formally is $\mu^2\phi^2$. On the right hand side, this is equivalent to $p^2 \to \infty$. Hence we find

$$\lim_{p^2 \to \infty} g(p^2/\mu^2) \propto (p^2/\mu^2)^{d-1} \quad . \tag{7.32}$$

Since $d = 1$ for Boson fields, we further conclude that the Boson propagator goes as $1/p^2$ for large p^2.

This result is manifestly false in perturbation theory where it is known that logarithmic terms are present in the asymptotic domain. Thus we must abandon the steps which lead from the true (by definition) Eq. (7.31) to the false result. Specifically we cannot conclude that $d = 1$ and that F vanishes with the mass.

E. True Theorems

Detailed calculation in perturbation theory *in lowest non trivial order of the interaction* yield the following conclusions. It remains possible to assert that F vanishes with the mass. However, d changes from its canonical value of 1. To exhibit the change in d, we consider the definition of that object.

$$i[D(0),\phi(0)] = i\int d^3x\, x_i[\theta^{0i}(0,\vec{x}),\phi(0)] = d\phi(0) \tag{7.33}$$

The commutator is evaluated by the BJL prescription. Specifically an application of this technique to $F^{\mu\nu}(p,q)$ gives, by definition,

$$\lim_{q_0 \to \infty} \frac{\partial}{\partial q^i} q_0 F^{0i}(p,q)\Big|_{\vec{q}=0} = idG(p) \quad . \tag{7.34}$$

Hence the true value of d may be computed from the high-energy behavior of $F^{\mu\nu}$. Explicit calculation in lowest order gives

(7.35) $$d = 1 + c\lambda^2$$

where c is a well defined positive numerical constant. Substituting this value of d into Eq. (7.32) (which remains valid to lowest order, since F *does* vanish with the masses), we find

(7.36) $$\lim_{p^2 \to \infty} g(p^2/\mu^2) \propto (p^2/\mu^2)^{c\lambda^2} \approx 1 + c\lambda^2 \log p^2/\mu^2 \ .$$

Explicit calculation of the propagator to the same order, verifies (7.36) with precisely the same coefficient. Perturbative calculations for several models have been performed, and the conclusion is always the same, *in lowest order:* although the scale breaking term vanishes with the masses, the dimension changes, and the resulting theorem about high energy behavior is verified by comparison with explicit calculation of the propagator.[7]

The situation to higher orders has been studied both by explicit calculations and by general analysis of the structure of Feynman graphs.[8] The conclusion is that d continues to migrate from the canonical value. However a new phenomenon sets in: F does not vanish with the masses; i.e., scale symmetry is broken by terms other than masses. This "anomalous", non-canonical scale breaking can be understood in the following way. In calculating matrix elements of $\theta^{\mu\nu}$ it is necessary that they be conserved. This is the requirement of Poincaré covariance. However, in specific calculations these matrix elements are *not* conserved, and conservation is achieved for example, by Pauli-Villars regularization. One defines $\theta_{Reg}{}^{\mu\nu} = \theta^{\mu\nu} - \theta_M{}^{\mu\nu}$, where $\theta_M{}^{\mu\nu}$ is formed from regulator fields $\bar{\phi}$ carrying mass M. Physical, conserved matrix elements are obtained by letting $M \to \infty$. Consider now the trace of $\theta_{Reg}{}^{\mu\nu}$, which breaks scale invariance. Evidently we have

(7.37) $$g_{\mu\nu}\theta_{Reg}{}^{\mu\nu} = \mu^2\phi^2 - M^2\bar{\phi}^2 \ .$$

Thus if matrix elements of $\tilde{\phi}^2$ behave as M^{-2} for large M, the regulator contribution to (7.37) survives, even in the physical limit $M \to \infty$. Specific calculation shows that $\tilde{\phi}^2$ does, indeed, behave in this fashion. Therefore even when μ^2 is zero, $g_{\mu\nu}\theta_{Reg}^{\mu\nu}$ does not vanish. (The analogy with the anomalous divergence of the axial current is clear.)

At the present time attention is being directed to the question of determining the precise form of the non-canonical scale breaking terms. It is believed that they are absent if the coupling constant renormalization factor is finite. In perturbation theory this object is infinite; it is not yet known whether in the complete theory it can be finite. If it becomes established that local quantum field theory does not scale at high energy, one will have to turn to non-Lagrangian models to implement this physically attractive idea.

REFERENCES

[1] The treatment here follows C. G. Callan, Jr., S. Coleman and R. Jackiw, *Ann. Phys. (N.Y.)* 59, 42 (1970) and S. Coleman and R. Jackiw, *Ann. Phys. (N.Y.)* 67, 552 (1971). For a less general discussion, and references to older literature, see G. Mack and A. Salam, *Ann. Phys. (N.Y.)* 53, 174 (1969).

[2] C. G. Callan, Jr., S. Coleman and R. Jackiw, Ref. 1.

[3] It is perhaps possible that these symmetries are exact on the Lagrangian level; yet a non-trivial mass spectrum is present because the vacuum is degenerate. This example of Goldstone phenomenon has been studied by A. Salam and J. Strathdee, *Phys. Rev.* 184, 1750 and 1760 (1969). We shall not be considering this point of view.

[4] C. G. Callan, Jr., S. Coleman and R. Jackiw, Ref. 2; see also F. Gursey, *Ann. Phys. (N.Y.)* 24, 211 (1963) and R. Penrose, *Proc. Roy. Soc. (London)* 284A, 159 (1965).

[5] This approach has been taken by M. Gell-Mann, Hawaii Summer School lectures (1969), and P. Carruthers, *Phys. Rev.* D2, 2265 (1970). The main obstacle to rapid advance in this direction seems to be the experimental fact that there are no low-lying scalar mesons.

[6] A detailed derivation of (7.23) and (7.24) is given in the first two papers of Ref. 1.

[7] These calculations are given by S. Coleman and R. Jackiw, Ref. 1. Similar calculations for the $\lambda\phi^4$ theory have been performed by K. Wilson, *Phys. Rev.* D2, 1478 (1970).

[8] S. Coleman and R. Jackiw, Ref. 1; K. Wilson, Ref. 7; C. G. Callan, Jr., *Phys. Rev.* D2, 1541 (1970); K. Symanzik, *Commun. Math. Phys.* 18, 227 (1970).

APPENDIX

A EXERCISES

2–1. By using only the canonical commutation relations (2.1), show that the space time "charges", P^{α}, associated with the translation current $\theta_c^{\mu\alpha}$, $P^{\alpha} = \int d^3x \theta_c^{0\alpha}(t,\vec{x})$, generate the correct transformation on the fields.

$$i[P^{\alpha},\phi(x)] = \partial^{\alpha}\phi(x) = \delta_T^{\alpha}\phi(x)$$

2–2. Consider Lorentz transformations.

$$\delta_L^{\alpha\beta}\phi(x) = (x^{\alpha}\partial^{\beta}-x^{\beta}\partial^{\alpha} + \Sigma^{\alpha\beta})\phi(x)$$

Here $\Sigma^{\alpha\beta}$ is the spin matrix appropriate to the field $\phi(x)$. It specifies the representation of Lorentz group under which $\phi(x)$ transforms. $\sigma^{\alpha\beta} = 0$ for spin zero Bosons; $\Sigma_{(ij)}{}^{\alpha\beta} = \frac{i}{2}\sigma_{ij}{}^{\alpha\beta}$ for spin ½ Fermions; (ij are the Dirac indices appropriate to a 4 component Fermion field); $\Sigma_{(\mu\nu)}{}^{\alpha\beta} = g_{\mu}{}^{\alpha}g_{\nu}{}^{\beta}-g_{\nu}{}^{\alpha}g_{\mu}{}^{\beta}$ for vector bosons ($\mu\nu$ are the space-time indices appropriate to a vector field). Under what conditions, on a translationally invariant $\mathcal{L}$, is the above a symmetry transformation of the theory? Show that the conserved, canonical, space-time current appropriate to Lorentz transformation is

$$M_c^{\mu\,\alpha\beta}(x) = x^{\alpha}\theta_c^{\mu\beta}(x)-x^{\beta}\theta_c^{\mu\alpha}(x)+ \pi^{\mu}(x)\Sigma^{\alpha\beta}\phi(x) \ .$$

By use of the canonical commutators (2.1), verify that the "charges"

$$M^{\alpha\beta} = \int d^3x\, M_c^{0\,\alpha\beta}(t,\vec{x}),$$

generate the correct transformations on the fields.

$$i[M^{\alpha\beta},\phi(x)] = \delta_L^{\alpha\beta}\,\phi(x)$$

2–3. Consider the Belinfante tensor $\theta_B^{\mu\alpha}$ defined in (2.17). Show that the Belinfante tensor is conserved when $\theta_c^{\mu\alpha}$ is. Next show that the charge P^α, defined in Exercise II–1, are the same regardless whether they are constructed from $\theta_B^{\mu\alpha}$ or $\theta_c^{\mu\alpha}$. Therefore for purposes of describing translations, $\theta_B^{\mu\alpha}$ may be used instead of $\theta_c^{\mu\alpha}$. $\theta_B^{\mu\alpha}$ has additional advantages. With the help of the equations of motion, as well as the condition for Lorentz covariance, derived in Exercise 2–2, show that $\theta_B^{\mu\alpha}$ is symmetric in μ and α, while $\theta_c^{\mu\alpha}$ has this property only for spinless fields $\Sigma^{\alpha\beta} = 0$.

The Belinfante tensor takes on a significance over the canonical tensor in connection with the space-time current associated with Lorentz transformation. Consider

$$M_B^{\mu\,\alpha\beta}(x) = x^\alpha \theta_B^{\mu\beta}(x) - x^\beta \theta_\beta^{\mu\alpha}(x) \quad .$$

Show that $M_B^{\mu\,\alpha\beta}(x)$ is conserved when $M_c^{\mu\,\alpha\beta}(x)$, defined in Exercise 2–2, possesses this property. Also show that $M_B^{\mu\,\alpha\beta}(x)$, leads to the same charges $M^{\alpha\beta}$ as $M_c^{\mu\,\alpha\beta}(x)$ does. Hence $M_B^{\mu\,\alpha\beta}$ can be used instead of $M_c^{\mu\,\alpha\beta}$ as the space time current for Lorentz transformations; however $M_B^{\mu\,\alpha\beta}$ is much simpler than $M_c^{\mu\,\alpha\beta}$, since the former contains no explicit spin term $\pi^\mu\,\Sigma^{\alpha\beta}\phi$. The full significance of the Belinfante tensor emerges when one asserts that it is this energy-momentum tensor (rather than any other one) to which gravitons couple in the Einstein theory of gravity, general relativity.

2–4. By use of canonical expressions for θ_c^{00} and J_0^a, as well as the canonical commutators, derive the equal time commutator between these two operators, given in (2.18). Verify that the answer remains unchanged when θ_c^{00} is replaced by θ_B^{00}.

2–5. Consider the scalar field Lagrangian.

$$\mathcal{L}(x) = \tfrac{1}{2}\,\phi^\mu(x)\phi_\mu(x) + gx^2\phi(x)$$

Construct the energy momentum tensor from the formula (2.11), and show that it is not conserved. Evaluate the commutator $[P^0(t), P^i(t)]$ and verify that it does not vanish. (When translation invariance holds this commutator vanishes.) Construct the Lorentz current for this model, and verify that it is conserved.

2–6. Verify that (2.24) is the most general solution of (2.23b). Hint: Multiply (2.23b) by y_i and integrate over $\vec{y}$.

2–7. Consider the ETC of electromagnetic currents which hold in scalar electrodynamics.

$$[J^0(0,\vec{x}),J^0(0)] = 0$$

$$[J^0(0,\vec{x}),J^i(0)] = S(0)\partial^i\delta(\vec{x})$$

$$[J^i(0,\vec{x}),J^i(0)] = 0$$

Here $S(y)$ is a scalar operator. Rewrite this commutator in a Lorentz covariant, but frame dependent fashion by introducing the unit time like vector n^μ. Consider also

$$TJ^\mu(x)J^\nu(0) \equiv T^{\mu\nu}(x;n) \quad .$$

Show that this is not covariant, and construct a covariantizing seagull $\tau^{\mu\nu}(x;n)$. Determine the seagull by requiring that $T^{*\mu\nu}(x) = T^{\mu\nu}(x;n) + \tau^{\mu\nu}(x;n)$ be conserved.

3–1. Consider the free Boson propagator $D(q) = \frac{i}{q^2 - m^2}$. By use of the BJL theorem verify the canonical commutation relations.

$$i[\phi(0,\vec{x}),\phi(0)] = i[\dot{\phi}(0,\vec{x}),\dot{\phi}(0)] = 0$$

$$i[\dot{\phi}(0,\vec{x}),\phi(0)] = \delta(\vec{x})$$

Next consider the full propagator of renormalized fields. It may be written in the form

$$G(q) = i\int_0^\infty da^2 \frac{\rho(a^2)}{q^2 - a^2} .$$

The spectral function ρ can be shown to be non-negative. (A renormalized field $\tilde{\phi}$ is proportionally related to the unrenormalized field ϕ by $\phi = Z^{1/2}\tilde{\phi}$. In perturbation theory Z is cutoff dependent.) What can you deduce about the vacuum expectation of the canonical commutators in the complete theory?

3–2. Consider the vacuum polarization tensor, which (formally) can be written as

$$T^{*\mu\nu}(q) = \int d^4x e^{iqx} < 0|T^* J^\mu(x) J^\nu(0)|0> =$$

$$(g^{\mu\nu}q^2 - q^\mu q^\nu)\int_0^\infty da^2 \frac{\sigma(a^2)}{q^2 - a^2} .$$

It can be shown that $\sigma(a^2)$ has a definite sign. Extract the T product from $T^{*\mu\nu}(q)$ and calculate $< 0|[J^0(0,\vec{x}),J^i(0)]|0>$ in terms of $\sigma(a^2)$. (Do not concern yourselves here with problems of convergence; these will be discussed in Exercise 6–1.)

4–1. Show that

$$\Delta^\mu(a) = i\int \frac{d^4r}{(2\pi)^4}\left[\frac{r^\mu+a^\mu}{([r+a]^2-m^2)^2} - \frac{r^\mu}{(r^2-m^2)^2}\right] = \frac{-a^\mu}{32\pi^2}\,.$$

This verifies (4.19b).

4–2. Show that

$$\Delta^{\alpha\mu\nu}(p,q|a) = i\int \frac{d^4r}{(2\pi)^4}\,\mathrm{Tr}\gamma^5\gamma^\alpha\Big[[r_\beta\gamma^\beta+a_\beta\gamma^\beta+p_\beta\gamma^\beta-m]^{-1}$$

$$\gamma^\mu[r_\beta\gamma^\beta+a_\beta\gamma^\beta-m]^{-1}\gamma^\nu[r_\beta\gamma^\beta+a_\beta\gamma^\beta+q_\beta\gamma^\beta-m]^{-1}$$

$$-[r_\beta\gamma^\beta+p_\beta\gamma^\beta-m]^{-1}\gamma^\mu[r_\beta\gamma^\beta-m]^{-1}\gamma^\nu[r_\beta\gamma^\beta-q_\beta\gamma^\beta-m]^{-1}\Big]$$

$$= \frac{-1}{8\pi^2}\,\epsilon^{\alpha\mu\nu\beta}\,a_\beta\,.$$

This verifies (4.21).

4–3. Show that

$$i\int \frac{d^4r}{(2\pi)^4}\,\mathrm{Tr}\gamma^5\gamma^\alpha\Big[[r_\beta\gamma^\beta+q_\beta\gamma^\beta-m]^{-1}\gamma^\nu[r_\beta\gamma^\beta-p_\beta\gamma^\beta-m]^{-1}$$

$$-[r_\beta\gamma^\beta+p_\beta\gamma^\beta-m]^{-1}\gamma^\nu[r_\beta\gamma^\beta-q_\beta\gamma^\beta-m]^{-1}\Big]$$

$$= \frac{1}{4\pi^2}\,\epsilon^{\alpha\mu\nu\beta}\,p_\mu q_\beta\,.$$

This verifies (4.25c).

5–1. Compute canonically the symmetric part of

$$\int d^3x[\dot{J}^i(0,\vec{x}),J^i(0)] \quad ,$$

in the quark-vector gluon model, where

$$i\gamma_\mu \partial^\mu \psi = m\psi - g\gamma^\mu \psi B_\mu \quad .$$

Verify that the spin averaged, matrix element of that commutator, between diagonal proton states is of the form

$$A(\delta^{ij}\vec{p}^2 - p^i p^j) + B\delta^{ij}.$$

5–2. Assume that the dispersion relation for T_L needs one subtraction, performed at $\nu = 0$, $(\omega = \infty)$. Show that the ST sum rule reduces in that instance to an uninformative relation between the subtraction term and the ST.

6–1. When the spectral function $\sigma(a^2)$, relevant to the vacuum polarization tensor $T^{*\mu\nu}$, does not vanish as $a^2 \to \infty$, the dispersive representation for $T^{*\mu\nu}$ given in Exercise 3–2 will not converge. Assuming that

$$\lim_{a^2 \to \infty} \sigma(a^2) = A$$

$$\lim_{a^2 \to \infty} a^2 \sigma(a^2) = B \quad ,$$

the following subtracted form for $T^{*\mu\nu}$ may be established

$$T^{*\mu\nu}(q) = \int d^4x \, e^{iqx} < 0|T^* J^\mu(x) J^\nu(0)|0 >$$

$$= (g^{\mu\nu}q^2 - q^\mu q^\nu)\left[C + q^2 \int_{4m^2}^{\infty} da^2 \, \frac{\sigma(a^2)}{a^2(q^2 - a^2)} \right] .$$

Here C is a subtraction constant, and $4m^2$ is the appropriate threshold for the dispersive integral. Calculate the vacuum expectation value of the $[J^0,J^i]$ ETC, and show that a triple derivative of the δ function is present. Hint: Let $\tilde{\sigma}(a^2) = \sigma(a^2) - A - B/a^2$.

6–2. Consider T(p,q), defined by

$$T(p,q) = \int d^4x d^4y\, e^{ipx} e^{iqy} < \alpha | TA(x)B(y)C(0) | \beta > ,$$

where α and β are arbitrary states. Show that

$$\lim_{q_0 \to \infty} \lim_{p_0 \to \infty} -p_0 q_0 T(p,q) =$$

$$= \int d^3x d^3y\, e^{-i\vec{p}\cdot\vec{x}} e^{-i\vec{q}\cdot\vec{y}} < \alpha | [B(0,\vec{y}),[A(0,\vec{x}),C(0)]] | \beta > .$$

What is the result if the limit is performed on opposite order,
$\lim_{p_0 \to \infty} \lim_{q_0 \to \infty} -p_0 q_0 T(p,q)$?

7–1. Show that the invariant functions $\tilde{F}_i$, defined from

$$C^{\mu\nu}(q,p) = \int \frac{d^4x}{(2\pi)^4} e^{iqx} < p | [J^\mu(x),J^\nu(0)] | p > =$$

$$- (g^{\mu\nu} - \frac{q^\mu q^\nu}{q^2})\tilde{F}_1 + (p^\mu - q^\mu \frac{p\cdot q}{q^2})(p^\nu - q^\nu \frac{p\cdot q}{q^2}) \frac{\tilde{F}_2}{p\cdot q}$$

are dimensionless. The state $|p>$ is normalized covariantly.

7–2. Compute $i[D(t),P^\mu]$ and $i[D(t),M^{\mu\nu}]$, where D(t) is the dilatation "charge".

7–3. By use of canonical ETC, show that

$$i[D(t),\phi(x)] = \delta_D \phi(x)$$

$$i[K^\alpha(t),\phi(x)] = \delta_C{}^\alpha\phi(x) \ .$$

Here D(t) and $K^\alpha(t)$ are the dilatation and conformal "charges" respectively. Hint: Use the canonical formulas for the currents.

7–4. Consider a scalar field Lagrangian $\mathcal{L}$ depending on ϕ and $\partial^\mu\phi$. What is the most general form for $\mathcal{L}$ which is scale invariant? What is the most general form that is conformly invariant? Derive $\theta_c{}^{\mu\nu}$ for this conformaly invariant theory. Since this is a spinless theory, $\theta_c{}^{\mu\nu} = \theta_B{}^{\mu\nu}$. What is $\theta^{\mu\nu}$?

B SOLUTIONS

2–1. It is clear that the argument of $\phi(x)$ is inessential. Hence we consider

$$i[P^\alpha,\phi(0)] = \int d^3y\ i[\theta_c{}^{0\alpha}(0,y),\phi(0)] \ .$$

(a) $\alpha = i$

$$\begin{aligned} i[P^i,\phi(0)] &= \int d^3y\ i[\pi^0(0,\vec{y})\ \phi^i(0,\vec{y}),\phi(0)] \\ &= \int d^3y\ i[\pi^0(0,\vec{y})\ \phi(0)]\phi^i(0,\vec{y}) \\ &= \int d^3y\ \delta(\vec{y})\phi^i(0,\vec{y}) \\ &= \phi^i(0) \end{aligned}$$

(b) $\alpha = 0$

$$\begin{aligned} i[P^0,\phi(0)] &= \int d^3y\ i[\pi^0(0,\vec{y}),\phi^0(0,\vec{y}) - \mathcal{L}(0,\vec{y}),\phi(0)] \\ &= \phi^0(0) \\ &\quad + \int d^3y\ (\pi^0(0,\vec{y})i[\phi^0(0,\vec{y}),\phi(0)] - i[\mathcal{L}(0,\vec{y}),\phi(0)]) \end{aligned}$$

We now show that the second term in the above is zero. To do this, we introduce the technique of functional differentiation with respect to $\phi(0)$ and $\pi^0(0)$. By definition,

$$\frac{\delta'\phi(0,\vec{x})}{\delta'\phi(0)} = \delta(\vec{x}) \ , \qquad \frac{\delta'\phi^i(0,\vec{x})}{\delta'\phi(0)} = \partial^i\delta(\vec{x})$$

$$\frac{\delta'\pi^0(0,\vec{x})}{\delta'\phi(0)} = 0 \ , \qquad \frac{\delta'\phi(0,\vec{x})}{\delta'\pi^0(0)} = 0 \ ,$$

$$\frac{\delta'\partial^i\phi(0,\vec{x})}{\delta'\pi^0(0)} = 0 \ , \qquad \frac{\delta'\pi^0(0,\vec{x})}{\delta'\pi^0(0)} = \delta(\vec{x}) \ .$$

The quantity ϕ^0 is considered to be a functional of both π^0 and ϕ. For any functional F which depends on both ϕ and ϕ^μ, the chain rule holds

$$\frac{\delta' F(0,\vec{x})}{\delta'\phi(0)} = \frac{\delta F(0,\vec{x})}{\delta\phi}\delta(\vec{x}) + \frac{\delta F(0,\vec{x})}{\delta\phi^i}\partial^i\delta(\vec{x}) + \frac{\delta F(0,\vec{x})}{\delta\phi^0}\frac{\delta'\phi^0(0,\vec{x})}{\delta'\phi(0)}$$

$$\frac{\delta' F(0,\vec{y})}{\delta'\pi^0(0)} = \frac{\delta F(0,\vec{x})}{\delta\phi^0}\frac{\delta'\phi^0(0,\vec{x})}{\delta'\pi^0(0)} \ .$$

Here δ indicates the ordinary variational derivative; δ' is the functional derivative. Applying this formalism, gives

$$\pi^0(0,\vec{y})i[\phi^0(0,\vec{y}),\ \phi(0)] - i[\mathfrak{L}(0,\vec{y}),\phi(0)]$$

$$= \pi^0(0,\vec{y})\frac{\delta'\phi^0(0,\vec{y})}{\delta'\pi^0(0)} - \frac{\delta'\mathfrak{L}(0,\vec{y})}{\delta'\pi^0(0)}$$

$$= \pi^0(0,\vec{y})\frac{\delta'\phi^0(0,\vec{y})}{\delta'\pi^0(0)} - \frac{\delta\mathfrak{L}(0,\vec{y})}{\delta\phi^0}\frac{\delta'\phi^0(0,\vec{y})}{\delta'\pi^0(0)} = 0 \ .$$

The definition of π^0 was used: $\pi^0 = \dfrac{\delta\mathcal{L}}{\delta\phi^0}$.

(c) Note that we have established the stronger result

$$i[\theta_c{}^{0\alpha}(t,\vec{y}),\phi(t,\vec{x})] = \phi^\alpha(x)\delta(\vec{x}-\vec{y})$$

2–2.

(a)
$$\delta_L^{\alpha\beta}\mathcal{L} = \pi_\mu\delta_L^{\alpha\beta}\phi^\mu + \frac{\delta\mathcal{L}}{\delta\phi}\delta_L^{\alpha\beta}\phi$$

$$= \pi_\mu\partial^\mu\delta_L^{\alpha\beta}\phi + \frac{\delta\mathcal{L}}{\delta\phi}\delta_L^{\alpha\beta}\phi$$

$$= \pi_\mu(g^{\mu\alpha}\phi^\beta - g^{\mu\beta}\phi^\alpha + (x^\alpha\partial^\beta - x^\beta\partial^\alpha)\phi^\mu + \Sigma^{\alpha\beta}\phi^\mu)$$

$$+ \frac{\delta\mathcal{L}}{\delta\phi}(x^\alpha\partial^\beta\phi - x^\beta\partial^\alpha\phi) + \frac{\delta\mathcal{L}}{\delta\phi}\Sigma^{\alpha\beta}\phi$$

By translation invariance, the above may be written as

$$\delta^{\alpha\beta}\mathcal{L} = x^\alpha\partial^\beta\mathcal{L} - x^\beta\partial^\alpha\mathcal{L} + \pi_\mu\,\Sigma^{\alpha\beta}\phi^\mu + \frac{\delta\mathcal{L}}{\delta\phi}\Sigma^{\alpha\beta}\,\phi + \pi^\alpha\phi^\beta$$
$$- \pi^\beta\phi^\alpha$$
$$= \partial_\mu[g^{\mu\beta}x^\alpha\mathcal{L} - g^{\mu\alpha}x^\beta\mathcal{L}] + \pi_\mu\,\Sigma^{\alpha\beta}\phi^\mu + \frac{\delta\mathcal{L}}{\delta\phi}\Sigma^{\alpha\beta} +$$
$$\pi^\alpha\phi^\beta - \pi^\beta\phi^\alpha \ .$$

Hence the condition for Lorentz covariance of the theory is

$$\pi_\mu\Sigma^{\alpha\beta}\phi^\mu + \frac{\delta\mathcal{L}}{\delta\phi}\Sigma^{\alpha\beta}\phi = \pi^\beta\phi^\alpha - \pi^\alpha\phi^\beta \ .$$

(b) In the notation of Chapter II,

$$\Lambda^{\mu\ \alpha\beta} = x^\alpha g^{\mu\beta}\mathcal{L} - x^\beta g^{\mu\alpha}\mathcal{L} \quad ,$$

and the canonical conserved current is

$$M_c^{\mu\ \alpha\beta} = \pi^\mu \delta_L^{\alpha\beta}\phi - \Lambda^{\mu\ \alpha\beta}$$

$$= \pi^\mu(x^\alpha\phi^\beta - x^\beta\phi^\alpha + \Sigma^{\alpha\beta}\phi) - x^\alpha g^{\mu\beta}\mathcal{L} - x^\beta g^{\mu\alpha}\mathcal{L}$$

$$= x^\alpha\theta_c{}^{\mu\beta} - x^\beta\theta_c{}^{\mu\alpha} + \pi^\mu\Sigma^{\alpha\beta}\phi \quad .$$

(c) Consider

$$i[M_c^{0\ \alpha\beta}(t,\vec{y}),\phi(t,\vec{x})] =$$

$$= iy^\alpha[\theta_c{}^{0\beta}(t,\vec{y}),\phi(t,\vec{x})] - iy^\beta[\theta_c{}^{0\alpha}(t,\vec{y}),\phi(t,\vec{x})]$$

$$+ i[\pi^0(t,\vec{y})\,\Sigma^{\alpha\beta}\phi(t,\vec{y}),\phi(t,\vec{x})] \ .$$

The commutators with $\theta_c{}^{0\alpha}$ are evaluated in Exercise 2-1. Hence

$$i[M_c^{0\ \alpha\beta}(t,\vec{y}),\phi(t,\vec{x})] = (x^\alpha\partial^\beta\phi(x) - x^\beta\partial^\alpha\phi(x) + \Sigma^{\alpha\beta}\phi(x)\delta(\vec{x}-\vec{y})$$

$$= \delta_L^{\alpha\beta}\phi(x)\delta(\vec{x}-\vec{y}) \quad .$$

The desired result now follows.

2-3.

(a) $$\theta_B^{\mu\alpha} = \theta_c{}^{\mu\alpha} + \tfrac{1}{2}\,\partial_\lambda X^{\lambda\mu\alpha}$$

$$X^{\lambda\mu\alpha} = -X^{\mu\lambda\alpha} = \pi^\lambda\Sigma^{\mu\alpha}\phi - \pi^\mu\Sigma^{\lambda\alpha}\phi - \pi^\alpha\Sigma^{\lambda\mu}\phi$$

Since $X^{\lambda\mu a}$ is explicitly anti-symmetric in $\lambda\mu$, $\partial_\lambda X^{\lambda\mu a}$ does not contribute to the divergence of $\theta_B{}^{\mu a}$ in μ, nor does it contribute to the charges $\int d^3x\theta_B{}^{0a}(x)$.

(b) $$\theta_B{}^{\mu a} = \pi^\mu\phi^a - g^{\mu a}\mathcal{L} + \tfrac{1}{2}\partial_\lambda[\pi^\lambda\Sigma^{\mu a}\phi - \pi^\mu\Sigma^{\lambda a}\phi - \pi^a\Sigma^{\lambda\mu}\phi]$$

The only terms that are not explicitly symmetric in μa are

$$\pi^\mu\phi^a + \tfrac{1}{2}\partial_\lambda[\pi^\lambda\,\Sigma^{\mu a}\phi] = \pi^\mu\phi^a + \tfrac{1}{2}\partial_\lambda\pi^\lambda\,\Sigma^{\mu a}\phi + \tfrac{1}{2}\pi^\lambda\,\Sigma^{\mu a}\phi_\lambda$$

$$= \pi^\mu\phi^a + \tfrac{1}{2}\frac{\delta\mathcal{L}}{\delta\phi}\Sigma^{\mu a}\phi + \tfrac{1}{2}\pi^\lambda\,\Sigma^{\mu a}\phi_\lambda\ .$$

The equations of motion have been used. Next use the constraint on $\mathcal{L}$ imposed by Lorentz covariance; see Exercise 2–2. We have

$$\pi^\mu\phi^a + \tfrac{1}{2}\partial_\lambda[\pi^\lambda\,\Sigma^{\mu a}\phi] = \pi^\mu\phi^a - \tfrac{1}{2}\pi^\mu\phi^a + \tfrac{1}{2}\pi^a\phi^\mu$$

$$= \tfrac{1}{2}\pi^\mu\phi^a + \tfrac{1}{2}\pi^a\phi^\mu.$$

(c) $$M_B{}^{\mu a\beta} = x^a\theta_B{}^{\mu\beta} - x^\beta\theta_B{}^{\mu a}$$

$$= x^a\theta_c{}^{\mu\beta} - x^\beta\theta_c{}^{\mu a} + x^a\,\tfrac{1}{2}\partial_\lambda\,X^{\lambda\mu\beta} - x^\beta\,\tfrac{1}{2}\partial_\lambda X^{\lambda\mu a}$$

$$= M_c{}^{\mu a\beta} + \tfrac{1}{2}\partial_\lambda[X^{\lambda\mu\beta}x^a - X^{\lambda\mu a}x^\beta]$$

$$+ \tfrac{1}{2}X^{\beta\mu a} - \tfrac{1}{2}X^{a\mu\beta} - \pi^\mu\Sigma^{a\beta}\phi$$

From the explicit form for $X^{\lambda\mu a}$, we see that the last three terms cancel. The total derivative term is the divergence in λ of a tensor which is anti-symmetric in $\mu\lambda$; hence that object does not contribute to the divergence in μ, nor to charges.

2–4.

(a) $$i[\theta_c^{00}(t,\vec{x}),J_0^{a}(t,\vec{y})] = i[\theta_c^{00}(t,\vec{x}),\ \pi^0(t,\vec{y})T^a\phi(t,\vec{y})]$$

$$= \pi^0(t,\vec{y})T^a\ i[\theta_c^{00}(t,\vec{x}),\phi(t,\vec{y})] + i[\theta_c^{00}(t,\vec{x}),\ \pi^0(t,\vec{y})]T^a\phi(t,\vec{y})$$

The first commutator is evaluated as in Exercise 2–1. The second is expressed in terms of functional derivatives. The result is

$$i[\theta_c^{00}(t,\vec{x}),J_0^{a}(t,\vec{y})] = \pi^0(t,\vec{y})\,T^a\phi^0(t,\vec{y})\delta(\vec{x}-\vec{y})$$

$$-\frac{\delta'\theta_c^{00}(t,\vec{x})}{\delta'\phi(t,\vec{y})}\ T^a\phi(t,\vec{y})\ .$$

The functional derivative is now calculated from the formula for $\theta_c^{00}(t,\vec{x})$.

$$\frac{\delta'\theta_c^{00}(t,\vec{x})}{\delta'\phi(t,\vec{y})} = \pi^0(t,\vec{x})\ \frac{\delta'\phi^0(t,\vec{x})}{\delta'\phi(t,\vec{y})} - \frac{\delta'\mathcal{L}(t,\vec{x})}{\delta'\phi(t,y)}$$

$$\frac{\delta'\mathcal{L}(t,\vec{x})}{\delta'\phi(t,\vec{y})} = \frac{\delta\mathcal{L}(t,\vec{x})}{\delta\phi}\ \delta(\vec{x}-\vec{y}) + \frac{\delta\mathcal{L}(t,\vec{x})}{\delta\phi^0}\ \frac{\delta'\phi^0(t,\vec{x})}{\delta'\phi(t,\vec{y})}$$

$$+\frac{\delta\mathcal{L}(t,\vec{x})}{\delta\phi^i}\ \partial^i\delta(\vec{x}-\vec{y})$$

$$= \partial_\lambda\pi^\lambda(t,\vec{x})\delta(\vec{x}-\vec{y}) + \pi^0(t,\vec{x})\,\frac{\delta'\phi^0(t,\vec{x})}{\delta'\phi(t,\vec{y})} + \pi_i(t,\vec{x})\partial^i\delta(\vec{x}-\vec{y})$$

We have used the equations of motion and the definition of π^μ. It now follows that

$$\frac{\delta'\theta_c^{00}(t,\vec{x})}{\delta'\phi\,(t,\vec{y})} = -\partial_\lambda\pi^\lambda(t,\vec{x})\ \delta(\vec{x}-\vec{y}) - \pi_i(t,\vec{x})\partial^i\delta(x-\vec{y})$$

$$i[\theta_c^{\,00}(t,\vec{x}),J_0^{\,a}(t,\vec{y})] = \pi^0(t,\vec{x})T^a\phi^0(t,\vec{x})\delta(\vec{x},\vec{y})$$

$$+ \partial_\lambda\pi^\lambda(t,\vec{x})T^a\phi(t,\vec{x})\delta(\vec{x}-\vec{y}) + \pi^i(t,\vec{x})T^a\phi(t,\vec{y})\partial_i\delta(\vec{x}-\vec{y})$$

$$= \partial_0[\pi^0(t,\vec{x})T^a\phi(t,\vec{x})]\delta(\vec{x}-\vec{y})$$

$$+ \partial_i[\pi^i(t,\vec{x})T^a\phi(t,\vec{x})]\delta(\vec{x}-\vec{y}) + \pi^i(t,\vec{x})T^a\phi(t,\vec{x})\partial_i\delta(\vec{x}-\vec{y})$$

$$= \partial^\mu J_\mu^{\,a}(x)\delta(\vec{x}-\vec{y}) + J_i^{\,a}(x)\partial^i\delta(\vec{x}-\vec{y}) \ .$$

In the last formula we have used the definition for the current.

$$J_\mu^{\,a} = \pi_\mu T^a\,\phi$$

(b) The difference between $\theta_B^{\,00}$ and $\theta_c^{\,00}$ is

$$\tfrac{1}{2}\,\partial_\lambda\,X^{\lambda 00} = \tfrac{1}{2}\,\partial_i X^{i00} = \partial_i[\pi^0\Sigma^{0i}\phi] \ .$$

Since Σ^{0i} and T^a commute, (they operate in different spaces) J_0^a commutes with $\tfrac{1}{2}\,\partial_i X^{i00}$.

2–5.

(a)

$$\theta_c^{\,\mu\nu} = \pi^\mu\phi^\nu - g^{\mu\nu}\mathcal{L}$$

$$= \phi^\mu\phi^\nu - g^{\mu\nu}\mathcal{L}$$

$$\partial_\mu\theta_c^{\,\mu\nu} = \partial_\mu\phi^\mu\phi^\nu + \phi^\mu\partial_\mu\phi^\nu - \partial^\nu\mathcal{L}$$

$$= gx^2\phi^\nu + \phi^\mu\partial_\mu\phi^\nu - \phi^\mu\partial^\nu\phi_\mu$$

$$- 2g\,x^\nu\phi - gx^2\phi^\nu = -2gx^\nu\phi \ .$$

We have used the equation of motion: $\partial_\mu \phi^\mu = gx^2$.

(b) $$i[P^0(t),P^i(t)] = \int d^3x\, d^3y\; i[\theta_c^{\,00}(t,\vec{x}),\theta_c^{\,0i}(t,\vec{y})]$$

$$= \int d^3x\, d^3y\; i[\tfrac{1}{2}\phi^0(t,\vec{x})\,\phi^0(t,\vec{x}) - \tfrac{1}{2}\phi^j(t,\vec{x})\,\phi_j(t,\vec{x}) - gx^2\phi(t,\vec{x}),\; \phi^0(t,\vec{y})\,\phi^i(t,\vec{y})]$$

$$= \int d^3x\, d^3y \left\{ \phi^0(t,\vec{x})\phi^0(t,\vec{y}) \frac{\partial}{\partial y_i}\delta(\vec{x}-\vec{y}) + \phi^j(t,\vec{x})\phi^i(t,\vec{y}) \frac{\partial}{\partial x^j}\delta(\vec{x}-\vec{y}) + gx^2\phi^i(t,\vec{x})\delta(\vec{x}-\vec{y}) \right\}$$

$$= \int d^3x\, \{ -\phi^0(t,\vec{x})\partial^i\phi^0(t,\vec{x}) + \phi^j(t,\vec{x})\partial_j\phi^i(t,\vec{x}) + gx^2\phi^i(t,\vec{x}) \}$$

$$= -\tfrac{1}{2}\int d^3x\,\partial^i\, \{ \phi^0(t,\vec{x})\phi^0(t,\vec{x}) - \phi^j(t,\vec{x})\phi_j(t,\vec{x}) - 2gx^2\phi(t,\vec{x}) \} - g\int d^3x\; x^i\phi(t,\vec{x})$$

The first term is a surface integral; it may be dropped. We are left with

$$i[P^0(t),P^i(t)] = -g\int d^3x\; x^i\phi(t,\vec{x}) \;.$$

(c) Since $\Sigma^{\alpha\beta}$ is zero for spin zero fields,

$$M_c^{\mu\,\alpha\beta}(x) = x^\alpha\theta_c^{\,\mu\beta} - x^\beta\theta_c^{\,\mu\alpha}$$

$$\partial_\mu M_c^{\,\alpha\beta}(x) = \theta_c^{\,\alpha\beta} + x^\alpha\partial_\mu\theta_c^{\,\alpha\beta} - \theta_c^{\,\beta\alpha} - x^\beta\partial_\mu\theta_c^{\,\mu\alpha}$$

$$= x^\alpha\partial_\mu\theta_c^{\,\mu\beta} - x^\beta\partial_\mu\theta_c^{\,\mu\alpha} \;.$$

We have used the symmetry of $\theta_c^{\mu\nu}$. From (a) we have

$$\partial_\mu M_c^{\mu\,\alpha\beta}(x) = 2g(x^\beta x^\alpha - x^\alpha x^\beta) = 0 \ .$$

2–6. $$[J_0^{\,b}(t,\vec{y}),J_k^{\,a}(t,\vec{z})]\partial^k\delta(\vec{x}-\vec{z}) + [J_0^{\,a}(t,\vec{x}),J_k^{\,b}(t,\vec{z})]\partial^k\delta(\vec{z}-\vec{y})$$

$$= -\,f_{abc}\delta(\vec{x}-\vec{y})\,J_k^{\,c}(t,\vec{z})\partial^k\delta(\vec{z}-\vec{x})$$

Multiplying by y_i and integrating over $\vec{y}$ gives

$$[J_0^{\,a}(t,\vec{x}),J_i^{\,b}(t,z)] = -\,f_{abc}\;x_iJ_k^{\,c}(t,\vec{z})\partial^k\delta(\vec{z}-\vec{x})$$

$$-\int d^3y\;y_i[J_0^{\,b}(t,\vec{y}),J_k^{\,a}(t,\vec{z})]\partial^k\delta(\vec{x}-\vec{z})$$

$$= -\,f_{abc}\;J_i^{\,c}(t,\vec{z})\delta(\vec{z}-\vec{x}) - f_{abc}z_iJ_k^{\,c}(t,\vec{z})\;\partial^k\delta(\vec{z}-\vec{x})$$

$$-\int d^3y\;y_i[J_0^{\,b}(t,\vec{y}),J_k^{\,a}(t,\vec{z})]\partial^k\delta(\vec{x}-\vec{z})$$

Clearly this is of the form

$$[J_0^{\,a}(t,\vec{x}),J_i^{\,b}(t,\vec{z})] = -\,f_{abc}J_i^{\,c}(t,\vec{z})\delta(\vec{z}-\vec{x})$$

$$+\,S_{ij}^{\,ab}(t,\vec{z})\partial^j\delta(\vec{y}-\vec{z}) \ .$$

Reinserting this solution into the starting equation, we have

$$f_{abc}J_k^{\,c}(t,\vec{z})\delta(\vec{y}-\vec{z})\partial^k\delta(\vec{x}-\vec{z}) + S_{k\ell}^{\,ba}(t,\vec{z})\partial^\ell\delta(\vec{y}-\vec{z})\partial^k\delta(\vec{x}-\vec{z})$$

$$-\,f_{abc}\;J_k^{\,c}(t,\vec{z})\delta(\vec{x}-\vec{z})\partial^k\delta(\vec{z}-\vec{y}) + S_{k\ell}^{\,ab}(t,\vec{z})\partial^\ell\delta(\vec{x}-\vec{z})\partial^k\delta(\vec{z}-\vec{y})$$

$$= -\,f_{abc}J_k^{\,c}(t,\vec{z})\delta(\vec{x}-\vec{y})\partial^k\delta(\vec{z}-\vec{x}) \ .$$

Since

$$-\delta(\vec{y}-\vec{z})\partial^k\delta(\vec{x}-\vec{z}) + \delta(\vec{x}-\vec{z})\partial^k\delta(\vec{z}-\vec{y}) = \delta(\vec{x}-\vec{y})\partial^k\delta(\vec{z}-\vec{x}) ,$$

we are left with

$$S_{k\ell}{}^{ba}(t,\vec{z})\partial^\ell\delta(\vec{y}-\vec{z})\partial^k\delta(\vec{x}-\vec{z}) + S_{\ell k}{}^{ab}(t,\vec{z})\partial^k\delta(\vec{x}-\vec{z})\partial^\ell\delta(\vec{z}-\vec{y}) = 0$$

or

$$S_{k\ell}{}^{ba} = S_{\ell k}{}^{ab} .$$

❧ • ❧

2–7. The covariant, but frame dependent formula, for the ETC is

$$[J^\mu(x), J^\nu(0)]\delta(x\cdot n) = S(0)(n^\mu g^{\nu\alpha} + n^\nu g^{\mu\alpha})P_{\alpha\beta}\partial^\beta\delta^4(x) .$$

$$-\delta^4(x)n^\nu P^{\mu\alpha}\partial_\alpha S(0) .$$

Since there is a ST, $S^{\mu\nu\alpha}(n) = (n^\mu g^{\nu\alpha} + n^\nu g^{\mu\alpha})S(0)$, the T product is not covariant. Note that

$$S^{\mu\nu\alpha}(n) = S(0)\frac{\delta}{\delta n^\alpha}(n^\mu n^\nu) .$$

Hence a covariantizing seagull is $\tau^{\mu\nu}(x;n) = S(0)n^\mu n^\nu\delta^4(x)$, and

$$T^{*\mu\nu}(x) = TJ^\mu(x)J^\nu(0) + n^\mu n^\nu S(0)\delta^4(x) + \tau^{\mu\nu}(x)$$

is covariant. The covariant seagull $\tau^{\mu\nu}(x)$ can be chosen to satisfy conservation of $T^{*\mu\nu}(x)$. We have

$$\partial_\mu T^{*\mu\nu}(x) = 0 = \delta(x\cdot n)n_\mu[J^\mu(x),J^\nu(0)] + n^\mu n^\nu S(0)\partial_\mu\delta^4(x)+\partial_\mu\tau^{\mu\nu}(x)$$

$$-\partial_\mu\tau^{\mu\nu}(x) = S(0)P^{\nu\beta}\partial_\beta\delta^4(x) + S(0)n^\nu n^\beta\partial_\beta\delta^4(x) = S(0)\partial^\nu\delta^4(x) .$$

Therefore the choice $\tau^{\mu\nu}(x) = -g^{\mu\nu}S(0)\delta^4(x)$ assures conservation.

$$T^{*\mu\nu}(x) = T^{\mu\nu}(x) - P^{\mu\nu} S(0) \delta^4(x)$$

Note that the seagull is present only in the ij component of $T^{*\mu\nu}$.

3–1.

(a) $$D(q) = \int d^4x\, e^{iqx} < 0|T\phi(x)\phi(0)|0 >$$

According to the BJL theorem and the canonical ETC, we have

$$\lim_{q_0 \to \infty} q_0 D(q) = i \int d^3x\, e^{-i\vec{q}\cdot\vec{x}} < 0|[\phi(0,\vec{x}),\phi(0)]|0 >$$
$$= 0$$

$$\lim_{q_0 \to \infty} q_0^2 D(q) = - \int d^3x\, e^{-i\vec{q}\cdot\vec{x}} < 0|[\dot{\phi}(0,\vec{x}),\phi(0)]|0 >$$
$$= i$$

$$\lim_{q_0 \to \infty} q_0^3 [D(q) - \frac{i}{q_0^2}] = -i \int d^3x\, e^{-i\vec{q}\cdot\vec{x}} <0|[\dot{\phi}(0,x),\phi(0)]|0 >$$

$$= -i \int d^3x\, e^{-i\vec{q}\cdot\vec{x}} \{\partial_0 < 0|[\dot{\phi}(0,x),\phi(0)]|0 > -$$
$$< 0|[\phi(0,x),\dot{\phi}(0)]|0 > \} = 0 \ .$$

Since the explicit formula for D(q) has the expansion $D(q) = \frac{i}{q_0^2} + 0\frac{1}{q_0^4}$, all the BJL limits are satisfied.

(b) G(q) formally has the expansion

$$G(q) = \frac{i}{q_0^2} \int_0^\infty da^2 \rho(a^2) + 0\frac{i}{q_0^4} \ .$$

Hence we deduce that

$$< 0|[\tilde{\phi}(0,\vec{x}),\tilde{\phi}(0)]|0> = 0$$

$$< 0|[\dot{\tilde{\phi}}(0,\vec{x}),\dot{\tilde{\phi}}(0)]|0> = 0$$

$$i < 0|[\dot{\tilde{\phi}}(0,\vec{x}),\phi(0)]|0> = \int_0^\infty da^2\rho(a^2)\delta(\vec{x}) \ .$$

If $\int_0^\infty da^2\rho(a^2)$ is divergent, as it is in perturbation theory, one cannot properly be speaking of the ETC for renormalized fields. On the other hand, if that quantity is finite, we find that

$$i < 0|[\dot{\phi}(0,\vec{x}),\phi(0)]|0> = \delta(\vec{x})$$

$$\tilde{\phi} = Z^{-1/2}\phi$$

$$Z^{-1} = \int_0^\infty da^2\rho(a^2) \ .$$

$\rho(a^2)$ is non-negative. It has a contribution from the single particle state given by $\delta(a^2-m^2)$. Hence $\rho(a^2) = \delta(a^2-m^2) + \tilde{\rho}(a^2), \tilde{\rho}(a^2) \geq 0$. Therefore

$$0 < Z^{-1} \leq 1 \ .$$

3–2. $$T^{*\mu\nu}(q) = (q^{\mu\nu}q^2 - q^\mu q^\nu)\int \frac{da^2\sigma(a^2)}{q^2-a^2}$$

$$T^{*00}(q) = -\vec{q}^{\,2}\int \frac{da^2\sigma(a^2)}{q^2-a^2} \qquad T^{*00}(q) \xrightarrow[q_0 \to \infty]{} 0$$

$$T^{*0i}(q) = -q^0q^i\int \frac{da^2\sigma(a^2)}{q^2-a^2} \qquad T^{*0i}(q) \xrightarrow[q_0 \to \infty]{} 0$$

$$T^{*ij}(q) = (g^{ij}q^2 - q^i q^j)\int \frac{da^2\sigma(a^2)}{q^2-a^2} \qquad T^{*ij}(q) \xrightarrow[q_0 \to \infty]{} g^{ij}\int da^2\sigma(a^2)$$

Hence the seagull is only in the ij component. Therefore

$$T^{\mu\nu}(q) = T^{*\mu\nu}(q) - P^{\mu\nu}\int da^2\sigma(a^2) \ .$$

Finally

$$\lim_{q_0 \to \infty} q_0 T^{0i}(q) = i\int d^3x\, e^{-i\vec{q}\cdot\vec{x}} < 0|[J^0(0,\vec{x}),J^i(0)]|0 >$$

$$= \lim_{q_0 \to \infty} -q_0{}^2 q^i \int \frac{da^2\sigma(a^2)}{q^2-a^2} = -q^i\int da^2\sigma(a^2).$$

This shows that

$$< 0|[J^0(0,x),J^i(0)]|0 > = i\partial^i\delta(\vec{x})\int da^2\sigma(a^2) \ .$$

Since $\sigma(a^2)$ has definite sign, the vacuum expectation value of the ST is a non-vanishing. This provides an alternate argument for the necessary existence of ST. Note, however, that the derivation here is formal, in the sense that there is no guarantee that any of our steps are justified. Divergences in the integral over the spectral function will modify our result. In particular, there is no guarantee that only *one* derivative of the δ function is present; see Exercise 6–1. Since $\sigma(a^2)$ is proportional to the total lepton annihilation cross section, the present argument shows that the vacuum expectation value of the ST is a measureable quantity.

4–1. $$\Delta^\mu(a) = i\int \frac{d^4r}{(2\pi)^4}\left[\frac{r^\mu + a^\mu}{([r+a]^2-m^2)^2} - \frac{r^\mu}{(r^2-m^2)^2}\right]$$

$$= i\int \frac{d^4r}{(2\pi)^4}\left\{\exp a^\nu \frac{\partial}{\partial r^\nu} - 1\right\}\frac{r^\mu}{(r^2-m^2)^2}$$

$$= i\int \frac{d^4r}{(2\pi)^4}\ a^\nu \frac{\partial}{\partial r^\nu}\left\{\left[1+0(r^{-1})\right]\frac{r^\mu}{(r^2-m^2)^2}\right\}$$

The integrand is a total derivative. If the integrand were sufficiently damped at large r, one would conclude by Gauss' theorem that the integral is zero. In the present instance, let us integrate over a surface r = R, by symmetric integration. Using the fact that

$$\int_{r=R} d^4r\ \partial^\nu f(r) = i2\pi^2\ R^\nu R^2 f(R),$$

we have

$$\Delta^\mu(a) = -\frac{1}{8\pi^2}\ a^\nu \lim_{R\to\infty} R_\nu R^2\left[1+0(R^{-1})\right]\frac{R^\mu}{(R^2-m^2)^2}$$

The limit is to be performed symmetrically.

$$\lim_{R\to\infty}\ \{\text{odd number of factors of } R\} = 0$$

$$\lim_{R\to\infty}\frac{R^\mu R^\nu}{R^2} = \frac{1}{4}\, g^{\mu\nu}$$

Hence

$$\Delta^\mu(a) = -\frac{a^\mu}{32\pi^2}\ .$$

❧ • ❧

4–2. $$\Delta^{\alpha\mu\nu}(p,q|a) = i\int\frac{d^4r}{(2\pi)^4}\ \mathrm{Tr}\gamma^5\gamma^\alpha\left\{\exp a^\beta\frac{\partial}{\partial r^\beta} - 1\right\}$$

$$(\gamma_\delta r^\delta+\gamma_\delta p^\delta-m)^{-1}\gamma^\mu(\gamma_\delta r^\delta-m)^{-1}\gamma^\nu(\gamma_\delta r^\delta+\gamma_\delta q^\delta-m)^{-1}$$

$$= i\int \frac{d^4r}{(2\pi)^4}\, a^\beta \frac{\partial}{\partial r^\beta}\left\{[1+0(r^{-1})]\mathrm{Tr}\gamma^5\gamma^\alpha(\gamma_\delta r^\delta+\gamma_\delta p^\delta-m)^{-1}\gamma^\mu\right.$$

$$\left.(\gamma_\delta r^\delta-m)^{-1}\gamma^\nu(\gamma_\delta r^\delta+\gamma_\delta q^\delta-m)^{-1}\right\}$$

$$= -\frac{a^\beta}{8\pi^2}\lim_{R\to\infty} R_\beta R^2[1+0(R^{-1})][\mathrm{Tr}\gamma^5\gamma^\alpha\gamma_\delta R^\delta\gamma_\epsilon R^\epsilon$$

$$\gamma^\nu\gamma_\phi R^\phi R^{-6} + 0(R^{-4})\,]$$

The trace is simply $4\epsilon^{\alpha\omega\mu\nu} R_\omega/R^4$. Hence

$$\Delta^{\alpha\,\mu\nu}(p,q|a) = -\frac{a_\beta}{8\pi^2}\epsilon^{\alpha\beta\,\mu\nu} \quad .$$

❧ • ☙

4–3.
$$i\int \frac{d^4r}{(2\pi)^4}\,\mathrm{Tr}\gamma^5\gamma^\alpha\left\{[\gamma_\delta r^\delta+\gamma_\delta q^\delta-m]^{-1}\gamma^\nu[\gamma_\delta r^\delta-\gamma_\delta p^\delta-m]^{-1} -\right.$$

$$\left.-[\gamma_\delta r^\delta+\gamma_\delta p^\delta-m]^{-1}\gamma^\nu[\gamma_\delta r^\delta-\gamma_\delta q^\delta-m]^{-1}\right\}$$

$$= i\int \frac{d^4r}{(2\pi)^4}\,\mathrm{Tr}\gamma^5\gamma^\alpha\left\{\exp(q-p)^\beta\frac{\partial}{\partial r^\beta} -1\right\}[\gamma_\delta r^\delta+\gamma_\delta p^\delta-m]^{-1}\gamma^\nu$$

$$[\gamma_\delta r^\delta-\gamma_\delta q^\delta-m]^{-1}$$

$$= i\int \frac{d^4r}{(2\pi)^4}(q-p)^\beta\frac{\partial}{\partial r^\beta}\left\{[1+0(r^{-1})]\mathrm{Tr}\gamma^5\gamma^\alpha[\gamma_\delta r^\delta+\gamma_\delta p^\delta-m]^{-1}\gamma^\nu\right.$$

$$\left.[\gamma_\delta r^\delta-\gamma_\delta q^\delta-m]^{-1}\right\}$$

$$= -\frac{(q-p)^\beta}{8\pi^2} \lim_{R\to\infty} R_\beta R^2[1+0(R^{-1})] \frac{Tr\gamma^5\gamma^\alpha[\gamma_\delta R^\delta+\gamma_\delta p^\delta+m]\gamma^\nu[\gamma_\epsilon R^\epsilon-\gamma_\epsilon q^\epsilon-m]}{([R+p]^2-m^2)([R-q]^2-m^2)}$$

The trace is $-4\epsilon^{\alpha\phi\nu\omega} R_\omega(q_\phi+p_\phi) + 0$ (constant). Therefore the limit of the above is

$$\frac{(q-p)_\omega}{8\pi^2} \epsilon^{\alpha\phi\nu\omega}(q_\phi+p_\phi) = \frac{1}{4\pi^2} \epsilon^{\alpha\mu\nu\beta} p_\mu q_\beta \quad .$$

❧ • ❧

5–1. $$J^i = \bar{\psi}\gamma^i\psi$$

$$\dot{J}^i = \bar{\psi}\gamma^i\dot{\psi} + \text{h.c.}$$

$$= -i\bar{\psi}\gamma^i\gamma^0 i\gamma^0\dot{\psi} + \text{h.c.}$$

From equation of motion, $i\gamma_\mu\partial^\mu\psi = m\psi - gB_\mu\gamma^\mu\psi$, it follows that

$$i\gamma^0\dot{\psi} = -i\gamma^k\partial_k\psi + m\psi - gB_\mu\gamma^\mu\psi \quad .$$

Hence

$$\dot{J}^i = -\bar{\psi}\gamma^i\gamma^0\gamma^k\partial_k\psi - im\bar{\psi}\gamma^i\gamma^0\psi + igB_\mu\bar{\psi}\gamma^i\gamma^0\gamma^\mu\psi + \text{h.c.} \quad .$$

The commutator may now be evaluated with the help of the canonical commutators between the fields. A long computation yields for the symmetric part of the commutator the following.

$$\int d^3x[\dot{J}^i(0,\mathbf{x}),J^j(0)] = \bar{\psi}(0)[\gamma^i \overleftrightarrow{\partial^j} + \gamma^j \overleftrightarrow{\partial^i} - 2g^{ij}\gamma^k \overleftrightarrow{\partial_k}$$

$$- 2ig(\gamma^i B^j+\gamma^j B^i-2g^{ij}\gamma^k B_k) - 4img^{ij}]\psi(0)$$

$$+ \text{ term anti-symmetric in } (i,j)$$

To calculate the diagonal, spin overaged proton matrix element of the above, we define

$$< p|\bar{\psi}\gamma^{\mu} \overleftrightarrow{\partial}^{\nu}\psi|p > = g^{\mu\nu}iA_1 + p^{\mu}p^{\nu}iA_2$$

$$< p|\bar{\psi}\gamma^{\mu}B^{\nu}\psi|p > = g^{\mu\nu}B_1 + p^{\mu}p^{\nu}B_2$$

$$< p|\bar{\psi}\psi|p > = C \ .$$

It then follows that

$$\int d^3x < p|[\dot{J}^i(0,x),J^j(0)]|p > = A(\delta^{ij}\vec{p}^2-p^ip^j)+B\ \delta^{ij}$$

$$A = -2i(A_2-2gB_2)$$

$$B = 4i(A_1-2gB_1+mC) \ .$$

⁂ • ⁂

5–2. When the fixed q^2 dispersion relation in ν for T_L needs a subtraction, the dispersion relation in ω has the form

$$T_L(\omega,q^2) = T_L(q^2) + \frac{1}{2\pi}\int_0^1 d\omega' \frac{\tilde{F}_L(\omega',q^2)}{\omega'^2-\omega^2}$$

Here $T_L(q^2)$ is the subtraction, performed at $\nu = 0$ ($\omega = \infty$). This equation replaces (5.15a) in the text. It is evident that one may still establish that

$$-\int d^3x\, e^{-i\vec{q}\cdot\vec{x}} < p|[J^0(0,\vec{x}),J^i(0)]|p >$$

$$= \lim_{q_0 \to \infty} q^iT_L \ .$$

However, from the dispersion relation, one now concludes only the uninformative relation

$$\lim_{q_0 \to \infty} T_L(q^2,\omega) = T_L(\infty) \ .$$

⁂ • ⁂

6–1. $$T^{*\mu\nu}(q) = (g^{\mu\nu}q^2 - q^\mu q^\nu)T(q^2)$$

$$T(q^2) = C + q^2 \int_{4m^2}^{\infty} da^2 \, \frac{\sigma(a^2)}{a^2(q^2-a^2)}$$

Define $\tilde{\sigma}(a^2) = \sigma(a^2) - A - B/a^2$, $\lim_{a^2\to\infty} a^2\tilde{\sigma}(a^2) = 0$. Therefore $T(q^2)$ can be represented by

$$T(q^2) = C + q^2 \int_{4m^2}^{\infty} da^2 \, \frac{\tilde{\sigma}(a^2)}{a^2(q^2-a^2)} + Aq^2 \int_{4m^2}^{\infty} da^2 \, \frac{1}{a^2(q^2-a^2)}$$

$$+ Bq^2 \int_{4m^2}^{\infty} da^2 \, \frac{1}{a^4(q^2-a^2)}$$

$$= \tilde{C} + \int_{4m^2}^{\infty} da^2 \, \frac{\tilde{\sigma}(a^2)}{q^2-a^2} + A \log\left(1 - \frac{q^2}{4m^2}\right) + \frac{B}{q^2} \log\left(1 - \frac{q^2}{4m^2}\right) ,$$

$$\tilde{C} = C + \int_{4m^2}^{\infty} da^2 \, \frac{\tilde{\sigma}(a^2)}{a^2} + \frac{B}{4m^2} .$$

For large q_0

$$T(q^2) \xrightarrow[q_0\to\infty]{} \tilde{C} + A\log\left(-\frac{q_0^{\,2}}{4m^2}\right) + \frac{B}{q_0^{\,2}} \log\left(-\frac{q_0^{\,2}}{4m^2}\right)$$

$$- \frac{A}{q_0^{\,2}} (4m^2 + \vec{q}^{\,2}) + \frac{1}{q_0^{\,2}} \int_{4m^2}^{\infty} da^2 \tilde{\sigma}(a^2) .$$

Clearly $\tilde{C}$ is a covariant seagull; it may be dropped. Therefore it follows that the T product has the following form at large q_0.

$$q^0 T^{0i}(q) = -q_0{}^2 q^i T(q^2) \xrightarrow[q_0{}^2 \to \infty]{} -q^i \left[A\, q_0{}^2 \log\left(-\frac{q_0{}^2}{4m^2}\right) + B \log\left(-\frac{q_0{}^2}{4m^2}\right) - A(4m^2+\vec{q}^{\,2}) + \int_{4m^2}^{\infty} da^2 \tilde{\sigma}(a^2) \right]$$

It is thus seen that the terms proportional to one derivative of the δ function (q^i in momentum space) are the following: a quadratically divergent term, $Aq_0{}^2\log(-q_0{}^2/4m^2)$; a logarithmically divergent term, $B\log(-q_0{}^2/4m^2)$; and finite contributions;

$-4m^2 A + \int_{4m^2}^{\infty} da^2\, \tilde{\sigma}(a^2)$. In addition to these, there is a *triple* derivative of a δ function ($q^i \vec{q}^{\,2}$ in momentum space) given by the well defined coefficient $-A$. Thus the existence of a quadratically divergent term in the $[J^0, J^i]$ ETC proportional to $\partial^i \delta(\vec{x})$, necessarily implies that an object proportional to $\partial^i \nabla^2 \delta(\vec{x})$ will be present with a finite coefficient.

❧ • ☙

6–2. $$T(p,q) = \int d^4x d^4y\, e^{ipx} e^{iqy} \langle \alpha | T\, A(x)B(y)C(0) | \beta \rangle$$

$$\lim_{p_0 \to \infty} p_0 T(p,q) = i \int d^3x d^4y\, e^{-i\vec{p}\cdot\vec{x}} e^{iqy} \langle \alpha | T\, B(y)[A(0,\vec{x}), C(0)] | \beta \rangle$$

$$\lim_{q_0 \to \infty} q_0 \lim_{p_0 \to \infty} p_0 T(p,q) = -\int d^3x d^3y\, e^{-i\vec{p}\cdot\vec{x}}\, e^{-i\vec{q}\cdot\vec{y}}$$

$$< \alpha|[B(0,\vec{y}),[A(0,\vec{x}),C(0)]]|\beta >$$

$$\lim_{q_0 \to \infty} q_0 T(p,q) = i \int d^4x d^3y\, e^{ipx} e^{-i\vec{q}\cdot\vec{y}}$$

$$< \alpha|T\ A(x),[B(0,\vec{y}),C(0)]|\beta >$$

$$\lim_{p_0 \to \infty} p_0 \lim_{q_0 \to \infty} q_0 T(p,q) = -\int d^3x d^3y\, e^{-i\vec{p}\cdot\vec{x}} e^{-i\vec{q}\cdot\vec{y}}$$

$$< \alpha|[A(0,\vec{x}),[B(0,\vec{y}),C(0)]]|\beta >$$

7–1.
$$C^{\mu\nu}(q,p) = \int \frac{d^4x}{(2\pi)^4}\, e^{iqx} < p|[J^\mu(x),J^\nu(0)]|p >$$

The states are covariantly normalized.

$$< p|p'> = 2p_0\delta(\vec{p}-\vec{p}\,')$$

Hence the dimension of the states is m^{-2}. The dimension of $[J^\mu(x),J^\nu(0)]$ is m^6. Finally d^4x is of dimension m^{-4}. Therefore $C^{\mu\nu}$ is dimensionless. Since the tensors which appear in the decomposition of $C^{\mu\nu}$ are also dimensionless, this property is shared by the $\tilde{F}_i$.

7–2.

(a)
$$i[D(t),P^\mu] = i\int d^3x\ x_\alpha[\theta^{\alpha 0}(x),P^\mu]$$

$$= -\int d^3x\ x_\alpha \partial^\mu \theta^{\alpha 0}(x)$$

Take $\mu = 0$.

$$i[D(t), P^0] = -\int d^3x\, x_\alpha \partial^0 \theta^{\alpha 0}(x) = \int d^3x\, x_\alpha \partial_i \theta^{\alpha i}(x)$$

$$= \int d^3x\, x_j \partial_i \theta^{ji}(x)$$

$$= -\int d^3x\, \theta_i{}^i(x) = \int d^3x\, \theta^{00}(x) - \int d^3x\, \theta_\mu{}^\mu(x)$$

$$= P^0 - \int d^3x\, \theta_\mu{}^\mu(x)$$

We have used conservation of $\theta^{\mu\nu}$ and Gauss' theorem. Now take $\mu = i$.

$$i[D(t),P^i] = -\int d^3x\, x_\alpha \partial^i \theta^{\alpha 0}(x) = -\int d^3x\, x_j \partial^i \theta^{j0}(x)$$

$$= \int d^3x\, \theta^{i0}(x) = P^i$$

It is seen that the algebra $i[D,P^\mu]$ is not satisfied for $\mu = 0$, when the dilatation current is not conserved, $\theta_\mu{}^\mu \neq 0$.

(b) $$i[D(t),M^{\mu\nu}] = i\int d^3x\, x_\alpha [\theta^{\alpha 0}(x), M^{\mu\nu}]$$

$$= -\int d^3x\, x_\alpha (x^\mu \partial^\nu \theta^{\alpha 0} + g^{\alpha\mu}\theta^{\nu 0} + g^{0\mu}\theta^{\alpha\nu}) - \{\mu \leftrightarrow \nu$$

The commutator with $M^{\mu\nu}$ has been explicitly evaluated, since it is known that $\theta^{\alpha\beta}$ transforms as a second rank tensor under Lorentz transformations. Take first $\mu = 0$, $\nu = i$.

$$i[D(t),M^{0i}] = -\int d^3x\, \{x_\alpha x^0 \partial^i \theta^{\alpha 0}(x) + x^0 \theta^{i0}(x)$$

$$+ x_\alpha \theta^{\alpha i}(x) - x_\alpha x^i \partial_0 \theta^{\alpha 0}(x) - x^i \theta^{00}(x)\}$$

$$= -\int d^3x\, \{x_j x^0 \partial^i \theta^{j0}(x) + 2x^0 \theta^{i0}(x)$$

$$+ x_j \theta^{ji}(x) + x_\alpha x^i \partial_j \theta^{\alpha j}(x) - x^i \theta^{00}(x)\}$$

$$= - \int d^3x \, \{x^0\theta^{i0}(x) + x_j\theta^{ji}(x)$$

$$- x^0\theta^{0i}(x) - x_j\theta^{ij}(x) - x^i\theta_j{}^j(x) - x^i\theta^{00}(x)\}$$

$$= \int d^3x \, x^i\theta_\mu{}^\mu(x)$$

We have used conservation of $\theta^{\mu\nu}$ and Gauss' theorem. Next we consider $\mu = i$, $\nu = j$.

$$i[D(t), M^{ij}] = - \int d^3x[x_\alpha x^i\partial^i\theta^{\alpha 0}(x) + x^i\theta^{j0}(x) - \{i \leftrightarrow j\}]$$

$$= - \int d^3x[-x^i\theta^{j0}(x) + x^i\theta^{j0}(x) - \{i \leftrightarrow j\}]$$

$$= 0$$

As expected from general principles, we find that D(t) is not a Lorentz scalar when $\theta_\mu{}^\mu \neq 0$ scalar, though it is a rotational invariant.

7–3.

(a) $$i[D(t), \phi(x)] = i\int d^3y y_\alpha[\theta_c{}^{0\alpha}(t,\vec{y}), \phi(t,\vec{x})]$$

$$+ i\int d^3y[\pi^0(t,\vec{y})d\phi(t,\vec{y}), \phi(t,\vec{x})]$$

We have used the canonical formula for $D_c{}^\mu$; of course D(t) is insensitive to which current is used. The $[\theta_c{}^{0\alpha}, \phi]$ ETC has been evaluated in Exercise 2–1. We therefore have

$$i[D(t), \phi(x)] = \int d^3y y_\alpha \phi^\alpha(t,\vec{y})\delta(\vec{x}-\vec{y}) + d\phi(t,\vec{x})$$

$$= x_\alpha\phi^\alpha(x) + d\phi(x) = \delta_D\phi(x)$$

(b) $$i[K^{\alpha}(t),\phi(x)] = i\int d^3y[(2y^{\alpha}y_{\nu}-g^{\alpha}_{\nu}y^2)\theta_c^{\;0\nu}(t,\vec{y}) + 2y_{\nu}\pi^0(t,\vec{y})(g^{\nu\alpha}d-\Sigma^{\nu\alpha})\phi(t,\vec{y})-2\sigma^{\alpha 0}(t,\vec{y}),\phi(t,\vec{x})]$$

Again the canonical formula for the conformed current has been used. The commutators of ϕ with the first two terms in the expression for $K_c^{\;\alpha 0}$ are readily evaluated. To calculate the commutator of $\sigma^{\alpha 0}$ with ϕ, we observe that $\sigma^{\mu\nu}$ does not depend on derivatives of fields. The reason for this is that if $\sigma^{\mu\nu}$ were to depend on derivatives of fields then $\partial_{\nu}\sigma^{\mu\nu}$ would depend on second derivatives of the fields. This is impossible when $\mathcal{L}$ has no dependence on second derivatives, since

$$\partial_{\nu}\sigma^{\mu\nu} = V^{\mu} = \frac{\delta\mathcal{L}}{\delta\phi^{\beta}}[g^{\beta\alpha}d - \Sigma^{\beta\alpha}]\phi \;.$$

Therefore the commutator of ϕ with $\sigma^{\alpha 0}$ vanishes, and we are left with

$$\begin{aligned} i[K^{\alpha}(t),\phi(x)] &= \int d^3y[(2y^{\alpha}y_{\nu}-g_{\nu}^{\;\alpha}y^2)\phi^{\nu}(t,\vec{y}) \\ &\quad + 2y_{\nu}(g^{\nu\alpha}d-\Sigma^{\nu\alpha})\phi(t,\vec{y})] \\ &= (2x^{\alpha}x_{\nu}-g_{\nu}^{\;\alpha}x^2)\phi^{\nu}(x)-2x_{\nu}(g^{\nu\alpha}d-\Sigma^{\nu\alpha})\phi(x) \\ &= \delta_C^{\;\alpha}\phi(x) \;. \end{aligned}$$

7–4.

(a) For the scale invariance we need $4\mathcal{L} = \frac{\delta\mathcal{L}}{\delta\phi^{\mu}}(d+1)\phi^{\mu} + \frac{\delta\mathcal{L}}{\delta\phi}d\phi$, $d = 1$. Since $\mathcal{L}$ is Poincaré invariant, it can depend on ϕ^{μ} only through the combination $\phi^{\mu}\phi_{\mu}$. Therefore we set, without loss of generality, $\mathcal{L}(\phi^{\mu},\phi) = f(\frac{\phi^{\mu}\phi_{\mu}}{\phi^4},\phi)\,\phi^4$. The above condition now gives a differential equation for f: $\frac{\partial}{\partial y} f(x,y) = 0$. Therefore the most general scale invariance Lagrangian is of the form $\mathcal{L}(\phi^\mu, \phi) = f(\frac{\phi^\mu \phi_\mu}{\phi^4})\, \phi^4$, where f is an arbitrary function.

(b) For the Lagrangian to be conformally invariant, the field virial, $V^\mu = \pi^\mu \phi$, must be a total divergence. In the present case this requires

$$2\phi^\mu \phi\, f'(\frac{\phi^\alpha \phi_\alpha}{\phi^4}) = \partial_\alpha \sigma^{\alpha\mu} \quad .$$

The quantity $\sigma^{\alpha\mu}$ is a function of ϕ and ϕ^μ. However, it cannot depend on ϕ^μ, for then the indicated differentiation on the right hand side would produce second derivatives of ϕ, which are not present on the left hand side. Thus

$$\sigma^{\alpha\mu} = g^{\alpha\mu}\sigma(\phi^2)$$

$$\partial_\alpha \phi^{\alpha\mu} = 2\sigma'(\phi^2)\phi^\mu \phi \quad ,$$

and the condition on the field virial becomes

$$f'(\frac{\phi^\alpha \phi_\alpha}{\phi^4}) = \sigma'(\phi^2) \quad .$$

It is now seen that f' must be constant, since $\sigma'(\phi^2)$ cannot depend on ϕ^α. We conclude therefore that the most general conformally invariant single Boson Lagrangian is

$$\mathcal{L} = \tfrac{1}{2}\, \phi^\mu \phi_\mu - \frac{\lambda}{4!}\, \phi^4$$

$$\theta_c^{\mu\nu} = \phi^\mu \phi^\nu - g^{\mu\nu}\, \mathcal{L}$$

$$\sigma^{\mu\nu} = \frac{g^{\mu\nu}}{2}\phi^2$$

$$\theta^{\mu\nu} = \phi^\mu\phi^\nu - g^{\mu\nu}\mathfrak{L} + \frac{1}{6}(g^{\mu\nu}\Box - \partial^\mu\partial^\nu)\phi^2$$

Massachusetts Institute of Technology

THE HIGH ENERGY BEHAVIOR OF WEAK AND ELECTROMAGNETIC PROCESSES

David J. Gross

I. INTRODUCTION AND KINEMATICS

a) Introduction

The subject matter to be covered in these lectures is the high energy behavior of weak and electromagnetic processes. Most of the emphasis will be placed on the highly inelastic region of lepton-hadron scattering. This is an exciting and undeveloped area of research which can be highly recommended to young theorists, particularly since most of the experimental surprises and the important theoretical advances are yet to come. In addition it is a field which demands an interplay of varied theoretical methods and intuition. Finally, there is some reason to hope that the use of the weak and electromagnetic interactions as a probe of the short distance structure of hadrons will provide critical information concerning the structure of the latter.

The great advantage of scattering leptons off hadrons is that we believe that we know (at least to lowest order) the weak and electromagnetic interactions of the leptons. Consider the scattering of an electron (or a positron, muon) of a nucleon target. To lowest order in α (the fine structure constant) the amplitude is given by the exchange of one photon (Fig. 1).

We can, therefore, separate the scattering amplitude into two parts, the purely leptonic vertex and the hadronic vertex. In practice higher order

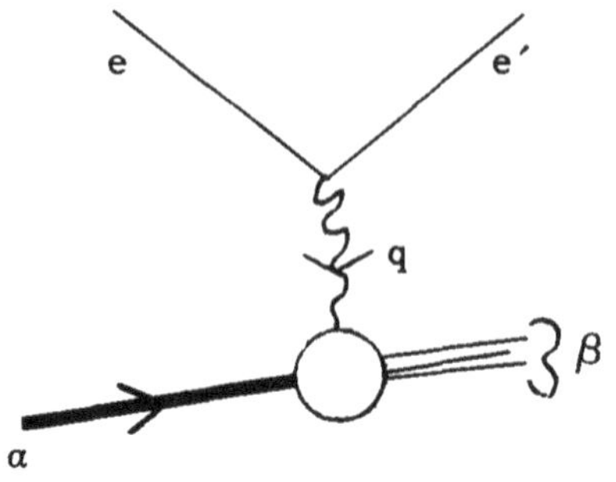

Fig. 1

radiative corrections must be taken into account; however, the important radiative corrections only involve the leptons and can therefore be calculated. As long as we assume that one photon exchange dominates the amplitude, and that we indeed have an accurate description of the leptonic vertex (these assumptions can, of course, be tested experimentally) we can deduce directly from experiment the amplitude for the scattering of a photon of momentum q (see Fig. 1) off the hadronic target.

The same holds for the scattering of neutrinos off hadrons, where to lowest order in the Fermi coupling constant we believe the amplitude to be described by the effective Lagrangian

$$\mathcal{L} = \frac{G}{(2)^{1/2}}\, j_\mu^{\,\text{leptonic}}\, J^{\mu\,(\text{hadronic})} + \text{h.c.}$$

Therefore neutrino scattering experiments determine the amplitudes for the "scattering" of the weak current, $J_\mu^{\text{had.}}$, off hadrons (Fig. 2). Here the analysis depends upon whether or not the intermediate vector boson (the W meson) exists. We shall, for lack of contrary evidence, assume that it does not.

In both of the above cases, i.e., electron or neutrino scattering, we are measuring the matrix elements of a current operator between hadron

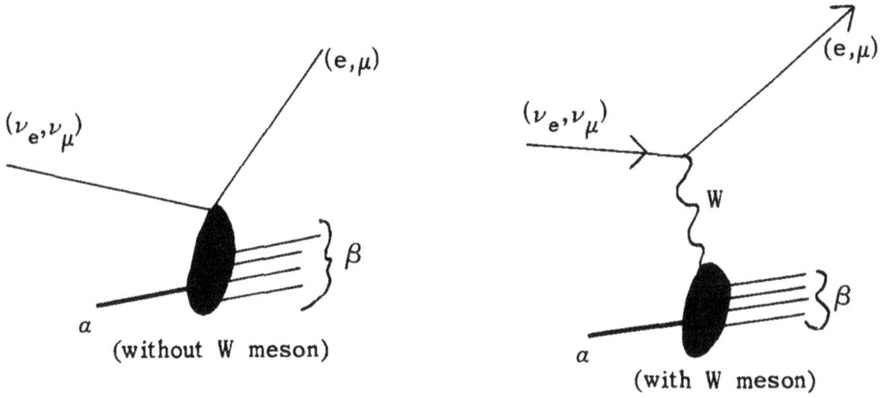

Fig. 2

states, $<\alpha|J_\mu^{E.M.}|\beta>$ or $<\alpha|J_\mu^{WEAK}|\beta>$, where $|\alpha>$ ($|\beta>$) denotes the initial (final) hadronic state. The distinguishing feature of these experiments is that the knowledge of the leptonic vertices allows us to vary the virtual "mass" of the current (i.e., the photon or W-meson). The assumption of local weak couplings [i.e., that J_μ^{lept} couples locally to the local hadronic electromagnetic current] is of crucial importance, allowing us operationally to define a local hadronic current operator and probe the small distance structure of hadrons. This is to be contrasted with pure hadronic scattering, where one cannot, in a model independent fashion factorize amplitudes in such a fashion or measure local operators off the mass shell.

Although these are weak and electromagnetic processes, most of our attention will be focused on the strong interactions. To be sure one can use these processes to test our hypothesis regarding the weak and electromagnetic interactions. Thus one can look for corrections to the photon propagator, test for T invariance of leptonic couplings, test CVC, etc. . . . However, our primary interest will be the strong interactions.

b) A List of Relevant Experiments

It is instructive to review the various kinds of experiments that one can in practice perform to investigate the high energy behaviour of weak amplitudes. Roughly speaking these can be divided into 2 groups, on shell photon scattering and lepton-hadron scattering.

1) On shell photon scattering

These experiments include photoproduction and Compton scattering, $\gamma+N \to N\pi$, $N^*\pi$, $N\rho$, etc. or $\gamma+N \to \gamma+$hadron. Since the photon is on its mass shell in these experiments we might expect the large energy behavior of these processes to be similar to that of purely hadronic processes. There are, however, some important differences which shall be discussed in the following.

2) Lepton-Hadron Scattering

In these experiments one can investigate both the large energy and the large virtual mass behavior of "current" hadron scattering. Such processes include:

a) Elastic Form Factors

Here one measures the large momentum transfer behavior of hadronic electromagnetic (or weak) form factors by observing elastic electron (or neutrino) scattering off a hadronic target (Fig. 3).

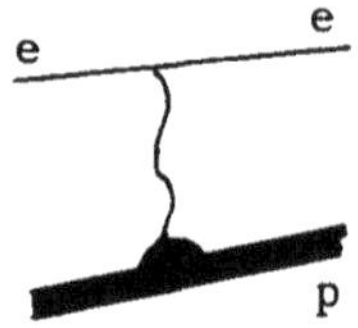

Fig. 3

b) Total "Current"-Hadron Cross Sections

If one measures the cross section for lepton + hadron → lepton + anything, one can determine the total absorption cross section for the scattering virtual photon (or W-meson) off a hadron, as a function of the energy and the "mass" of the photon (or W-meson) (Fig. 4).

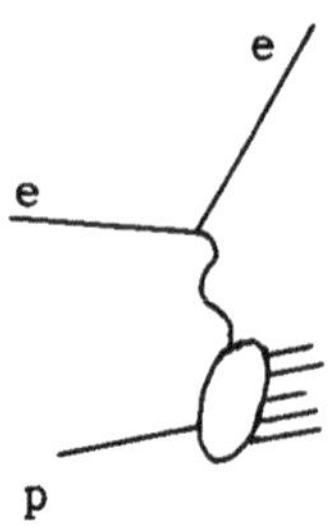

Fig. 4

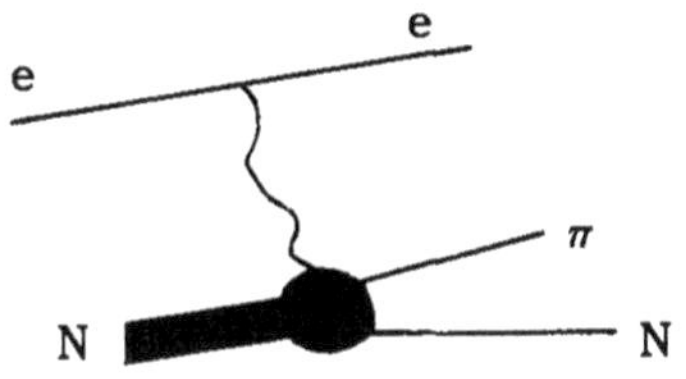

Fig. 5

c) Inelastic Form Factors

If one measures the cross-section for, say, lepton + N → lepton + [N*, or (Nπ), or (Nρ), etc. . .] one determines a form factor of the appropriate current between a one nucleon state and the given final state (Fig. 5).

d) Annihilation Processes

These experiments include $e^+e^- \to \pi^+\pi^-$; $N\bar{N}$, anything; and determine electromagnetic form factors for timelike momentum transfer (Fig. 6).

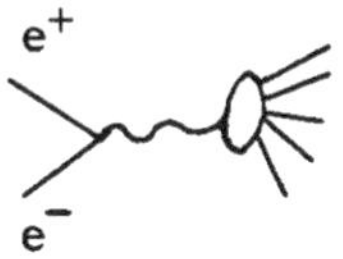

Fig. 6

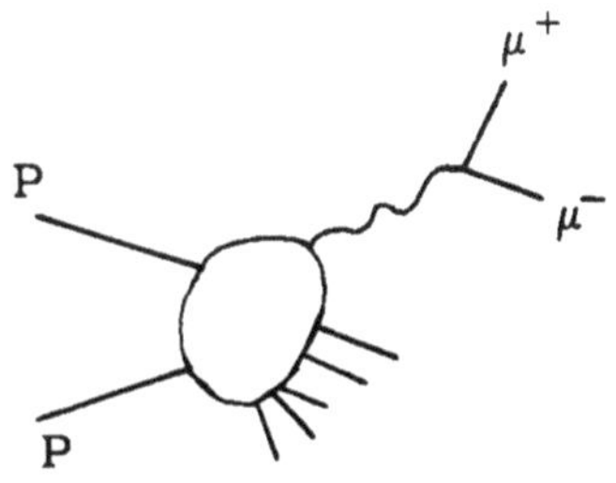

Fig. 7

e) Production of Virtual Photons

By observing the process, say, P+P → anything + $(\mu^+\mu^-)$, one can determine electromagnetic form factors for timelike momentum transfer (Fig. 7).

c) **The Kinematics of Lepton-Hadron Scattering**

Consider the scattering of a lepton from a nucleon to an arbitrary final state. The variables we use to describe the scattering amplitude are:

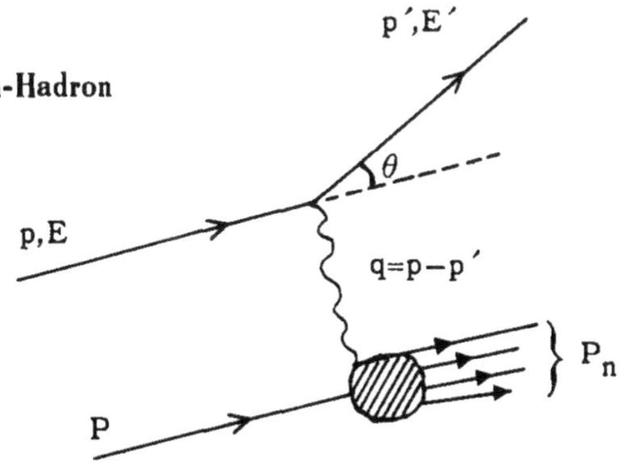

Fig. 8

$p(p')$ = initial (final) momentum of the lepton
$P(P_n)$ = initial (final) momentum of the hadron
$q = p-p'$
$\nu = q\cdot P = M(E-E')$ in the lab.
θ = scattering angle of the lepton in the lab.
$Q^2 = -q^2 = 4EE'\sin^2\theta/2$ (in the limit $m_e \approx 0$)
$S = (q+P)^2 = 2\nu + q^2 + M^2$ and a scale variable $\omega = \dfrac{Q^2}{2\nu}$. The hadronic vertex depends upon P, q and P_n. The structure functions to be defined later depend upon ν and Q^2.

Notation: Our metric is

$$g_{\mu\nu} = \begin{pmatrix} 1 & & & \\ & -1 & & \\ & & -1 & \\ & & & -1 \end{pmatrix}, \text{ such that } P^2 = P_0{}^2 - \vec{P}^2 = M^2 ;$$

spinors are normalized to $\bar{u}u = 1$; and fermion states are normalized so that

$$< \underset{\sim}{p}\sigma | \underset{\sim}{p}'\sigma' > = (\frac{E}{M})\, (2\pi)^3\, \delta^{(3)}\, (\underset{\sim}{p}-\underset{\sim}{p}') \; .$$

The amplitude for this process to a given final state $|n>$ is described by the matrix element

$$M = \bar{u}(p'\lambda')\gamma_\mu u(p,\lambda)\, \frac{e^2}{Q^2} < n|\, J^\mu(0)^{E.M.} |P\sigma > \quad \text{for eP scattering}$$

(1.1)

$$M = \bar{u}(p'\lambda')\gamma_\mu(1-\gamma_5)u(p\lambda)\, \frac{G}{(2)^{1/2}} < n|\, J^\mu(0)^{wk} |P\sigma > \quad \text{for } \nu P$$

scattering where:

$$J_\mu{}^{EM} = V_\mu{}^3 + \frac{1}{(3)^{1/2}}\, V_\mu{}^8$$

(1.2) $$J_\mu{}^{wk} = (V_\mu{}^{1+i2} - A_\mu{}^{1+i2})\cos\theta_c + (V_\mu{}^{4+i5} - A_\mu{}^{4+i5})\sin\theta_c \; .$$

The Cabibbo angle, θ_c, is approximately 15° and in most applications we will take $\theta_c \approx 0$.

Most of the experiments in the highly inelastic region do not observe the final hadronic state. The differential cross section for this type of inclusive experiment will be given by the probability for the above process summed over all hadron states and averaged over initial spins. This will be proportional to

$$\frac{1}{2}\sum_{\lambda,\lambda'} [\bar{u}(p'\lambda')0_\mu u(p\lambda)] * [\bar{u}(p'\lambda')0_\nu u(p,\lambda)] \times$$

$$\times \frac{1}{2}\sum_{\sigma,n} < P\sigma | J_\mu{}^\dagger(0) | n > < n | J_\nu(0) | P\sigma >$$

where $0_\mu = \gamma_\mu$ for the electromagnetic interaction (ep), $\gamma_\mu(1-\gamma_5)$ for the weak interaction (νp).

The spin sum and average of the lepton currents yields the tensor

$$\omega_{\mu\nu} = \frac{1}{2}\,\mathrm{Tr}[\not{p}'0_\mu \not{p} 0_\nu]$$

$$= 4[p_\mu p_\nu{}' + p_\mu{}' p_\nu{}' - g_{\mu\nu} p\cdot p' + i\epsilon_{\mu\nu\sigma\lambda} p_\sigma p_\lambda{}'] \text{ for the weak interaction}$$

$$(1.3) \qquad = 2[p_\mu p_\nu{}' + p_\mu{}' p_\nu - g_{\mu\nu} p\cdot p'] \text{ for the electromagnetic interaction.}$$

In this expression we have neglected the lepton mass, in which case the weak leptonic current is also conserved. The $\epsilon_{\mu\nu\sigma\lambda}$ term violates parity conservation and is due to the interference between the vector and the axial vector currents (hence it is not present in the electromagnetic case).

Similarly, we define the hadronic tensors

$$(1.4)\qquad W_{\mu\nu}{}^{ab} = \sum_{n,(\text{spin average})} \langle P|J_\mu{}^a(0)|n\rangle\langle n|J_\nu{}^b(0)|P\rangle\,(2\pi)^3\delta^4(P+q-P_n)$$

for arbitrary SU_3 currents, labelled by SU_3 indices a and b. We shall now express $W_{\mu\nu}{}^{ab}$ in terms of the commutator of the two currents. To this end we replace the δ-function, in (1.4), by $\frac{1}{(2\pi)^4}\int e^{i(P_n-P-q)\cdot x}\,d^4x$ and use translational invariance to write

$$\begin{aligned}\langle P|J_\mu{}^a(x)|n\rangle &= \langle P|e^{-i\mathcal{P}\cdot x}J_\mu{}^a(0)\,e^{i\mathcal{P}\cdot x}|n\rangle\\ &= e^{i(P_n-P)\cdot x}\langle P|J_\mu{}^a(0)|n\rangle\ .\end{aligned}$$

Then, using the completeness of the states $|n\rangle$ we can write (1.4) in the form:

$$(1.5)\qquad W_{\mu\nu}{}^{ab}(P,q) = \int\frac{d^4x}{2\pi}\,e^{-iq\cdot x}\langle P|J_\mu{}^a(x)J_\nu{}^b(0)|P\rangle\ ,$$

where an average over spins is understood.

It is sometimes convenient to rewrite this in terms of the commutator of the two currents. This is possible since

$$\begin{aligned}&\int\frac{d^4x}{2\pi}\,e^{-iq\cdot x}\langle P|J_\nu{}^b(0)J_\mu{}^a(x)|P\rangle\\ &\qquad=\sum_n\langle P|J_\nu{}^b(0)|n\rangle\langle n|J_\mu{}^a(0)|P\rangle\,(2\pi)^3\delta(P-q-P_n)\end{aligned}$$

vanishes, since in the lab frame $E_n = M-q_0 \le M$ and no baryon state exists with mass less than the nucleon mass. For $q_0 > 0$ we can therefore subtract this term from (1.5)

$$(1.6) \qquad W_{\mu\nu}{}^{ab}(P,q) = \int \frac{d^4x}{2\pi} e^{-iq\cdot x} < P|[J_\mu{}^a(x), J_\nu{}^b(0)]|P > .$$

The tensor $W_{\mu\nu}{}^{ab}$ has the following general properties:

1) *Crossing:*

$$(1.7) \qquad W_{\mu\nu}{}^{ab}(P,q) = -W_{\nu\mu}{}^{ba}(P,-q) .$$

This is obvious from (1.6).

2) *Positivity:* If J^a is a hermitean current, then $W_{\mu\nu}{}^{aa}$ is a positive semi-definite form, i.e.,

$$(1.8) \qquad \eta_\mu{}^* W_{\mu\nu}{}^{aa} \eta_\nu \geq 0 ,$$

for an arbitrary 4-vector η_μ. This follows immediately from (1.4)

3) *Conservation:* If J^a (J^b) is a conserved current, e.g., the electromagnetic current, then $q^\mu W_{\mu\nu}{}^{ab}$ ($q^\nu W_{\mu\nu}{}^{ab}$) vanishes.

We now construct the invariant amplitudes for the two cases of interest. In the electromagnetic case:

$$W_{\mu\nu}(P,q) = \frac{1}{M^2}(P_\mu - \frac{\nu}{q^2} q_\mu)(P_\nu - \frac{\nu}{q^2} q_\nu) W_2(q^2,\nu) -$$

$$- (g_{\mu\nu} - \frac{q_\mu q_\nu}{q^2}) W_1(q^2,\nu)$$

with

$$(1.9) \qquad W_i(q^2,\nu) = -W_i(q^2,-\nu) .$$

For the weak interactions:

$$W_{\mu\nu}^{(\mp\pm)}(P,q) = \frac{1}{M^2} P_\mu P_\nu W_2^{\binom{\nu}{\bar\nu}}(q^2,\nu) - g_{\mu\nu} W_1^{\binom{\nu}{\bar\nu}}(q^2,\nu)$$

$$(1.10) \qquad -i\epsilon_{\mu\nu\sigma\lambda} \frac{P_\sigma q_\lambda}{2M^2} W_3^{\binom{\nu}{\bar\nu}}(q^2,\nu) + \ldots\ldots$$

where (−) and (+) refer to 1−i2 and 1+i2 indices for a and b in the general definition of $W_{\mu\nu}{}^{ab}$. In this case, where the current is not conserved, there will be six structure functions. The deleted terms contain a factor of q_μ or q_ν which, when contracted with the leptonic tensor $\omega_{\mu\nu}$, give zero in the approximation of zero lepton mass. Crossing symmetry, (1.7), implies that:

$$W_i^{\nu}(q^2,\nu) = -W_i^{\bar\nu}(q^2,-\nu) \quad .$$

It is straightforward to calculate the differential cross section, in angle and energy of the outgoing lepton, in terms of the structure functions. The result is

$$\frac{d^2\sigma}{d\Omega' dE'} = \frac{EE'}{\pi} \frac{d^2\sigma}{dQ^2 d\nu} =$$

$$\text{for ep} \qquad \frac{\alpha^2 \cos^2\theta/2}{4E^2 \sin^4\theta/2} \left[W_2^{\text{em.}}(q^2,\nu) + 2\tan^2\theta/2\, W_1^{\text{em.}}(q^2,\nu)\right]$$

$$\text{for } \binom{\nu}{\bar\nu}p \qquad \frac{G^2(E')^2}{2\pi^2} \cos^2\theta/2 \left[W_2^{\binom{\nu}{\bar\nu}}(q^2,\nu) + 2\tan^2\theta/2 \left(W_1^{\binom{\nu}{\bar\nu}}(q^2,\nu)\right.\right.$$

$$(1.11) \qquad \left.\left. \mp \frac{E+E'}{2M} W_3^{\binom{\nu}{\bar\nu}}(q^2,\nu)\right)\right] \quad .$$

Note that the expression outside of the brackets is the differential cross section off a structureless target, in the case of ep scattering the Mott cross section. For fixed Q^2 and ν and large E, $\sin^2 \theta/2 = -\frac{Q^2}{4EE'} \to 0$ and $\cos^2 \theta/2 \to 1$ so that

$$(1.12) \qquad \frac{d^2\sigma^{ep}}{d\Omega' dE'} \xrightarrow[E \to \infty]{} \left(\frac{\alpha^2}{4E^2 \sin^2 \theta/2}\right) \cdot W_2(q^2,\nu) \ldots\ldots .$$

Let us note the following properties of (1.11):

1) *Locality:* The explicit dependence of the cross section (1.11) on θ is a consequence of the *local* coupling of the weak or electromagnetic currents to the leptons, and the assumption of one photon exchange. Therefore a test of whether

$$\frac{\frac{d\sigma}{d\Omega dE'}}{\left(\frac{d\sigma}{d\Omega dE'}\right)_{MOTT}} = W_1(q^2,\nu) + 2\tan^2 \theta/2 \; W_2(q^2,\nu)$$

is indeed linear in $\tan^2 \theta/2$ for fixed q^2 and ν (the Rosenbluth plot) is a test of these assumptions.

2) *Positivity:* The positivity conditions, (1.8), lead to various inequalities for W_i. Physically these inequalities express the positivity of the total absorption cross section for virtual photons (or W-mesons) of different helicities.

They are thus most easily derived by taking for η the polarization vectors of the exchanged photon or W-meson.

An off mass shell photon can be described by either a transverse or scalar polarization vector $\underset{\sim}{\epsilon}$. For transverse photons $\epsilon \cdot q = 0$, $\underset{\sim}{\epsilon}^2 = 1$ and (evaluated in the lab. where $P = (M,0)$) $\sigma_T \sim \epsilon_i{}^* W_{ij} \epsilon_j = W_1$. For scalar photons $\underset{\sim}{\epsilon} || \underset{\sim}{q}$ and

$$\sigma_S \sim q_i W_{ij} q_j = \frac{1}{M^2} \frac{\nu^2}{(Q^2)^2} W_2 |\underset{\sim}{q}|^2 + (|\underset{\sim}{q}|^2 - \frac{|\underset{\sim}{q}|^2}{Q^2}) W_1 .$$

Now $|\underset{\sim}{q}|^2 = Q^2 + \frac{\nu^2}{M^2}$ so that $\sigma_S \approx (1 + \frac{\nu^2}{M^2 Q^2}) W_2 - W_1$. We choose the normalization adopted by L. Hand so that:

$$\sigma_T(\nu, q^2) = \frac{4\pi\alpha M}{\nu - \frac{1}{2} q^2} W_1(q^2, \nu) \xrightarrow[q^2 \to 0]{} \sigma_\gamma(\nu) \quad , \tag{1.13}$$

where $\sigma_\gamma(\nu)$ is the photo absorption cross-section for real photons, and

$$\sigma_S(\nu, q^2) = \frac{4\pi\alpha M}{\nu - \frac{1}{2} q^2} \left[(1 + \frac{\nu}{Q^2 M^2}) W_2 - W_1 \right] \xrightarrow[q^2 \to 0]{} 0 . \tag{1.14}$$

The vanishing of σ_S for real photons is not obvious from (1.14), one must go back to (1.9) and derive the kinematic zeroes in W_2 and W_1 that are implied by the finite value of $W_{\mu\nu}$ at $q^2 = 0$. The positivity of these cross sections implies that

$$0 \lesssim W_1 \lesssim (1 + \frac{\nu^2}{M^2 Q^2}) W_2 . \tag{1.15}$$

In the case of neutrinos we must distinguish between right and left hand transverse polarizations. The parity non-conserving term, W_3, in fact distinguishes between the two. The form of σ_S remains the same but

$$\begin{matrix} \sigma_R \\ \sigma_L \end{matrix} \sim \epsilon_i^* W_{ij} \epsilon_j = |\underset{\sim}{\epsilon}|^2 W_1 \mp i \frac{1}{2M} (\underset{\sim}{\epsilon}^* \times \underset{\sim}{\epsilon}) \cdot \underset{\sim}{q} W_3 = (W_1 \mp \frac{1}{2M} (\frac{\nu^2}{M^2} - q^2)^{1/2} W_3) .$$

Therefore

$$0 \leq \frac{1}{2M}(\nu^2+M^2Q^2)^{1/2}|W_3| \leq W_1 \leq (1+\frac{\nu^2}{M^2Q^2})W_2 \quad . \tag{1.16}$$

When $q^2 \to 0$ the lepton current is proportional to q_μ, which projects out the divergence of the axial vector current. If one dominates these divergences by the pion pole the cross-sections for $q^2 = 0$ can be related to pion cross sections. This yields a test of PCAC which is independent of assumptions concerning current commutators.

3) *Isotopic Spin:* Consider ν or $\bar{\nu}$ scattering off a proton (p) or a neutron (n). The neutrino interacts via J^{1-i2} (W^- meson) and the anti-neutrino via J^{1+i2} (W^+ meson). Charge symmetry requires that a given W-Nucleon cross section obey:

$$\sigma_i(W^{\pm}p) = \sigma_i(W^{\mp}n) \quad ,$$

where i refers to R, L or S. If we recall that in the expression for the ν-nucleon cross-sections (1.11), the contribution of W_3 changes sign when we switch from ν to $\bar{\nu}$, and that $W_3 \sim \sigma_R - \sigma_L$ we have schematically

$$\sigma(\nu p) = \alpha\sigma_S(W^- p) + \beta\sigma_R(W^- p) + \gamma\sigma_L(W^- p)$$

$$\sigma(\bar{\nu} p) = \alpha\sigma_S(W^+ p) + \beta\sigma_L(W^+ p) + \gamma\sigma_R(W^+ p)$$

$$\sigma(\nu n) = \alpha\sigma_S(W^+ p) + \beta\sigma_R(W^+ p) + \gamma\sigma_L(W^+ p)$$

$$\sigma(\bar{\nu} n) = \alpha\sigma_S(W^- p) + \beta\sigma_L(W^- p) + \gamma\sigma_R(W^- p) \quad . \tag{1.17}$$

Thus $\sigma(\nu p)-\sigma(\bar{\nu} n)$ measures $\sigma_R(W^- p)-\sigma_L(W^- p) \approx W_3^{\nu p}$. Similarly $\sigma(\nu n)-\sigma(\bar{\nu} p) \approx \sigma_R(W^+ p)-\sigma_L(W^+ p) \approx W_3^{\bar{\nu} p}$. Therefore the isoscalar part of W_3 can be determined by measuring the difference of the differential cross sections of ν and $\bar{\nu}$ off heavy nuclei (which have roughly equal amounts of p and n). Similarly $\sigma(\nu p)+\sigma(\nu n)+\sigma(\bar{\nu}p)+\sigma(\bar{\nu}n)$ is pure isoscalar and proportional to $\sigma_S+\sigma_R+\sigma_L$, i.e., independent of W_3.

II. ELASTIC FORM FACTORS AND PHOTON HADRON SCATTERING

a) A Kinematical Plot

This section consists of a discussion of the experimental and theoretical situation regarding elastic e–p scattering, photoproduction and Compton scattering. These processes lie on the boundaries of the physical region for inelastic lepton-hadron scattering, to which we shall devote most of our attention. This is illustrated in Fig. 9, which shows the various physical processes on a $(Q^2-\nu)$ plot.

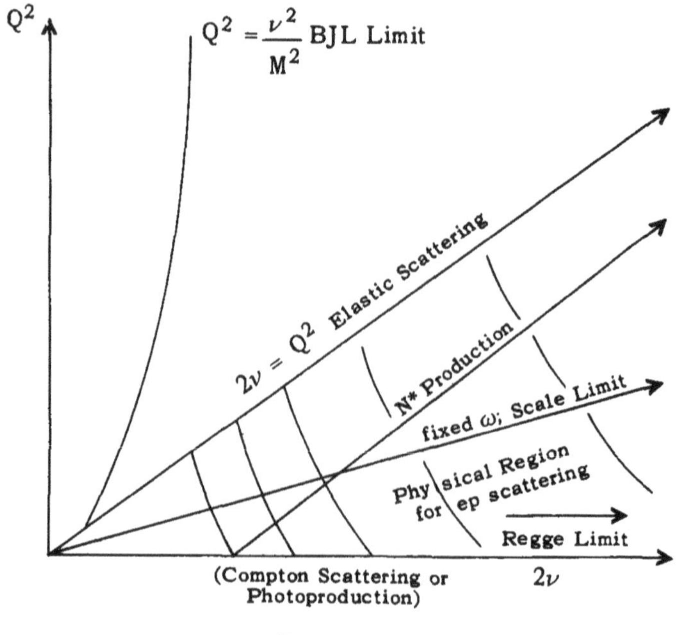

Fig. 9

The physical region for e–p scattering is determined by $S = (q+p)^2 \geq M^2$. The various processes illustrated in Fig. 9 are

1) Elastic Scattering corresponding to $2\nu = Q^2$. Along this curve the elastic form factors are measured as a function of Q^2.

2) Photoproduction and Compton scattering correspond to $Q^2 = 0$.

3) Production of Inelastic discrete states occurs along the lines of fixed $S = 2\nu - Q^2$.

The various asymptotic limits considered by theorists are:

1) The Regge limit, i.e., large ν and fixed external mass Q^2.

2) The scale (or Bjorken) limit which corresponds to large ν and Q^2 and fixed $\omega = \frac{Q^2}{2\nu}$.

3) The Bjorken-Johnson-Low (BJL) Limit corresponding to large q_0 and fixed $\vec{q}$ and p. In this limit

$$|q^2| \cong |q_0{}^2| \cong |\frac{\nu}{M}|^2 ,$$

and the resulting parabola lies outside the physical region. It is in this limit, however, that current algebra yields immediate results.

b) Electromagnetic Form Factors

The structure functions for elastic electron (or muon) nucleon scattering are determined by the charge and magnetic form factors:

$$W_2(q^2,\nu) = M\delta(\nu + \frac{q^2}{2}) \frac{[G_E(q^2)]^2 - q^2/4M^2[G_M(q^2)]^2}{1 - q^2/4M^2}$$

$$W_1(q^2,\nu) = -M\delta(\nu + \frac{q^2}{\nu}) \frac{q^2}{4M^2} [G_M(q^2)]^2 \quad . \tag{2.1}$$

The charge and magnetic form factors correspond respectively to scalar and transverse photon polarizations, and are given in terms of Dirac form factors by:

$$G_E(q^2) = F_1(q^2) - \frac{q^2}{4M^2}\,\mu\, F_2(q^2)$$

$$G_M(q^2) = F_1(q^2) + \mu\, F_2(q^2) \quad , \tag{2.2}$$

where $F_1(0) = F_2(0) = 1$ and μ is the anomalous magnetic moment.

The experimental data can be succinctly summarized by the dipole fit and the "scaling" laws:

$$G_E{}^p(q^2) = \frac{G_\mu{}^p(q^2)}{\mu_p} = \frac{G_\mu{}^n(q^2)}{\mu_n} = \frac{1}{\left[1 - \dfrac{q^2}{(0.71\ \mathrm{BeV}^2/c^2)}\right]^2}$$

$$G_E{}^n(q^2) = 0 \ . \tag{2.3}$$

These empirical fits agree well with the data, up to $-q^2 \sim 40\ \mathrm{BeV}^2/c^2$, although there seems to be some departure from both the scaling law and the dipole formula at the largest values of q^2.

Little detailed information is known about the other electromagnetic form factors. There is, however, recent evidence from SLAC that indicates that the transition form factors from the nucleon to the nucleon resonances have identical $(\frac{1}{q^2})^2$ behavior for large values of $-q^2$.

In the non-relativistic quantum mechanics the form factor is the Fourier transform of the spatial charge distribution. The rapid falloff in q^2 is thus, roughly speaking, some indication that the nucleon charge is smeared out, and perhaps can be taken as an indication of the composite nature of the nucleon.

Most of the theoretical attempts to explain the q^2-behavior of these form factors are based on unsubtracted dispersion relations. One writes

$$G(q^2) = \frac{1}{\pi} \int_{s_0}^{\infty} \frac{\rho(s)ds}{s-q^2-i\epsilon} ,$$

and assumes that the discontinuity, $\rho(s)$, is dominated by a few nearby singularities. Historically, in fact, the existence of the vector mesons was suggested by such attempts to fit the electromagnetic form factors. If one assumes "vector dominance", i.e., that the known vector meson (ρ, ω, Φ) poles dominate the dispersion relations (Fig. 10)

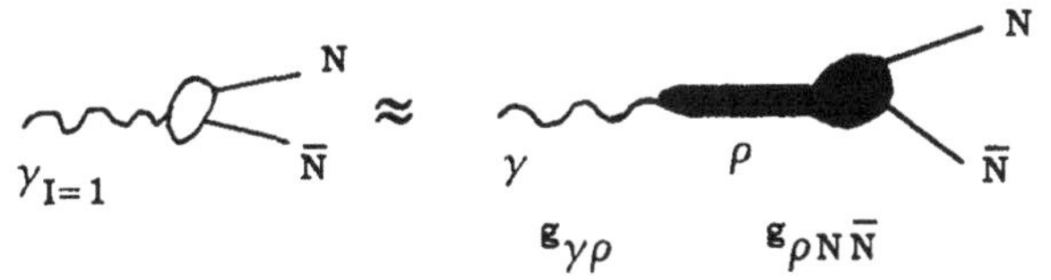

Fig. 10

then

$$G^{I=1}(q^2) \sim \frac{g_{\gamma\rho} g_{\rho NN}}{[1- q^2/m_\rho^{\ 2}]}$$

The rapid fall-off of the form factors is then understood to be the tail of the ρ-pole.

It is evident from the dipole formula that vector meson dominances does not provide a quantitative description of the q^2 dependence. Many attempts have been made to improve matters by adding additional poles. None of these are especially convincing. It is fair to say that, at present, there exists no adequate theory of electromagnetic form factors.

c) Photo-Hadron Scattering

1) *Photoproduction:* One might expect that two-body photoproduction cross sections would be quite similar to hadronic cross-sections. This, in fact, seems to be the case. High energy two body photoproduction reactions, in both the forward and backward directions, are diffractive and exhibit normal hadron like angular distributions.

These reactions show, in a crude sense, the following regularities:

a) For ρ, ω, ϕ photoproduction the cross section behaves as

$$\frac{d\sigma}{dt} \underset{s\to\infty}{\approx} f(t).$$ This is to be expected since vacuum quantum numbers are exchanged in the t-channel.

b) Forward photoproduction in the channels πN, $\pi\Delta$, ηN, $K\Sigma$, $K\Lambda$ exhibits a power behavior (Fig. 11)

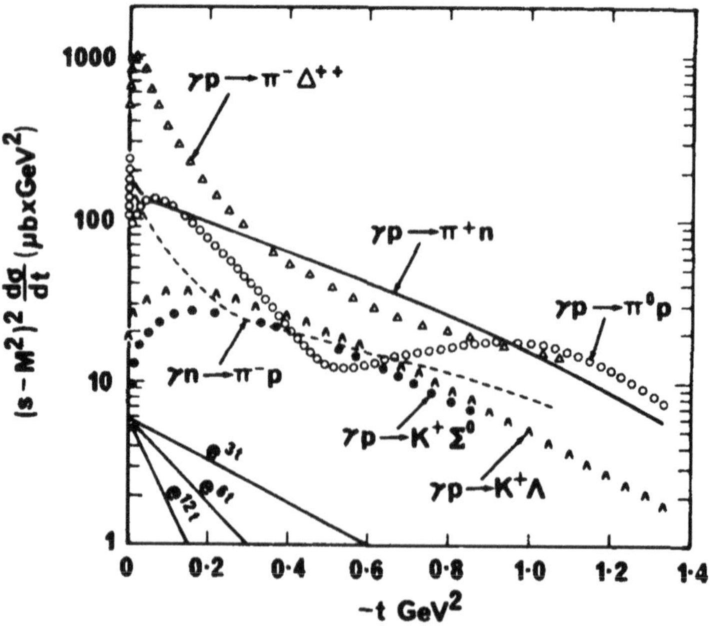

Fig. 11

General features of pseudoscalar meson photoproduction at forward angles as compiled by R. Diebold.

(2.4) $$\frac{d\sigma}{dt} \approx s^{\alpha-2} f(t) \qquad f(t) \sim e^{3t}$$

with $-0.2 \lesssim \alpha \lesssim 0.2$. In all these reactions a pion or another pseudoscalar meson can be exchanged in the t-channel. A $(\frac{1}{s})^2$ behavior would be the consequence of elementary meson exchange.

c) Backward photoproduction, e.g., $\gamma+N \to N+\pi$ behaves as

(2.5) $$\frac{d\sigma}{du} \approx \frac{1}{s^3} f(u) \ .$$

The power s^{-3}, in this case, is not what one would expect from the exchange of an elementary nucleon in the u-channel, which would give a s^{-1} behavior.

2) *Compton Scattering:* The total photoabsorption cross section has been measured up to photon energies of 20 GeV. Once again the result is in accord with the statement that photons of fixed mass have cross sections similar to hadronic cross sections. The total photoabsorption cross section, which is in the imaginary part of the forward Compton Scattering amplitude, appears to approach a constant value of roughly 125μb. (See Fig. 12)

d) Regge Poles and Fixed Poles in Weak Processes

There are many indications that the high energy behavior of the strong interactions can be described in terms of Regge poles in angular momentum. While it is not established experimentally that these poles do indeed explain *all* the features of the high energy behavior of hadronic amplitudes the following gross features are widely accepted:

1) *Inelastic Processes* in which there is an exchange of non-vacuum quantum numbers (e.g., $\pi+p \to \pi^0 n$) are dominated by moving Regge poles. Thus the differential cross section behaves as

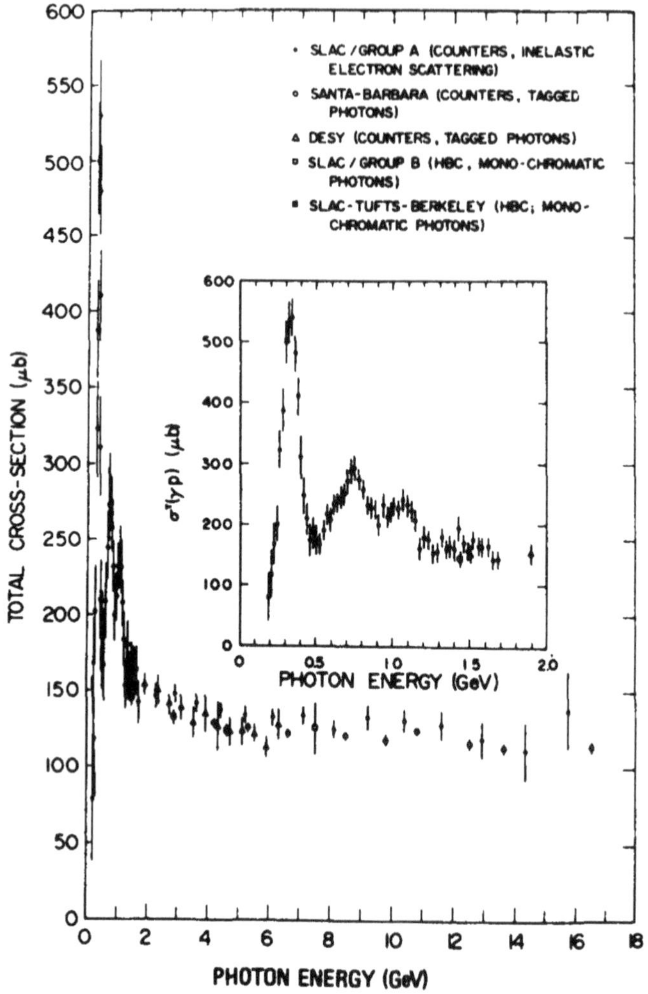

Fig. 4. The total photoabsorption cross section, $\sigma_T(\gamma p)$ measured in recent experiments[8].

Fig. 12

The total photoabsorption cross section, $\sigma_T(\gamma p)$ measured in recent experiments.

$$\frac{d\sigma}{dt} \underset{s \to \infty}{\sim} \beta(t) s^{2[\alpha(t)-1]} ,$$

where the Regge pole trajectory, $\alpha(t) = \alpha_0 + \alpha' t$, is roughly linear and α_0 is approximately equal to 1/2 for octet trajectories (the vector and tensor mesons lie on these trajectories) and is less than zero for processes in which the exchanged quantum numbers are exotic (i.e., I = 2 or $\underset{\sim}{27}$).

2) *Elastic Processes* in which the exchange quantum numbers are those of the vacuum (e.g., $\pi^+ p \to \pi^+ p$) have constant cross sections at large energies:

$$\left.\frac{d\sigma}{dt}\right|_{t=0} \underset{s \to \infty}{\sim} \text{const.}; \qquad \sigma_{tot} \xrightarrow[s \to \infty]{} \text{const.}$$

The nature of the vacuum singularity in angular momentum is still unclear.

Accepting the Regge pole model for hadronic processes we are led to inquire what asymptotic behavior one might except for weak processes, i.e., the scattering of weak or electromagnetic currents off hadrons. Do we have any reason to assume that $W_i(q^2,\nu)$, for fixed q^2 and large ν is dominated by Regge poles? Two answers can be given to this question. We do expect that if a pole is present in a hadronic partial wave amplitude at some $J = \alpha(t)$ the same pole will appear in all weak partial wave amplitudes with the same quantum numbers. On the other hand, since weak amplitudes (to lowest order in the coupling constant, α or G) are not restricted by a non-linear unitarity condition they can contain singularities in J which are forbidden in strong amplitudes.

Consider a weak process $W+A \to B+C$, where W is a photon or weak current with fixed virtual mass and A,B,C are hadrons. Our reason for believing that Regge poles will appear in the amplitude for this process is that the essential dynamics is governed by the strong interactions. For example, consider the Bethe-Salpeter equation, schematically illustrated in Fig. 13.

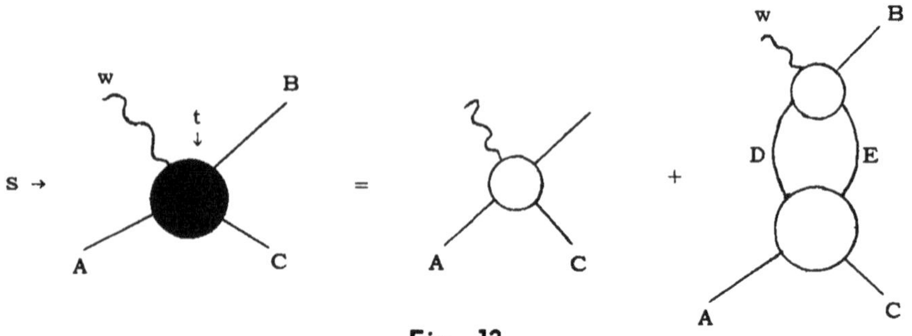

Fig. 13

The kernel of this integral equation is a strong amplitude and would appear in the Bethe-Salpeter equation for a strong process with the same quantum numbers in the t-channel. The singularities, in the angular momentum of the t-channel, of the strong amplitude arise due to the vanishing, at $J = \alpha(t)$, of the Fredholm determinent of this kernel. They will therefore, in general, occur also in the weak amplitude.

A more general argument follows from unitarity in the t-channel:

$$\text{Im}\, A_J(W+A \to B+C) = \sum_n A_J{}^*(W+A \to n) A_J(n \to B+X) \quad , \tag{2.6}$$

where n is a hadronic state. If we assume some region of analyticity, in the J plane, for the weak amplitude then this relation can be continued to complex J. If the strong amplitude $A_J(n \to B+C)$ has a pole at $J = \alpha(t)$ the weak amplitude $A_J(W+A \to B+C)$ will contain the same pole.

A weak amplitude can contain, however, singularities which are not present in strong amplitudes. In particular, the unitarity equation for a strong amplitude, e.g.,

$$a^{\pm}_{\text{had.}}(J,t) - a^{\pm}_{\text{had}}(J^*,t)^* = \rho(t) a^{\pm}_{\text{had}}(J,t) a^{\pm}_{\text{had}}(J^*,t)^* \tag{2.7}$$

is not consistent with a fixed pole at $J = J_0$ in the hadronic partial wave amplitude. Unitarity for weak amplitudes (to lowest order in the weak coupling constant) is a linear relation and does not include fixed poles in J. To see where such fixed poles might appear we will briefly review the Regge analysis for spinless particles.

The scattering amplitude, A(s,t), is expanded in a partial wave series in the t-channel and this is rewritten, using the Watson-Sommerfeld transformation, as:

$$A(s,t) = -\frac{1}{2\pi}\oint dJ\, \frac{2J+1}{\sin\pi J}\, \{a^{+}(J,t)[P_J(-Z_t)+P_J(Z_t)] +$$

$$a^{+}(J,t)[P_J(-Z_t)-P_J(Z_t)]\}\ , \tag{2.8}$$

where the contour encloses the positive integers and Z_t is the cosine of the t-channel center of mass scattering angle. The signatured partial wave amplitudes are given by

$$a^{\pm}(J,t) = \frac{1}{\pi}\int_{Z_0}^{\infty} dZ_t\ Q_J(Z_t)[A_s(t,Z_t) \pm A_u(t,Z_t)]\ , \tag{2.9}$$

where $A_s(A_u)$ is the absorptive part of A(s,t) in the s (u) channel. The above (Froissart-Gribov) definition of the partial wave amplitude seems to naturally contain poles at $J = -1, -2, -3, \ldots,$ since Q_J has simple poles at the negative integers. These unphysical values of J are called "nonsense" values. When spinning particles are involved these nonsense values of J can be promoted to positive integers.

The nonsense values actually occur for negative values of the orbital angular momentum, and if s is the maximum spin in the t-channel then $J = s-1, s-2, \ldots$ can be a nonsense value. A fixed pole cannot occur at a sense (physical) value of J in the correct signatured partial wave amplitude, since that would make a physical partial wave amplitude infinite. For example, in the case of vector + scalar → vector + scalar, where the maximum spin in the t-channel is 2, one could have a fixed pole at $J = 1, 0, -1,$

Since we have no good reason for excluding such fixed poles let us consider their consequences. Two types might occur:

1) *Correct Signature:* Such a fixed pole, for example in $a^-(s,t)$ at $J = -1$, gives rise to a fixed power behavior of the amplitude for large s:

$$A(s,t) \xrightarrow[s \to \infty]{} \frac{\beta(t)}{s} \; .$$

It does not contribute to the asymptotic behavior of the absorptive part, unless it coincides with a moving Regge pole at some value of t.

2) *Incorrect Signature:* Such a fixed pole, for example in $a^+(J,t)$ at $J = -1$, does not contribute to the asymptotic behavior of the amplitude for large s.

Do we have any reason to suspect that either of these fixed poles exist? The classic example of a theoretical necessity for fixed poles arises in the analysis of isotopic spin current scattering. Consider the "scattering" of the isospin current J^a (with virtual mass q_1^2) off a spinless hadron (Fig. 14). From the equal time

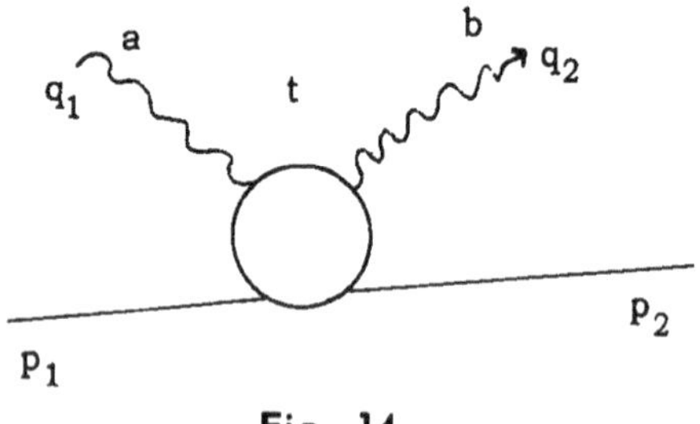

Fig. 14

commutation relations of the isotopic charge densities one derives the Adler-Fubini-Dashen-Gell-Mann sum rule

$$\begin{aligned} \int_{-\infty}^{+\infty} d\nu \, \mathrm{Im}\, A^{ab}(q_1^2, q_2^2, \nu, t) &= \epsilon_{abc} \langle p_1 | J^c(0) | p_2 \rangle \\ &= \epsilon_{abc} \, f^c(t) \end{aligned} \tag{2.10}$$

where the absorptive part of the above amplitude is given by $P_\mu P_\nu \mathrm{Im} A^{ab} + .$ $p = p_1+p_2$, $\nu = (p_1+p_2)(q_1+q_2)$, $t = (p_1-p_2)^2$. From this sum rule we can deduce that the partial wave projection of the invariant amplitude A^{ab} has a fixed pole at $J = 1$. To see this note that the residue of the negative signatured partial wave amplitude Eq. (2.9) at $J = -1$, where

$$Q_J \underset{J \approx -1}{\approx} \frac{1}{J+1} \, ,$$

is given by a formula identical to (2.10). For the amplitude A^{ab}, which is a helicity flip two t-channel amplitude, $J = +1$ is a nonsense value corresponding to $\ell = -1$. Alternatively we can deduce the presence of a fixed pole in the correct signature amplitude at $J = 1$ by examining the asymptotic behavior of $A^{ab}(\nu,t,q_1^2,q_2^2)$ as $\nu \to \infty$. A singularity at $J = \alpha(t)$ in $A_J^{ab(\pm)}(t,q_1^2,q_2^2)$ contributes a term to A^{ab} behaving like $\nu^{\alpha(t)-2}$ (because of the two p's in the definition of A^{ab}). But

$$A^{ab}(\nu,t,q_1^2q_2) = \frac{1}{\pi}\int \frac{\mathrm{Im}A^{ab}(\nu',t,q_1^2q_2^2)d\nu}{\nu'-\nu} \to -\frac{1}{\pi\nu}$$

$$\int \mathrm{Im}A^{ab}(\nu',t,q_1^2q_2^2)d\nu = -\frac{\epsilon_{abc}f^c(t)}{\pi\nu} \quad \ldots \tag{2.11}$$

and thus we see the fixed power behavior which indicates a fixed pole in $A_J^{ab(-)}(t,q_1^2q_2^2)$ with residue proportional to $\epsilon_{abc}f^c(t)$.

We also note that the residue of the fixed pole, i.e., $f^c(t)$, is independent of q_1^2 and q_2^2. This is reasonable since the discontinuities of A^{ab} in q_1^2 and q_2^2 are related to hadronic scattering amplitues which cannot have fixed poles. Thus at $q_1^2 = q_2^2 = m_\rho^2$

$$A^{ab} = \frac{T^{ab}}{(q_1^2-m_\rho^2)(q_2^2-m_\rho^2)} \, ,$$

where T^{ab} is the ρ-hadron scattering amplitude.

Unfortunately one cannot observe this amplitude directly since there are no charged photons, and for Compton scattering, the right hand side of (2.10) vanishes, leaving us with no prediction as to the presence of fixed poles at $J = 1$.

The only place fixed poles may be forced in Compton scattering is the positive signatured amplitude at $J = 1$. If one wishes to couple the Pomeranchuk Regge pole to photons one needs either a wrong signatured fixed pole at $J = 1$ or a singular Regge residue in order to cancel a kinematical zero.

Other fixed poles might arise in:

1) *Compton Scattering.* Gilman claims, on the basis of a dispersion theory fit, that Compton scattering seems to require a fixed pole at $J = 0$ with a rather large residue. Evidence against the presence of a different fixed pole in a rather oscure helicity amplitude for photon nucleon scattering is provided by the success of the Drell-Hearn sum rule. This sum rule is, in fact, a super convergence relation, whose derivation rests on the assumption of no fixed pole at $J = 0$.

2) *Photo Production.* There are no theoretical arguments in favor of fixed poles for amplitudes involving only one current. Experimentally, as we have seen, there is some evidence for fixed power behavior in $\gamma N \to$ meson$+N$ at $J = 0$ (like a π exchange) and in backward $\gamma N \to N +$ mesc at $J = 3/2$ (not like a nucleon exchange).

In the application of current algebra and the derivation of sum rules, it will be necessary to make assumptions regarding the asymptotic behavior of weak amplitudes in order to decide whether one can write unsubtracted dispersion relations. It seems quite safe to assume, as we shall, that weak amplitudes are bounded by s for vacuum exchange, by $s^{1/2}$ for $I = 1$ or octet exchange and perhaps by $s^{-\epsilon}$ for $I=2$ or $\underset{\sim}{27}$ exchange.

III. INELASTIC ELECTRON-HADRON SCATTERING

a) Theoretical Expectations

In this section I will review the results of the recent SLAC-MIT experiments on highly inelastic electron proton scattering, and the inelastic neutrino scattering experiments performed at CERN. First, however, I will describe the theoretical expectations that one had *before* the experiments were performed. The purpose of this recounting of past failure is twofold. First it will enable us to understand why the SLAC data was so surprising to so many theorists. It was not what many would have expected based on the extrapolation of simple models. In addition, it is useful to compare the naive predictions (by this I mean those predictions that could be or actually were made before the experiments were performed) of various theoretical models. Many of these models, whose naive extrapolation was in conflict with the data, were able, after the fact, to accommodate themselves to the experiments, however a historical review will allow us to evaluate their predicitive power.

We are interested in the behavior of the structure functions $W_i(q^2,\nu)$ in the region where both q^2 and ν are large. The only previous *experimental* information we have as to the large q^2 behavior of these structure functions is the behavior of elastic form factors, i.e., at the point $2\nu + q^2 = 0$. These, as we have seen, decrease rapidly with increasing q^2. The structure functions W_i are sums of inelastic form factors

$$W_i(q^2,\nu) \sim \sum_n |\langle P|J|n\rangle|^2 \quad .. \tag{3.1}$$

Although each individual inelastic form factor might be expected to exhibit the same rapid falloff with increasing Q^2 as the elastic form factor, the sum might behave quite differently. Even so, many experimentalists and theorists were worried, at the planning stage of the SLAC experiment, that the structure functions would fall off so rapidly as to make highly inelastic events extremely rare.

1) *Regge Theory*

We have previously given arguments in favor of the assumption of Regge behavior for weak amplitudes. What does Regge theory tell us about the behavior of $W_i(q^2,\nu)$? The relevant variable for the Regge pole expansion is the cosine of the scattering angle in the center-of-mass of the t-channel, z_t. It is easy to see that, for large ν and large Q^2, z_t is given by

$$z_t = \frac{\nu}{M(Q^2)^{1/2}} \tag{3.2}$$

So that according to Regge theory:

$$W_i(Q^2,\nu) = \sum_i \beta_i(q^2) \left(\frac{\nu}{M(Q^2)^{1/2}}\right)^{a_i} + \text{background}, \tag{3.3}$$

where the sum extends over all relevant Regge poles. This is only useful in the region where $\nu \to \infty$ for fixed Q^2. To say anything about the region where both Q^2 and ν are large one must have information regarding the scale limit where ν/Q^2 is fixed. In this case, for large ν, z_t is indeed large; however, from the expansion

$$W_i(Q^2,\nu) = \sum_i \beta_i(Q^2)(Q^2)^{\frac{1}{2}a_i} \left(\frac{\nu}{MQ^2}\right)^{a_i} + \text{background}$$

we can say very little unless we know the behavior for large Q^2 of $\beta_i(Q^2)$, and the behavior of the background is this limit. I suspect that a Regge pole theorist would have said, before the experiments were performed, that $\beta_i(Q^2)$ would be a rapidly decreasing function of Q^2 in that it should behave like a form factor, and that therefore, at least for large and fixed ν/Q^2, W_1 would be a decreasing function of Q^2.

2) *Vector Meson Dominance*

Vector meson dominance is essentially the idea that the matrix elements of the electromagnetic current are dominated by their vector meson poles.

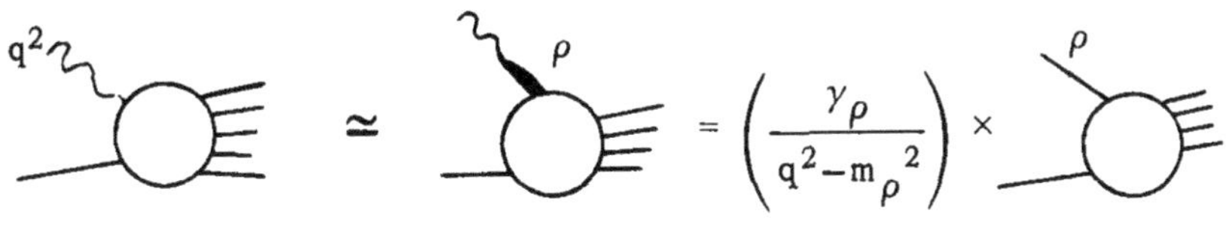

The electroproduction structure functions can, by this hypothesis, be calculated in terms of on-mass shell ρ-meson nucleon scattering amplitudes, and the only q^2 dependence arises from the vector meson poles. It is not at all obvious why this should be a good approximation away from the vector meson pole at $q^2 \cong m_\rho^{\,2}$. In fact, as we have seen this assumption is very bad for the elastic nucleon form factors. In any case the VMD model yields very definite predictions about the behavior of the photoabsorption cross sections $\sigma(q^2,\nu)$. Consider the cross section for transverse photons of mass q^2

$$\sigma_T(q^2,\nu) = \frac{M}{\nu+\tfrac{1}{2}Q^2}\, W_1(q^2,\nu) \ .$$

VMD tells us that this cross section is given by the on shell ρ-nucleon cross section:

$$\sigma_{\gamma N}^{(T)}(q^2,\nu) = \gamma_\rho^{\,2}\left(\frac{m_\rho^{\,2}}{m_\rho^{\,2}-q^2}\right)^2 \sigma_{\rho N}(2\nu+q^2+\mu^2) \ , \tag{3.4}$$

where the ρ-N cross section depends only upon the center of mass energy $s = M^2+2\nu+q^2$. For large ν and q^2 we then have

$$\sigma_T(q^2,\nu) \to \gamma_\rho^{\,2} \frac{m_\rho^{\,4}}{q^4} \sigma_{\rho N}(\infty) \qquad \text{or}$$

$$W_1(q^2,\nu) \to \frac{\gamma_\rho^{\,2} m_\rho^{\,4}(\nu - \tfrac{1}{2}q^2)}{Mq^4} \sigma_{\rho N}(\infty) \; . \tag{3.5}$$

It is slightly more problematic to predict the behavior of $\sigma_S(q^2,\nu)$, since this cross section must vanish at $q^2 = 0$. Sakurai has claimed, after the SLAC experiments, that the correct way to formulate VMD leads to essentially

$$\sigma_S(q^2,\nu) = \gamma_\rho^{\,2} \frac{m_\rho^{\,4}}{(m_\rho^{\,2} - q^2)^2} \frac{q^2}{m_\rho^{\,2}} \sigma_{\rho N} \sim \frac{\gamma_\rho^{\,2} m_\rho^{\,4}}{q^2} \sigma(\infty) \qquad \text{or}$$

$$\left(W_2 - \frac{Q^2M^2}{\nu^2} W_1\right) \to \frac{1}{\nu} \sigma(\infty) \; . \tag{3.6}$$

This additional factor of q^2 is then sufficient to give agreement with experiments insofar as W_2 is concerned, however the prediction that

$$R = \frac{\sigma_S(q^2,\nu)}{\sigma_T(q^2,\nu)} \xrightarrow[q^2 \to \infty,\ \frac{q^2}{\nu}\ \text{fixed}]{} \text{const.}\ q^2 \tag{3.7}$$

seems to be ruled out.

3) *Current Algebra*

Local current algebra yields direct predictions as to the asymptotic behavior of weak amplitudes. For example, Adler's sum rule for neutrino nucleon scattering (for a derivation see Treiman's lectures or Sec. IV) predicts that

$$\frac{d\sigma^{\bar{\nu}p}}{dq^2} - \frac{d\sigma^{\nu p}}{dq^2} \xrightarrow[E \to \infty]{} \frac{G^2}{2\pi}\int \frac{d\nu}{M}[W_2^{\bar{\nu}p}(\nu,q^2) - W_2^{\nu p}(\nu,q^2)]$$

(3.8) $$= \frac{G^2}{\pi}[\cos^2\theta_c + 2\sin^2\theta_c] \ .$$

The sum rule suggest that, in some sense, at large energies the neutrino-nucleon cross section is point like. For a bare, point like, particle of $I = I_3 = ½$ only $\bar{\nu}$ scatters, and for $I = -I_3 = ½$ only ν scatters, with a cross section of $\frac{G^2}{\pi}$. Therefore if at large q^2 we can treat the nucleon as a collection of point like, spin ½, I = ½ constituents (say quarks) we would expect,

(3.9) $$\frac{d\sigma^{\bar{\nu}p}}{dq^2} - \frac{d\sigma^{\nu p}}{dq^2} = \frac{G^2}{\pi} < N\uparrow - N\downarrow > = \frac{G^2}{\pi}$$

where $N\uparrow(N\downarrow)$ is the number of constituents with $I_3 = ½\ (-½)$. This agrees with (3.8) in the case of $\theta_c = 0$.

Bjorken considered the scale limit, $\nu, Q^2 \to \infty$ with $\omega = \frac{Q^2}{2\nu}$ fixed within the context of current algebra. He showed that current algebra together with what seemed to be reasonable assumptions as to smoothness of certain limits leads to the prediction that the following asymptotic limits exist

(3.10) $$F_1(Q^2,\omega) = MW_1(Q^2,\nu) \xrightarrow[Q^2 \to \infty]{} F_1(\omega)$$

(3.11) $$F_2(Q^2,\omega) = \frac{\nu}{M} W_2(Q^2,\nu) \xrightarrow[Q^2 \to \infty]{} F_2(\omega)$$

for either electron or neutrino nucleon scattering and

(3.12) $$F_3(Q^2,\omega) = \frac{\nu}{M} W_3(Q^2,\nu) \xrightarrow[Q^2 \to \infty]{} F_3(\omega)$$

for neutrino nucleon scattering. This scaling behavior is, of course, consistent with all current algebra sum rules and, if true, would explain why $\frac{d\sigma^{\bar{\nu}p}}{dq^2} - \frac{d\sigma^{\nu p}}{dq^2}$ is independent of q^2 for large q^2, since (3.8) can be written as:

(3.13) $$\frac{G^2}{2\pi} \int_0^1 d\omega[F_2^{\bar{\nu}p}(\omega,q^2) - F_2^{\nu p}(\omega,q^2) = \frac{G^2}{\pi}(\cos^2\theta_c + 2\sin^2\theta_c)$$

4) *Composite Particle Models*

Prior to the SLAC experiment there was some speculation along lines that developed afterwards into the "parton model". The basic idea here is that when one scatters a lepton off a composite object with large energy transfer one is measuring the instantaneous charge distribution within the object. Thus one can "observe" the charged constituents of the object.

There is precedent for this interpretation in electron scattering off atoms and off nuclei. Consider, for example, electron scattering off an atom (Fig. 15). The cross section is plotted as a function of channel number which is proportional to the final electron energy. We can see two distinct regions. First, the elastic peak and exitations of the atom as a whole to discrete atomic levels. Here the lepton sees the atom as a single coherent system with the sum of the charges of the constituents, ΣQ_i, and the cross section is proportional to $(\Sigma Q_i)^2$. As the energy transfer increases, (the channel number decreases), we arrive at the quasi-elastic peak where the incident electron scatters incoherently off the constituent electrons of the atom. Here the cross section is proportional to ΣQ_i^2.

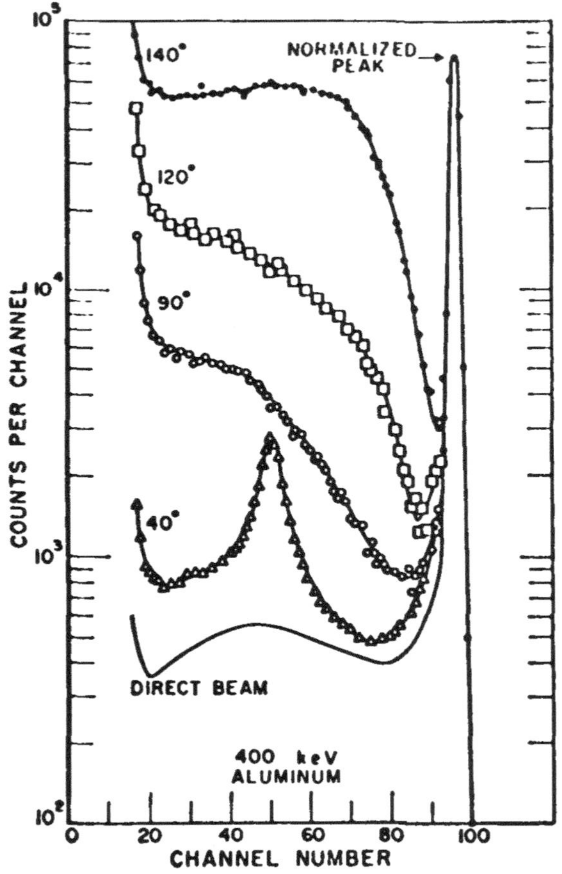

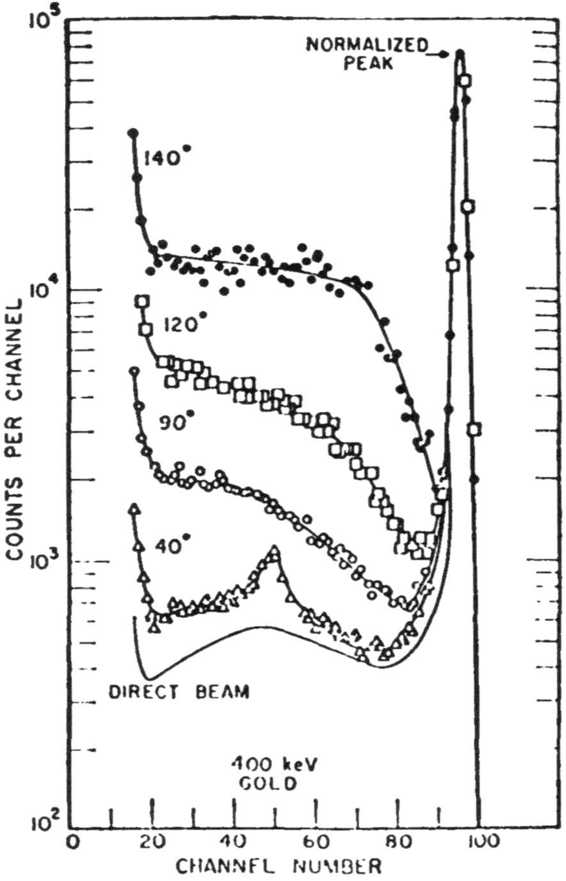

Fig. 15

Inelastic electron scattering from aluminum and gold atoms. The incident energy is 400 KeV and the final electron energy is directly proportional to the channel number.

The important exited states are those where the energy transfer, q_0, is related by classical kinematics to the momentum transfer

$q_0 \approx \frac{\vec{q}^{\,2}}{2m}$, where m is the mass of the constituent. Thus a naive quark model of the nucleon would suggest perhaps a quasi-elastic scattering,

for large q_0 and q^2, peaked about $\frac{\nu}{M_N} \approx \frac{q^2}{2M_Q}$ (M_Q = mass of the "quark").

II. EXPERIMENTAL RESULTS

1) Electron-Nucleon Scattering

The electrons at SLAC have a maximum energy of 18 GeV. Therefore in the SLAC-MIT experiment $\nu_{max} \approx 19\ (\mathrm{GeV}/c)^2$. Since the cross section falls rapidly with increasing momentum transfer to the nucleon (due to the photon propagator) large scattering angles are suppressed. The largest scattering angle in the experiment was 34°, ${q^2}_{max} \approx 14\ (\mathrm{GeV}/c)^2$. The cross section is measured at fixed scattering angle θ at various values of the incident electron energy, i.e.,

$\sin^2\theta = \dfrac{Q^2}{4E(E-\frac{\nu}{M})}$ is fixed. By varying the angle and the incident energy one can accumulate data points of fixed Q^2 and ν at various values of θ and thus separate W_2 and W_2. We recall that

$$\text{(3.14)} \qquad \frac{d^2\sigma}{d\Omega dE'} = \left(\frac{d^2\sigma}{d\Omega dE'}\right)_{\mathrm{MOTT}} [W_2(q^2,\nu)+2\tan^{2\theta}/2\ W_1(q^2,\nu)]$$

so that in order to determine

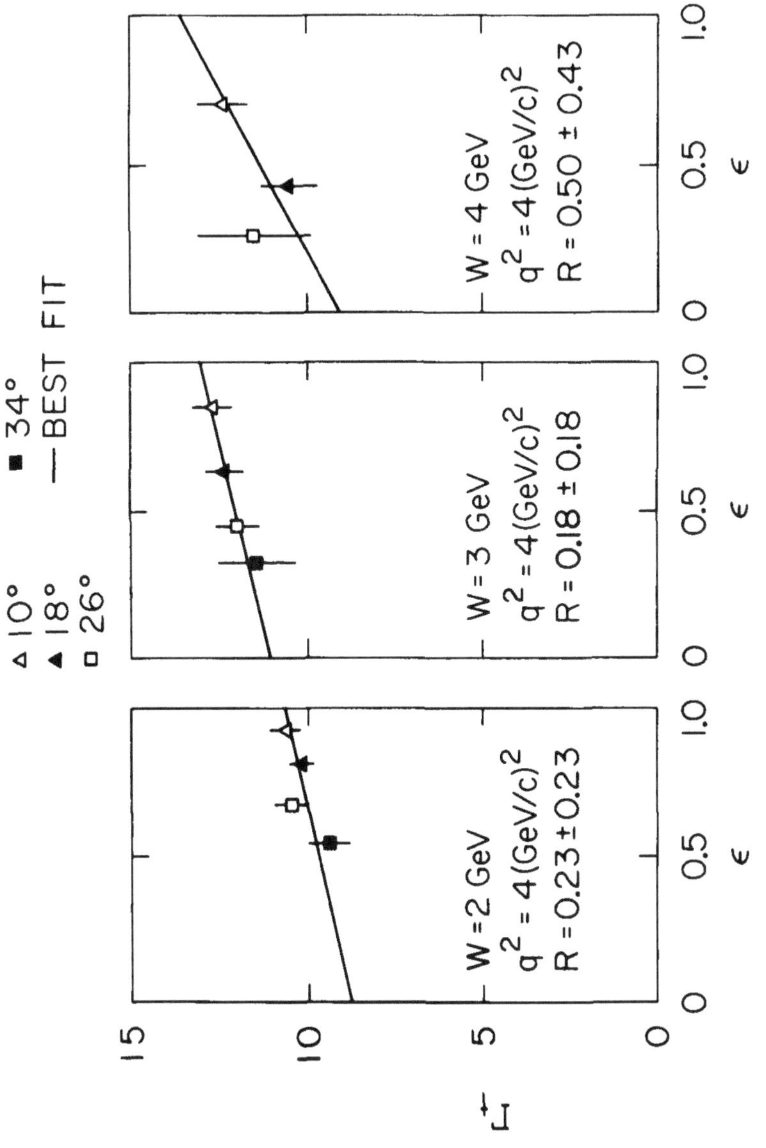

10°
18°
26°
34°
BEST FIT
W = 2 GeV
$q^2 = 4(\mathrm{GeV}/c)^2$
R = 0.23 ± 0.23
W = 3 GeV
$q^2 = 4(\mathrm{GeV}/c)^2$
R = 0.18 ± 0.18
W = 4 GeV
$q^2 = 4(\mathrm{GeV}/c)^2$
R = 0.50 ± 0.43
15
10
5
0
0
0.5
1.0
ε

Fig. 16

$$\text{(3.15)} \qquad \frac{W_1(q^2,\nu)}{W_2(q^2,\nu)} = \frac{q^2+\frac{\nu^2}{M^2}}{q^2}\,\frac{\sigma_T}{\sigma_T+\sigma_S} = \left(1+\frac{\nu^2}{M^2q^2}\right)\frac{1}{1+R}$$

one must achieve large values of $\tan^2\theta/2\,(1+\frac{\nu^2}{M^2q^2})$ (i.e., large angles or large values of $\frac{\nu^2}{M^2q^2}$). Such a separation is necessary before one can discuss the behavior of $W_i(q^2,\nu)$. Figure 16 illustrates the results on R, when the ordinate is equal to $\sigma_T(q^2,\nu)[1+\epsilon R]$, $\epsilon^{-1} = 1+2(1+\frac{\nu^2}{M^2q^2})\tan^2\theta/2$ and $W = (2\nu-Q^2+M^2)^{1/2}$. It is clear that R is small compared to one, and rather constant. The fits to the data at present are usually made under the assumption that $R = 0.18 \pm 0.18$.

A dramatic representation of the structure functions is presented in Fig. 17. Here essentially $W_2(q^2,\nu)$ is plotted vs. q^2 for fixed W, i.e., fixed mass for the produced hadrons. One notices immediately that in the region where the elastic form factor decreases rapidly, for $W = M$, the inelastic form factors for fixed $W > M$ vary rather slowly with increasing Q^2. This is consistent with scaling where

$$W_2(Q^2, W^2 = 2\nu-Q^2+M^2) = \frac{1}{\nu}F\left(\frac{Q^2}{2\nu}\right) = \frac{2}{W^2+Q^2-M^2}\,F\left(\frac{1}{1+\frac{W^2-M^2}{Q^2}}\right).$$

One therefore plots $\frac{\nu}{M}W_2$ and MW_1 as functions of $\frac{2\nu}{Q^2}$ and Q^2. Figures 18 and 19 show the $\omega = \frac{2\nu}{Q^2}$ behavior of the structure functions for various values of Q^2. The scaling seems to be quite striking – i.e., no Q^2 dependence is discernable. A more precise test of scaling is to plot νW_2 for fixed ω vs. Q^2, as in Fig. 20, where we see that the

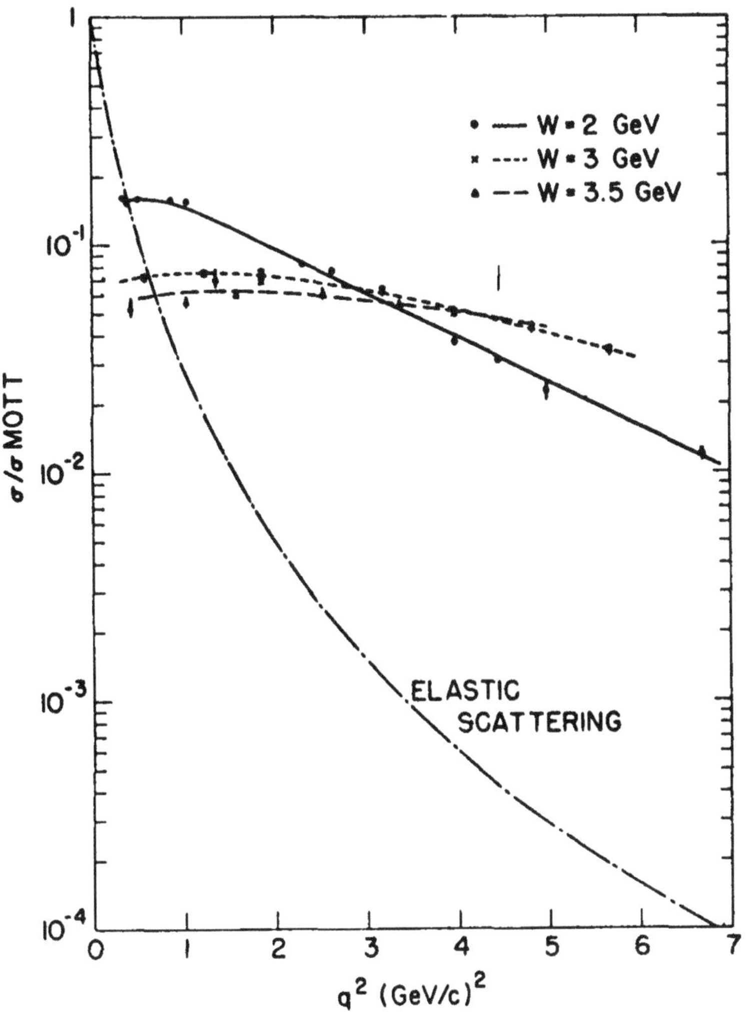

Fig. 17

$(d^2\sigma/d\Omega dE')/\sigma_{Mott}$, in GeV^{-1}, vs q^2 for $W=2$, 3, and 3.5 GeV. The lines drawn through the data are meant to guide the eye. Also shown is the cross section for elastic e–p scattering divided by σ_{Mott}, $(d\sigma/d\Omega)/\sigma_{Mott}$, calculated for $\theta = 10°$, using the dipole form factor. The relatively slow variation with q^2 of the inelastic cross section compared with the elastic cross section is clearly shown.

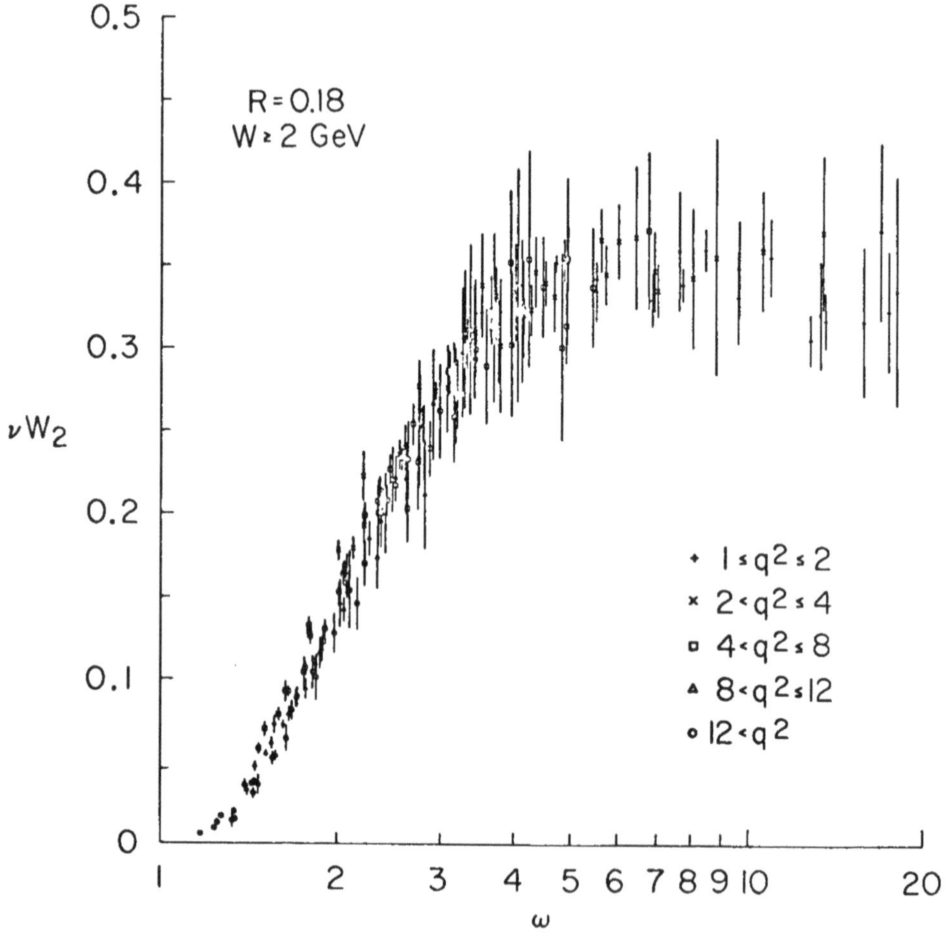
0.5
0.4
0.3
0.2
0.1
0
νW_2
R = 0.18
W ≥ 2 GeV
1 ≤ q² ≤ 2
2 < q² ≤ 4
4 < q² ≤ 8
8 < q² ≤ 12
12 < q²
1
2
3
4
5
6
7
8
9
10
20
ω

Fig. 18

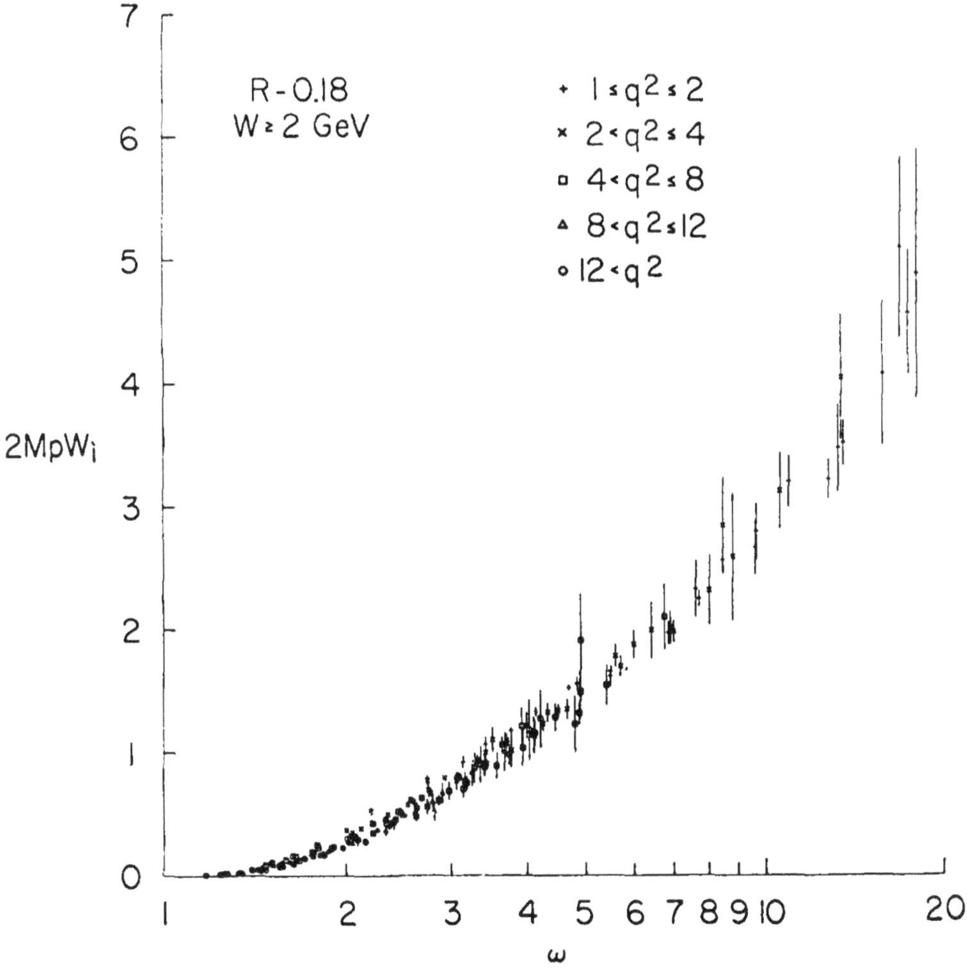
R-0.18
W ≥ 2 GeV
+ 1 ≤ q² ≤ 2
× 2 < q² ≤ 4
□ 4 < q² ≤ 8
△ 8 < q² ≤ 12
○ 12 < q²
2MpW₁
0
1
2
3
4
5
6
7
1
2
3
4
5
6
7
8
9
10
20
ω

Fig. 19

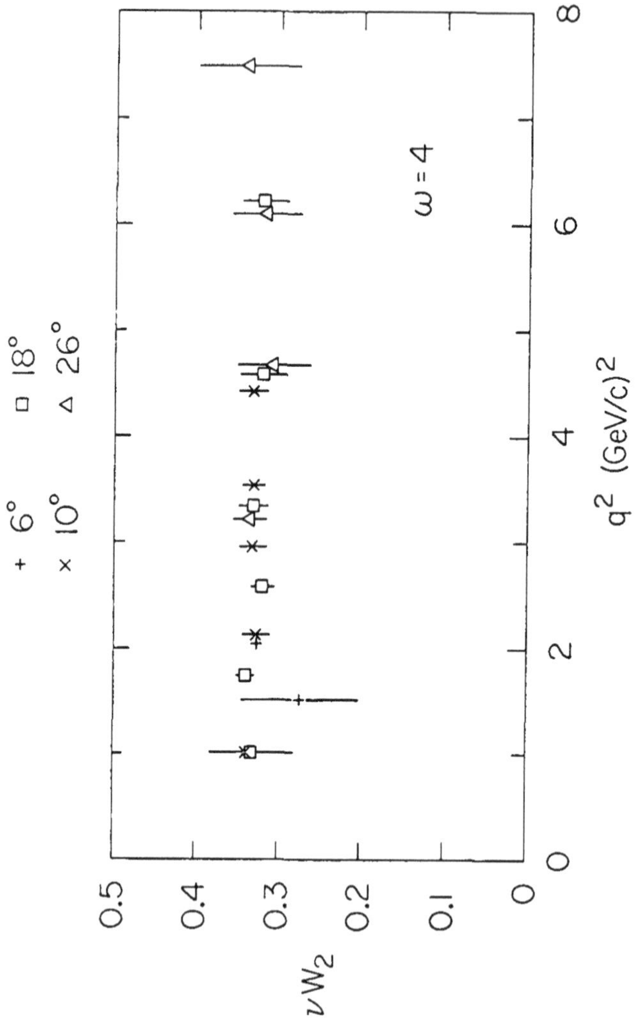

Fig. 20

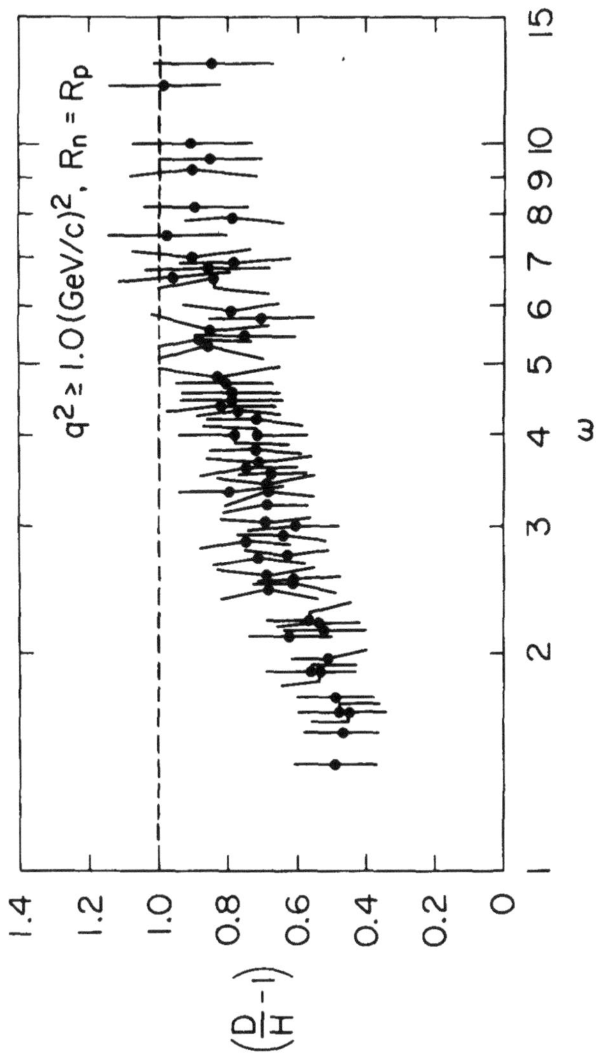
$q^2 \geq 1.0\,(\text{GeV}/c)^2$, $R_n = R_p$
$\left(\frac{D}{H} - 1\right)$
1.4
1.2
1.0
0.8
0.6
0.4
0.2
0
1
2
3
4
5
6
7
8
9
10
15
ω

Fig. 21

$\frac{2\nu}{Q^2} = 4$ there does not seem to be any Q^2 dependence for $Q^2 \geq 1\ (\text{GeV/c})^2$!
We can summarize the remarkable results of the SLAC-MIT experiment as follows:

1) Both $\frac{\nu}{M} W^2$ and MW_1 seems to be functions of ω alone already at surprisingly low values of Q^2. This scaling is very accurate for $3.5 \leq \frac{2\nu}{Q^2} \leq 10$ and $10 \geq Q^2 \geq 1\ (\text{GeV/c})^2$.

2) The "scalar to transverse" ratio, $R = \frac{\sigma_S}{\sigma_T}$, appears to be small, $R = 0.18 \pm 0.18$, and relatively constant, within the errors, for $Q^2 \geq 1 (\text{GeV/c})^2$. One cannot exclude the possibility that $R = 0$ or that R increases very slowly with Q^2.

3) The energy dependence of scaling structure functions has a diffractive nature. Thus $\frac{\nu}{M} W_2 = F_2 (\frac{2\nu}{Q^2})$ rises rapidly from threshold $(\frac{2\nu}{Q^2} = 1)$ to a value of ≈ 0.33 at $\frac{2\nu}{Q^2} = 4$ and thereafter remains constant. Since F_2 might be expected to behave for large $\frac{\nu}{Q^2}$ as $(\frac{2\nu}{Q^2})^{\alpha - 1}$ this behavior is consistent with $\alpha = \alpha_{\text{POMERON}} = 1$. Similarly $MW_1 = F_1(\frac{2\nu}{Q^2})$ is linear in $\frac{2\nu}{Q^2}$, consistent with $F_1(\frac{2\nu}{Q^2}) \approx (\frac{2\nu}{Q^2})^{\alpha_P}$. There is no evidence for a quasi-elastic peak.

4) Recently experiments involving electron deuterium inelastic scattering have yielded information on the isotopic spin dependence of the structure functions. The preliminary results indicate that both the neutron structure functions also scale and that the ratio of the neutron to the proton structure functions scale. In Fig. 21 this ratio is plotted vs $\omega = \frac{2\nu}{Q^2}$. We see that the ratio is close to $\frac{1}{2}$ near threshold $(2\nu = Q^2)$ and thereafter increases, perhaps approaching 1 for large $\frac{\nu}{Q^2}$ as would be

implied by Regge theory, where $(\nu W_2{}^p - \nu W_2{}^n) \approx \left(\frac{\nu}{Q^2}\right)^{\alpha_{A_2} - 1} \xrightarrow[\left(\frac{\nu}{Q^2}\right) \to \infty]{} 0.$

2) **Neutrino-Nucleon Scattering**

Neutrino scattering experiments are much more difficult to perform, the cross sections are smaller and the incident neutrino energy and flux is difficult to determine. The advantages of inelastic neutrino scattering are that one can easily measure the backward cross section, and that there is an additional interesting structure function W_3. The CERN experiment involved scattering neutrinos of energy 2–12 GeV off nuclei ($n \sim p$). The results of this experiment indicate, at least, that the structure functions of the weak current do not behave drastically different from those of the electromagnetic current. There is evidence for scaling and for diffractive behavior. The best evidence for scaling is the observed, Fig. 22, linear dependence of the total neutrino-nucleon cross section on the neutrino energy. This, as we shall see, is a direct consequence of the scaling of F_1, F_2 and F_3.

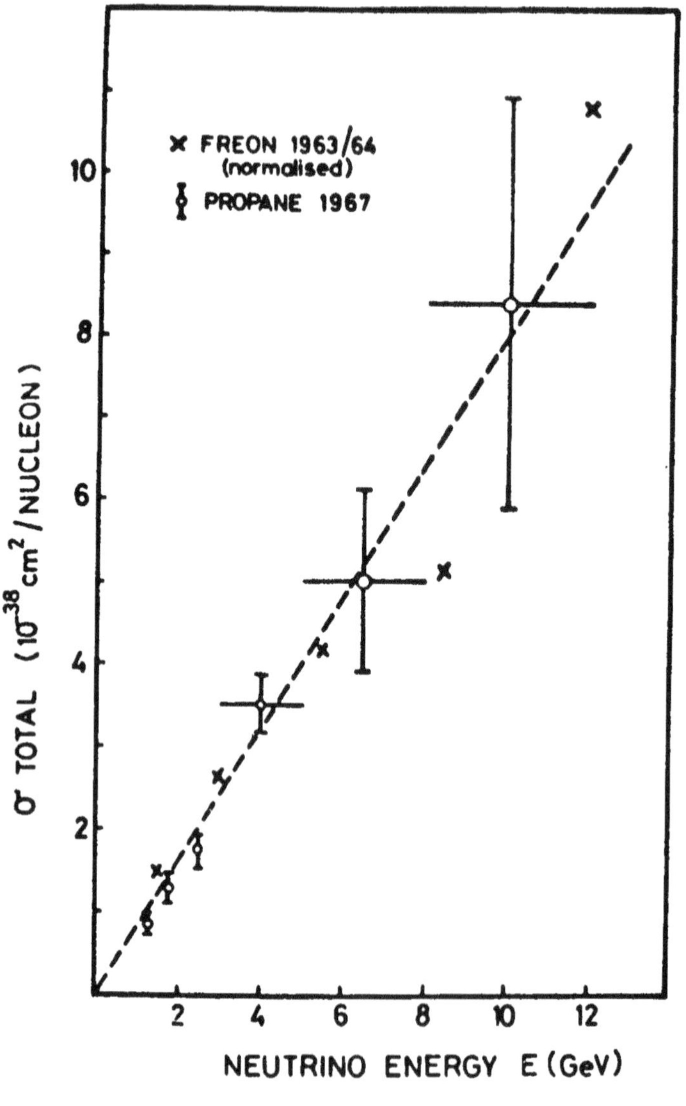

Fig. 22

Total neutrino-nucleon cross-sections as a function of neutrino energy E. The free cross sections have been multiplied by 1.35 to normalize them to the propane cross sections for $E > 2$ GeV. The errors shown are statistical only.

IV. *APPLICATIONS OF CURRENT ALGEBRA*

a) Assumptions about Commutators

It is intuitively obvious that the assumption of local equal time commutation relations (ETCR) between currents should have important implications regarding the high energy behavior of current amplitudes. The large momentum behavior of these amplitudes is clearly correlated with the small distance behavior of the relevant currents. The classic application of Current Algebra was the derivation of low energy theorems. Here one is not exploiting the locality of the ETCR's but rather the algebra of the charges which lead to Ward identities. Most of these Ward identities (except for those involving photons or gravitons) require the use of PCAC to make contact with the real world.

The applications we will be concerned with here rely upon the local commutation relations of currents, yield information about the asymptotic behavior of current amplitudes, and can be related to high energy lepton hadron scattering without the use of PCAC. However, the results we derive are on much shakier grounds than those which follow from Ward identities. This is, in part, because one must make rather strong assumptions about the ETCR's in order to get any information at all. It is not enough to assume the integrated charge algebra of Gell-Mann, i.e.,

$$(4.1) \qquad [\int d^3x J_0{}^{\alpha}(\underline{x},t), \int d^3y J_0{}^{\beta}(y,t)] = i f_{\alpha\beta\gamma} \int d^3x J_0{}^{\gamma}(\underline{z},t)$$

one must also require

$$(4.2) \qquad [J_0{}^{\alpha}(\underline{x},t), J_0{}^{\beta}(\underline{y},t)] = i f_{\alpha\beta\gamma} J_0{}^{\gamma}(\underline{x},t)\, \delta^{(3)}(\underline{x}-y) + \text{possible Schwinger terms}$$

and

$$[J_0^{\alpha}(\underline{x},t),J_i^{\beta}(\underline{y},t)] = if_{\alpha\beta\gamma}J_i^{\gamma}(\underline{x},t)\delta^{(3)}(\underline{x}-\underline{y}) + \text{(necessarily)}$$

$$S_{ik}^{\alpha\beta}(\underline{x})\partial^k\delta^{(3)}(\underline{x}-\underline{y}) \tag{4.3}$$

which are suggested, but not implied, by (4.1). In addition one sometimes needs to know the model dependent ETCR:

$$[J_i^{\alpha}(\underline{x},t),J_k^{\beta}(\underline{y},t)] = i[\delta_{ik}f_{\alpha\beta\gamma}J_0^{\gamma}\delta^{(3)}(\underline{x}-\underline{y}) -$$

$$-\epsilon_{ikj}d_{\alpha\beta\gamma}J_j^{\gamma}(\delta^{(3)}(\underline{x}-\underline{y})] \text{ (in the quark model)}$$

$$= 0 \text{ (in the algebra of fields)} . \tag{4.4}$$

The last set of ETCR are derived in the quark model, by constructing the currents as

$$J_{\mu}^{\alpha} = \bar{\psi}(x)\gamma_{\mu}\lambda^{\alpha}\binom{1}{\gamma_5}\psi(x) \qquad \alpha = \begin{cases} 1\ldots\ldots 8 \text{ for vector currents} \\ 9\ldots\ldots 16 \text{ for axial vector currents} \end{cases} \tag{4.5}$$

where $\psi(x)$ is the quark field and satisfies canonical ETCR's. In the algebra of fields J_i^{α} is identified with a vector or axial vector Yang Mills field.

Furthermore assumptions are sometimes made, or extracted from specific field theoretic models, about $[\partial_t J_{\mu}^{\alpha}(\underline{x},t),J_{\nu}^{\beta}(\underline{y},t)]$. These are even more model dependent since the time derivations of the currents involve the interaction Hamiltonian.

b) The Bjorken-Johnson-Low Limit

It was Bjorken, Johnson and Low who pointed out that the high energy behavior of time ordered (T) products of operators is governed by the

ETCR's of the operators. Johnson and Low used this fact, within the context of definite field theoretic models, to discuss the validity of canonical ETCR's. They calculated in perturbation theory (in particular they investigated the triangle graph in the gluon model) the high energy behavior of the T product to see whether it was determined by the ETCR which one would calculate canonically. Bjorken's motivation was quite different. He used canonical or assumed ETCR's to derive observable relations for high energy lepton scattering and to discuss the question of the finiteness of radiative corrections. My approach will be along the lines of Bjorken's.

Consider the matrix element for the scattering of two currents (operators) off arbitrary states. This is described by the Fourier transform of the T* product of the currents

$$< \alpha | T^*[J_1(x) J_2(y)] | \beta >$$

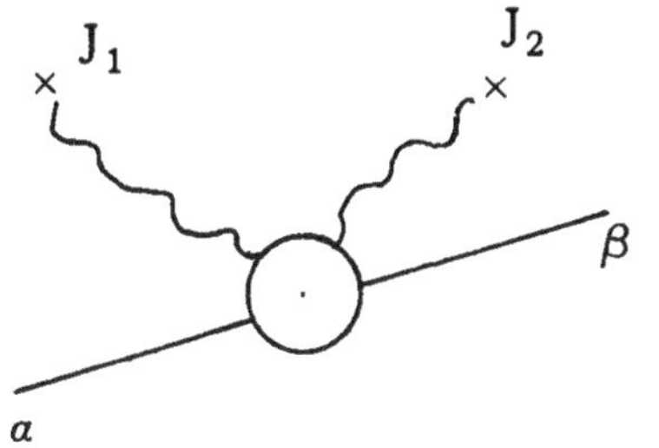

The reason that the T* product appears instead of the T product is that the latter is not in general Lorentz invariant. In fact, whenever the ETCR of the currents involve derivatives of delta functions (Schwinger terms) the time ordering operation is frame dependent. To reinstate Lorentz invariance one adds an appropriate local (proportional to $\delta^4(x-y)$ or derivatives of $\delta^4(x-y)$) "seagull" or contact term to T, thus converting it to T*. The contact term can be explicitly constructed in terms of the Schwinger terms which are responsible for its appearance. In order to be able to do this the Schwinger term must satisfy certain conditions. However, if the ETCR of the operators is the restriction of a casual commutator to equal times these conditions are automatically satisfied. As far as the matrix element in momentum space is concerned these terms merely add polynomials in the momenta of the currents to T so that

$$i\int e^{iqx}<\alpha|T^*(J_1(x)J_2(y))|\beta> \;=\; i\int e^{iqx}<\alpha|T(J_1(x)J_2(0))|\beta>$$

(4.6) $$+ \text{ polynomials in } q_\mu$$

or $T^*(q) = T(q) +$ polynomials in q. Thus the "seagulls" do not appear in the discontinuity of the T^* product, which is given by the Fourier transform of the commutator of the currents.

The BJL limit gives the large q_0 behavior of T(q). It can be derived from the Low equation for T(q). This Low equation can be derived by using the following representation of the step function:

$$e^{iq_0x}\theta(\pm x_0) = \int_{-\infty}^{\infty} \frac{dq_0'}{(2\pi i)} \frac{e^{\pm iq_0' \cdot x_0}}{q_0' \mp (q_0+i\epsilon)}$$

We then write T(q) as

$$T(q_0,\vec{q}) = i\int d^4x\, e^{iq\cdot x}[\theta(x_0)J_1(\underset{\sim}{x},x_0)J_2(0)+\theta(-x_0)J_2(0)J_1(x)]$$

$$= \frac{1}{2\pi}\int_{-\infty}^{\infty} d^4x \quad e^{-i\underset{\sim}{q}\cdot\underset{\sim}{x}} \left[J_1(x)J_2(0)\, \frac{e^{+iq_0'\cdot x_0}}{q_0'-q_0-i\epsilon} + \right.$$

$$\left. + \frac{e^{-iq_0'\cdot x_0}}{q_0'+q_0+i\epsilon} J_2(0)J_1(x) \right] dq_0'$$

(4.7) $$= \int_0^{\infty} dq_0' \left[\frac{W_{12}(q_0,\underset{\sim}{q})}{q_0'-q_0-i\epsilon} + \frac{\overline{W}_{21}(q_0,-\underset{\sim}{q})}{q_0'+q_0+i\epsilon} \right],$$

where $W_{12}(q_0,\underset{\sim}{q}) = \int \frac{d^4x}{2\pi} e^{iq\cdot x} < \alpha | J_1(x) J_2(0) | \beta >$

(=0 for $q_0 < 0$).

$$\overline{W}_{21}(q_0,\underset{\sim}{q}) = \int \frac{d^4x}{2\pi} e^{-iq\cdot x} < \alpha | J_2(0) J_1(x) | \beta > \; .$$

We now take the limit as $|q_0| \to \infty$, in any direction except along the positive real axis

$$T(q) \to -\frac{1}{q_0} \int_0^\infty dq_0{}'[W_{12}(q_0,\underset{\sim}{q}) - \overline{W}_{21}(q_0,-\underset{\sim}{q})]$$

$$- \frac{1}{q_0{}^2} \int_0^\infty dq_0{}'[q_0{}'][W_{12}(q_0,\underset{\sim}{q}) + \overline{W}_{21}(q_0,-\underset{\sim}{q})]$$

$$\to -\frac{1}{q_0} \int d^3x \, e^{-i\overline{q}\cdot\overline{x}} [J_1(x,0), J_2(0)] +$$

$$+ \frac{i}{q_0{}^2} \int d^3x \, e^{-i\overline{q}\cdot\overline{x}} [\dot{J}_1(x,0), J_2(0)] + \ldots$$

We have used $\int dq_0 \, e^{+iq_0x_0} = 2\pi\delta(x_0)$, $\int dq_0 q_0 \, e^{\pm iq_0x_0} = (\mp i) 2\pi\dot{\delta}(x_0)$.

Finally we have for the T^* product:

$$T^*(q) \to \text{polynomials in } q_\mu - \frac{1}{q_0} \sum_{n=0}^{\infty} \left(\frac{-i}{q_0}\right)^n$$

(4.8) $$\int d^3x \, e^{-i\overline{q}\cdot\overline{x}} [(\frac{\partial}{\partial t})^n J_1(\underset{\sim}{x},0), J_2(0)] \; .$$

We have been very cavalier in our assumptions. The main assumption we have made is that such an expansion in $\frac{1}{q_0}$ is possible. This requires, at the very least, that the coefficients in this expansion are finite. In particular the BJL theorem might make sense only for the first few terms in the expansion. In fact, it is known that in perturbation theory the BJL theorem does not, in general, hold. We shall discuss this point in detail in Sec. V.

How can we make use of the BJL theorem? Except for the special case of electron positron annihilation the BJL limit is highly unphysical, and one cannot directly measure the discontinuity in q_0 of the amplitude. Thus, for example, $W_{1,2}(q^2,\nu)$, as we have seen, can be measured in the physical region $-q^2 \leq 2\nu$; whereas a direct test of BJL would involve measuring $W_{1,2}(q^2,\nu)$ for $q_0 \to +\infty$, i.e., $q^2 \approx {q_0}^2 \simeq \frac{\nu^2}{{p_0}^2}$. Therefore the main applications of the BJL theorem have been indirect. The first is the investigation of the finiteness of electromagnetic mass splittings, other radiative corrections and second order weak interactions. We will illustrate these applications by considering the proton-neutron mass difference.

c) Application of BJL to Radiative Corrections and the Persistence of Divergences

$\delta M_N \quad =$ 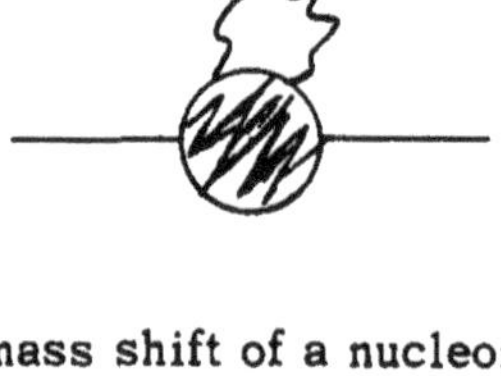 ,

Consider the electromagnetic mass shift of a nucleon. To lowest order in α this is given by the self energy graph; where the blob contains all the strong interactions and the (one) internal photon is exhibited. By cutting the photon line, one can write this mass shift in terms of the forward virtual Compton scattering amplitude:

$$\delta M_N = \alpha \int d^4q \; T_{\mu\nu}(q,p) \cdot \frac{g_{\mu\nu}}{q^2}$$

$$(4.9) \qquad = \alpha \int \frac{d^4q}{q^2} \int e^{iqx} \, d^4x < p| T\{J_\mu(x) J^\mu(0)\} | p > \; .$$

The proton-neutron mass difference is then given by

$$(4.10) \quad \Delta M_N = \delta M_p - \delta M_n = \alpha \int \frac{d^4q}{q^2} \, T_{\mu\nu}{}^{I=1}(q,p) g^{\mu\nu} \quad .$$

It is well known that in perturbation theory the electromagnetic mass splitting of a given particle is logarithmically divergent and therefore uncalculable. (In QED one renormalizes the mass and inserts the physical electron mass by hand). It was hoped, however, that this would not be the case for ΔM_N, which after all is experimentally observable (as contrasted with the mass shift of the electron or the proton) and is of order αM_N. One argued that the reason perturbation theory lead to divergences was the lack of form factors in the coupling of the photon to the particles and that the strong interactions would provide these damping form factors to give a convergent result.

In fact, by writing a dispersion relation for $T_{\mu\nu}$ in $\nu = q \cdot p$ (after Wick rotating the above formula so that q^2 became purely spacelike) one was able to reproduce certain mass differences by saturating the dispersion relations with nearby resonance poles. In particular such a calculation yields satisfactory results for the $\Delta I = 2$ $\pi^{\pm}-\pi^0$ and the $\Sigma^+ + \Sigma^- - 2\Sigma^0$ mass differences. The nucleon mass difference was unexplainable by these methods; however this failure was understandable. For $\Delta I = 1$, one of the invariant amplitudes in $T_{\mu\nu}$ behaves for large ν as $\nu^{1/2}$ and a subtraction is required for its dispersion relation. Unless one can calculate the subtraction constant the nucleon mass difference cannot be calculated – and it is no surprise that an unsubtracted dispersion relation gives wrong results.

The BJL theorem, however, makes it seem less likely that (4.10) is convergent. It informs us that as $|q_0| \to \infty$

$$(4.11) \qquad T_\mu{}^\mu(q,p) \to \frac{1}{q_0{}^2} \int e^{-i\bar{q}\bar{x}} < N|[\dot{J}_\mu(\underline{x},0), J^\mu(0)]|N > d^3x \quad .$$

(Note that there is no $\frac{1}{q^0}$ term since $T_\mu{}^\mu$ is a symmetric function of q_0). In almost any model of current commutators the $I = 1$ part of the above ETCR is non zero, and if the large q^2 behavior of $T_\mu{}^\mu(q,p)$ is uniform then

$$\delta M \sim \int \frac{d^4q}{q^4} \text{ diverges logarithmically.}$$

The local ETCR's of the currents therefore lead to point-like behavior at large energies and eliminate the possibility of a cutoff being provided by the strong interactions. Similar divergences are suggested by the BJL limit to other radiative corrections and for second order weak interactions (where the divergences are worse than logarithmic).

In order to eliminate the above divergence one might demand that $[\dot{J}_\mu, J^\mu]^{I=1} = 0$. This is, of course, possible but is an ad hoc assumption unsupported by any reasonable model. Furthermore all the divergences implied by BJL for higher order weak interactions cannot be eliminated in this fashion without relinquishing the Gell-Mann algebra.

Alternatively one can simply claim that the BJL limit (4.11) is incorrect. However, the breakdown of the BJL limit model field theories is due to the singular nature of perturbation theory. In these models the BJL theorem is, in fact, sometimes invalid; however the divergences persist.

Furthermore, if the structure functions $W_{1,2}$, as measured in SLAC, are indeed scale invariant then one can conclude that (4.10) diverges logarithmically (unless there is a miraculous cancellation) without the use of BJL. Scaling (point-like behavior) in fact implies (aside from subtractions) the same asymptotic behavior of current amplitudes as does BJL.

A radical suggestion to eliminate all such divergences is that of Lee and Wick. They propose a modification of the photon propagator by introducing negative metric states so as to replace $\frac{1}{q^2}$ by $\frac{1}{q^2} - \frac{1}{q^2-M^2}$.

In conclusion one should keep in mind the seriousness of the problem raised by these divergences. It is perhaps healthy to be somewhat sceptical about the extrapolation to arbitrarily high energy (or short distances) of the assumption of local ETCR's. While there is no obvious reason to expect trouble for electromagnetic interactions for energies below, say, 100 BeV/c^2, sooner or later the finiteness of higher order electromagnetic and weak corrections will have to be explained.

d) Applications of BJL to Lepton-Hadron Scattering

The BJL theorem allows us to derive sum rules for high energy lepton-hadron scattering. Consider, for example, the matrix element for virtual Compton scattering. Although the BJL limit is not in the physical region for this process, we can *calculate* the amplitude in this region by expressing it as a dispersion relation. To do this we must know the absorptive part of the amplitude (in the physical region). But the absorptive part of the virtual Compton scattering amplitude is precisely what is measured in lepton-hadron scattering. We then compare the large q_0 behavior of the dispersion relation with that implied by BJL, thereby deriving a sum rule for an integral of the measurable part in terms of ETCR's.

Consider the spin averaged amplitude for the forward scattering of a current J^a. For ep scattering J^a will be the electromagnetic current and for νp or $\bar{\nu} p$ scattering it will be the weak current as given by (1.2). We shall denote the hermitean conjugate of J^a by $J^{\bar{a}} = (J^a)^\dagger$.

$$T^{*a\bar{a}}_{\mu\nu}(q_0,\vec{q},p) = i\int d^4x\, e^{iq\cdot x} < p|T^*\{J_\mu{}^a(x)J_\nu{}^{\bar{a}}(0)\}|p> . \tag{4.12}$$

In most of the models that we shall consider the Schwinger term is a c-number. This is true by construction for the ETCR's of the algebra of fields and seems to be the case in perturbation theory for the quark model. In this case the Schwinger term does not contribute to the connected part of the above amplitude

$$\begin{aligned} < p|T^*(J^aJ^{+a})|p>_{\text{conn.}} &= < p|T^*(J^aJ^{a+})\,|p> \\ &\quad - < 0|T^*(J^aJ^{a+})\,|0> \ . \end{aligned}$$

since the contact or seagull terms are also c-numbers and therefore for connected matrix elements T^* is equal to T. Many of our conclusions are unaltered by the presence of operator Schwinger terms, but for convenience we shall assume that, for the moment, that the Schwinger terms are c-numbers.

The T-product can be decomposed into invariant amplitudes:

$$\begin{aligned} T_{\mu\nu}{}^{a\bar{a}}(q_0,\vec{q},p) &= \frac{1}{M^2}\,(p_\mu - \frac{\nu}{q^2}\,q_\mu)\,(p_\nu - \frac{\nu}{q^2}\,q_\nu)T_2{}^{a\bar{a}}(q^2,\nu) - \\ &\quad - (g_{\mu\nu} - \frac{q_\mu q_\nu}{q^2})T_1{}^{a\bar{a}}(q^2,\nu) - i\epsilon_{\mu\nu\sigma\lambda}\frac{p_\sigma q_\lambda}{2M^2}T_3{}^{a\bar{a}}(q^2,\nu) \end{aligned} \tag{4.13}$$

where, in the case of the weak current, we have deleted terms proportional to q_μ and T_3 vanishes in the case of the electromagnetic current. The absorptive parts of $T_i{}^{a\bar{a}}$, for $q^2 < 0$ and $2\nu + q^2 \geq 0$ are just the functions defined in (1.9) and (1.10):

$$\text{Im}\, T_i^{a\bar{a}}(q^2,\nu) = \pi\, W_i^{a\bar{a}}(q^2,\nu) \quad . \tag{4.14}$$

We now write dispersion relations in ν for fixed q^2 for T_i. For large ν the amplitudes and fixed q^2 we expect, as we have discussed previously, that $T_{\mu\nu} \sim \nu^{\alpha}$, where α is the appropriate Regge pole. If we keep q fixed and let $\nu = p \cdot q \to \infty$, then $p_{\mu} \approx \nu$ and thus

$$T_1 \approx \nu^{\alpha} \qquad T_2 \approx \nu^{\alpha-2} \qquad T_3 \approx \nu^{\alpha-1} \quad . \tag{4.15}$$

What power α do we expect? For T_1 and T_2 we can exchange vacuum quantum numbers so that $\alpha = 1$, whereas T_3 arises due to a vector-axial vector interference term and therefore has negative G-parity (ρ–A_1 interference) and thus has $\alpha \approx \frac{1}{2}$. Thus we might expect $T_1 \sim \nu$ $T_2 \sim \nu^{-1}$ $T \sim \nu^{-1/2}$ for large ν. Therefore T_2, T_3 satisfy unsubtracted dispersion relations, whereas T_1 requires one subtraction:

$$T_{2,3}^{a\bar{a}}(q^2,\nu) = \int_{-\infty}^{+\infty} \frac{d\nu'}{\nu'-\nu}\, W_{2,3}^{a\bar{a}}(q^2,\nu)$$

$$= \int_{-q^2/2}^{\infty} d\nu' \left[\frac{W_{2,3}^{a\bar{a}}(q^2\nu')}{\nu'-\nu} + \frac{W_{2,3}^{\bar{a}a}(q^2\nu')}{\nu'+\nu} \right] \tag{4.16}$$

$$T_1^{a\bar{a}}(q^2,\nu) = T_1^{a\bar{a}}(q^2,0) + \nu \int_{-\infty}^{+\infty} \frac{d\nu'}{(\nu'-\nu)\nu'}\, W_1^{a\bar{a}}(q^2,\nu') =$$

$$= T_1^{a\bar{a}}(q^2,0) + \nu \int_{-q^2/2}^{\infty} \frac{d\nu'}{\nu'} \left[\frac{W_1^{a\bar{a}}(q^2,\nu')}{\nu'-\nu} - \frac{W_1^{a\bar{a}}(q^2,\nu')}{\nu'+\nu} \right] \quad . \tag{4.17}$$

We have used the crossing properties of W_i:

$$W_i^{a\bar{a}}(q^2,\nu) = -W_i^{\bar{a}a}(q^2,-\nu) \quad . \tag{4.18}$$

We are interested in the behavior of $T_i^{a\bar{a}}(q^2,\nu)$ in the BJL limit, i.e., when $|q_0| \to \infty$ so that both q^2 and ν approach infinity. It is convenient to change variables, and work with q^2 and $\omega = -q^2/2\nu$, since the limits of integration are then fixed. Recalling our definitions: $F_1^{a\bar{a}} = MW_1^{a\bar{a}}$, $F_2^{a\bar{a}} = \frac{\nu}{M} W_2^{a\bar{a}}$, $F_3 = \frac{\nu}{M} W_3^{a\bar{a}}$, we have:

$$T_1(q^2,\omega) = T_1(q^2,\infty) - \frac{1}{M} \int_0^1 d\omega' \left[\frac{F_1^{a\bar{a}}(\omega',q^2)}{\omega'-\omega} + \frac{F_1^{\bar{a}a}(\omega'q^2)}{\omega'+\omega} \right]$$

$$T_{2,3}(q^2,\omega) = \frac{2\omega M}{q^2} \int_0^1 d\omega' \left[\frac{F_{2,3}^{a\bar{a}}(\omega'q^2)}{\omega'-\omega} - \frac{F_{2,3}^{\bar{a}a}(\omega'q^2)}{\omega'+\omega} \right] . \tag{4.19}$$

We are now in a position to derive sum rules. We evaluate the large q_0 behavior of (4.19) and compare with the BJL limit. It is particularly simple to work in the frame where $q = (q_0,\vec{0})$ $p = (E,\vec{p})$, and $\nu = Eq_0$, $q^2 = q_0^2$, $\omega = -q_0/2E$. In this frame $T_{0i}^{a\bar{a}} = T_{00}^{a\bar{a}} = 0$, (neglecting the terms in $T_{\mu\nu}$ proportional to q_μ) and

$$\begin{aligned} T_{ij}^{a\bar{a}}(q,p) \to & -\frac{1}{q^0}\int d^3x < p|[J_i^a(\underline{x},0),J_j^{\bar{a}}(0)]|p > \\ & +\frac{i}{q_0^2}\int d^3x < p|[\frac{\partial}{\partial t} J_i^a(\underline{x},0),J_j^{\bar{a}}(0)]|p > \quad . \end{aligned} \tag{4.20}$$

Alternatively, we have from (4.13) and (4.19) that

$$T_{ij}{}^{a\bar{a}}(q,p) \to \delta_{ij} \left\{ T_1{}^{a\bar{a}}(q^2,\infty) + \frac{2E}{Mq_0} \int_0^1 d\omega' [F_1{}^{a\bar{a}}(\omega',q^2) - \right.$$

$$\left. - F_1{}^{\bar{a}a}(\omega',q^2)] + \frac{4E^2}{Mq_0{}^2} \int_0^1 d\omega'\omega' [F_1{}^{a\bar{a}}(\omega' q^2) + F_1{}^{\bar{a}a}(\omega' q^2)] \right\}$$

$$- \frac{1}{M} p_i p_j \frac{2}{q_0{}^2} \int_0^1 d\omega' [F_2{}^{a\bar{a}}(\omega',q^2) + F_2{}^{\bar{a}a}(\omega' q^2)]$$

$$+ \frac{i\epsilon_{ijk}p_k}{M} \left\{ \frac{1}{q^0} \int_0^1 d\omega' [F_3{}^{a\bar{a}}(\omega' q^2) + F_3{}^{\bar{a}a}(\omega' q^2)] \right.$$

$$\left. - \frac{2E}{q_0{}^2} \int_0^1 d\omega'\omega' [F_3{}^{a\bar{a}}(\omega' q^2) - F_3{}^{a\bar{a}}(\omega' q^2)] \right\} + 0(\frac{1}{q_0{}^3}). \tag{4.21}$$

If we assume that $F_i{}^{a\bar{a}}(\omega,q^2)$ approach a finite limit as $q^2 \to -\infty$, then we can compare the coefficients of $1/q_0$ and $1/q_0{}^2$ in (4.21) and (4.20), thereby deriving sum rules for the moments of F_i. We shall now proceed to derive sum rules for the particular case of ep and $(\frac{\nu}{\bar{\nu}})$p scattering, for a variety of models for the ETCR's.

Before proceeding, however, let us consider the possible effects of an operator Schwinger term. If the ETCR involves an operator Schwinger term, then the BJL limit, (4.20), refers to the T product, whereas the dispersion limit (4.21) is, of course, valid for the covariant T* product. To compare (4.20) and (4.21) one must therefore add the appropriate seagull term to (4.20). If the Schwinger term is a Lorentz scalar, i.e.,

$$[J_0{}^{a}(\underset{\sim}{x},t),J_i{}^{\bar{a}}(0)] = \ldots . + S^{a\bar{a}}(0)\partial^i\delta^{(3)}(\underset{\sim}{x})$$

then the seagull term, constructed (by the methods of Gross and Jackiw) to render T* Lorentz invariant and gauge invariant, adds a term to T_{0j} of the form $q_j < p|S^{a\bar{a}}(0)|p >$. This term can be incorporated into the unknown, subtraction constant $T_1{}^{a\bar{a}}(q^2,\infty)$, and will not affect the resulting sum rules. This would not be the case for a spin-2 operator Schwinger term, $[J_0{}^{a}(\underset{\sim}{x}0)J_i{}^{a}(0)] = \ldots\ldots + [S_{ji}{}^{a\bar{a}}(0)]\partial^i\delta^{(3)}(\underset{\sim}{x})$, which would contribute to $T_{0j}{}^{a\bar{a}}$ a term proportional to $(\bar{q}\cdot\bar{p})p_j$. In fact, the only way to reconcile such a Schwinger term with an unsubtracted dispersion relation in ν for T_2 is that $F_2{}^{a\bar{a}}(\omega,q^2) \to q^2$ as $q^2 \to -\infty$.

e) **Sum Rules for Electron-Nucleon Scattering**

In the case of electron-nucleon scattering we have $J^{EM} = J^a = J^{\bar{a}}$. The first term in the BJL expansion vanishes when a spin average is performed, since the ETCR of $J_i{}^{EM}$ and $J_k{}^{EM}$ must be anti-symmetric in i and k and no such tensor can be formed from $\vec{p}$ alone. We therefore have:

$$\frac{8E^2}{M}\int_0^1 d\omega\left[\omega F_1(\omega)+\left(\delta_{ij}-\frac{p_ip_j}{E^2}\right)(F_2(\omega)-2\omega F_1(\omega))\right]$$

$$= -i\int d^3x < p|[\frac{\partial}{\partial t}J_i{}^{EM}(\underset{\sim}{x},0),J_j{}^{EM}(0)]|p > . \tag{4.22}$$

We have rearranged the terms so that, recalling (I.13-14), ωF_1 is proportional to σ_T, and $F_2-2\omega F_1$ is proportional to σ_S and therefore both combinations are positive definite.

In order to make definite predictions we must have information regarding the above ETCR. This ETCR is, of course, extremely model dependent We shall calculate it in the following models:

1) *Quark Model:*

In this model $J_\mu^{EM} = \bar{\psi}\gamma_\mu Q\psi$ where Q is the charge matrix of the quarks. Consider an interaction, $\mathcal{L}_I$, which binds the quarks together

$$\mathcal{L}_I = g_1\bar{\psi}\gamma_\mu\psi B^\mu + g_2\bar{\psi}\gamma_5\psi\phi + g_3\bar{\psi}\psi\sigma \quad , \tag{4.23}$$

where B^μ is a neutral vector meson field ("gluon"), $\phi(\sigma)$ is a pseudo-scalar (scalar) field. This interaction involves neutral mesons only and therefore does not affect the form or ETCR's of J^{EM}. It does, of course, alter the ETCR of the *time derivitive* of J with itself. We use the equations of motion to reduce time derivitives to spatial derivitives and then use the canonical ETCR's to derive:

$$-i < p| \int d^3x [\frac{\partial}{\partial t} J_i^{EM}(\underset{\sim}{x},0), J_j^{EM}(0)]|p >_{\substack{\text{SPIN}\\ \text{AVE.}}} = C_{ij}(\underset{\sim}{p})$$

$$= -i < p|\bar{\psi}(0)[\gamma_i\overleftrightarrow{\partial}_j + \gamma_j\overleftrightarrow{\partial}_i - 2\delta_{ij}\underset{\sim}{\gamma}\cdot\overleftrightarrow{\underset{\sim}{\partial}} + 2ig_1(\gamma_i B_j + \gamma_j B_i - 2\delta_{ij}\underset{\sim}{\gamma}\cdot\underset{\sim}{B})$$

$$-4iM\delta_{ij} + \delta_{ij}(g_2\gamma_5\phi + g_3\sigma)]Q^2\psi(0)|p >_{AVE} \tag{4.24}$$

where M is the mass matrix of the quarks, and

$$Q^2 = \frac{4}{9}B - \frac{1}{6}Y + \frac{1}{3}I_3 \quad . \tag{4.25}$$

This explicit form for the ETCR does not seem to give us any useful information since we have no independent knowledge of the matrix elements of, say, $\bar{\psi}\gamma\overleftrightarrow{\partial}\psi$. However, one can still extract something from (4.24), namely the tensor structure of $C_{ij}(\underset{\sim}{p})$. If we assume that the product of the operators in (4.24) is not too singular we must have that

$$i < p|\bar{\psi}(\gamma_\mu\partial_\nu - g_1\gamma_\mu B_\nu)\psi|p>_{AVE} = Ap_\mu p_\nu + Bg_{\mu\nu}$$

$$i < p|\bar{\psi}(-4iM+g_2\phi+q_3\sigma)\psi|p> = Cg_{\mu\nu} \quad ,$$

where A,B,C are Lorentz scalars; and therefore only functions of $p^2 = M^2$, i.e., constants. Therefore

$$C_{ij}(\underset{\sim}{p}) = 2A(p_i p_j - \delta_{ij}\underset{\sim}{p}^2) + \delta_{ij}(4B-C) \quad . \tag{4.26}$$

Dividing the sum rule (4.22) by E^2 and considering the limit $E \to \infty$, we have

$$\frac{8}{M}\int_0^1 d\omega\ \omega F_1(\omega) = 2A \tag{4.27}$$

$$p_i p_j \int_0^1 d\omega[F_2(\omega)-2\omega F_1(\omega)] = 0 \quad . \tag{4.28}$$

The sum rule (4.27) is useless, since A is unknown, but (4.28) tells us, since $F_2-2\omega F_1$ is positive definite, that

$$F_2(\omega)-2\omega F_1(\omega) = \lim_{q^2 \to -\infty} [F_2(\omega,q^2)-2\omega F_1(\omega,q^2)] = 0 \tag{4.29}$$

or recalling (1.13–14)

$$R = \frac{\sigma_S(q^2,\omega)}{\sigma_T(q^2,\omega)} \xrightarrow[q^2 \to -\infty]{} 0 \quad . \tag{4.30}$$

This result depends on the fact that the current is constructed from spin-½ constituents in a minimal fashion. If we add an anomalous magnetic moment to the current R will not necessarily vanish.

2) *Field Algebra*

In this model the SU(3) × SU(3) currents are given by Yang-Mills fields. The ETCR's of J with itself and of $\frac{\partial}{\partial t} J_i$ with J_j are canonical and are equal to

$$[J_i(\underset{\sim}{x},0),J_k(0)] = 0 \tag{4.31}$$

$$[\frac{\partial}{\partial t} J_i{}^a(\underset{\sim}{x},0),J_k{}^b(0)] = i\delta^3(\underset{\sim}{x})C_{ab}J_i{}^a(0)J_k{}^a(0) + \text{c-number.} \tag{4.32}$$

where C_{ab} is a numerical matrix, a and b run over the indices of SU(3) × SU(3). Once again the value of $C_{ij}(\underset{\sim}{p})$ is unknown, but its tensor structure is determined:

$$C_{ij}(\underset{\sim}{p}) = \langle p|C_{ab}J_i{}^aJ_k{}^b|p\rangle_{AVE} = A'p_ip_k+B'\delta_{ik} \ . \tag{4.33}$$

An identical tensor structure arises if the current is bilinear in scalar meson fields. Thus, if $J_\mu = \phi^*\overleftrightarrow{\partial}_\mu\phi$ then $\int d^3x[\frac{\partial}{\partial t} J_i(\underset{-}{x},0),J_k(0)] = \partial_i\phi^*(0)\partial_k\phi(0) + (i \leftrightarrow k)$, and $C_{ij}(\underset{\sim}{p})$ has the same form as (4.33). In this case (4.32) yields the sum rules:

$$\frac{4}{M}\int_0^1 d\omega[F_2(\omega)-2\omega F_1(\omega)] = A'$$

$$\int_0^1 d\omega\ \omega F_1(\omega) = 0 \ , \tag{4.34}$$

and therefore, for scalar mesons or field algebra we arrive at the opposite conclusion of (4.29) and (4.30), i.e.,

$$\omega F_1(\omega) = \lim_{q^2\to-\infty} \omega F_1(\omega,q^2) = 0$$

$$R = \frac{\sigma_S(q^2,\omega)}{\sigma_T(q^2,\omega)} \xrightarrow[q^2 \to -\infty]{} \infty \ . \tag{4.35}$$

Additional sum rules could be derived if one relinquishes the spin average or if one makes additional assumptions about the ETCR's. For electron scattering off a polarized target Bjorken has derived a sum rule for the spin dependent part of the cross-section in terms of

$$[J_i^{EM}(\underline{x},), J_k^{EM}(0)] = \left\{ \begin{array}{cl} 2i\epsilon_{ikj}A_j & \text{Quark Model} \\ 0 & \text{Field Algebra} \end{array} \right.$$

This sum rule unfortunately is extremely difficult to test.

A model for C_{ij} has been proposed by S. Ciccarrello, R. Gatto, G. Sartori, M. Tonin and by G. Mack. They, following arguments by K. Wilson, argue that the only operator of spin 2 and the appropriate dimension (4) to appear in C_{ij} is the energy momentum tensor $\theta_{\mu\nu}$ and therefore

$$[\frac{\partial}{\partial t} J_i^a(\underline{x},0), J_k^b(0)] = \delta^3(\underline{x})\delta_{ab}[a\theta_{ik}(0)+bg_{ik}\theta_{00}(0)] + \text{spin zero.}$$

This assumption is advantageous since the diagonal, spin averaged, matrix elements of θ_{ik} are, of course, known $< p|\theta_{\mu\nu}|p > = \frac{1}{M} p_\mu p_\nu$. An immediate consequence of this assumption is that the sum rules (4.27), (4.34) are isotopic spin scalars, since θ is an isoscalar. In other words, the integral of $(F_2^p - F_2^n)$ should vanish. This seems to be in contradiction with the latest SLAC results indicating that $F_2^p > F_2^n$ for all $0 \lesssim \omega \lesssim 1$. The quark model, on the other hand, has both iso-scalar and isovector components of C_{ij} as can be seen from (4.24) and (4.25).

f) Sum Rules for Neutrino-Nucleon Scattering

In the case of neutrino-nucleon scattering one can, in addition to deriving sum rules involving $[\frac{\partial}{\partial t} J_i, J_k]$, test the ETCR of the weak current with itself. First of all the time-time ETCR is non vanishing. This leads to the crucial Adler sum rule. Second, the presence of W_3 allows one to test the space-space ETCR.

The relevant ETCR's are, for $J_\mu{}^a = J_\mu{}^+ = V_\mu{}^+ - A_\mu{}^+$, where we have set $\theta_c = 0$ and $V^+ = V^{1+i2}$,

$$[J_0{}^+(\underset{\sim}{x},0), J_\mu{}^-(0)] = 4[V_\mu{}^3(0) - A_\mu{}^3(0)]\delta^3(\underset{\sim}{x}) \quad \ldots . \tag{4.36}$$

$$[J_i{}^+(x,p), J_k{}^-(0)] = \Big\{ 4\delta_{ik}[V_0{}^3(0) - A_0{}^3(0)] +$$

$$+ 4i\epsilon_{ikj}\left[(\tfrac{2}{3})^{1/2}(V_j{}^0(0) - A_j{}^0(0)) + \frac{1}{(3)^{1/2}}(V_j{}^8(0) - A_j{}^8(0)) \right] \Big\} \delta^3(x) \tag{4.37}$$

in the quark model. In the algebra of fields, (4.36) is, of course, unchanged, but (4.37) vanishes.

1) *Adler Sum Rule:*

The Adler sum rule tests (4.36) and can be derived without invoking the BJL limit. One way to derive it is to take the $\underset{\sim}{p} \to \infty$ limit of:

$$\int dq_0\, W_{00}{}^{+-}(q_0, \underset{\sim}{q}, \underset{\sim}{p}) = \int d^3x\, e^{-i\underset{\sim}{q}\cdot\underset{\sim}{x}} < N|[J_0{}^+(\underset{\sim}{x},0)J_0{}^-(0)]|N >_{AVE} .$$

In this limit, $\nu = q_0|\underset{\sim}{p}|$ and $W_{00} = p_0{}^2 W_2 + (1 - \frac{q_0{}^2}{q^2})W_2$. We have therefore

$$\lim_{|\underset{\sim}{p}| \to \infty} \int \frac{d\nu}{|\underset{\sim}{p}|} \left[\frac{|\underset{\sim}{p}|^2}{M^2} W_2^{+-}\left(\nu, \frac{\nu^2}{|\underset{\sim}{p}|^2} - q^2\right) - \frac{\underset{\sim}{q}^2}{\frac{\nu^2}{|\underset{\sim}{p}|^2} - q^2} \right.$$

$$\left. W_2^{+-}\left(\nu, \frac{\nu^2}{|\underset{\sim}{p}|^2} - \underset{\sim}{q}^2\right) \right] = \frac{4p_0}{M} < I_3 > .$$

If we are allowed to interchange the integral with the limit we derive:

$$(4.38) \quad \int_{-\infty}^{+\infty} \frac{d\nu}{M} [W_2^{+-}(\nu, -\underset{\sim}{q}^2)] = \int_{-q^2/2}^{\infty} \frac{d\nu}{M} [W_2^{\bar{\nu}}(\nu, q^2) - W_2^{\nu}(\nu, q^2)] = 4 < I_2 > ,$$

or, changing variables:

$$(4.39) \quad \int_0^1 \frac{d\omega}{\omega} [F_2^{\bar{\nu}}(\omega, q^2) - F_2^{\nu}(\omega, q^2)] = 4 < I_3 > = \begin{cases} 2 \text{ for p} \\ -2 \text{ for n} \end{cases} .$$

This sum rule is much more restrictive than the sum rules derived from the BJL limit, since it holds for all q^2. It requires measuring the difference of the inelastic cross sections for neutrino and anti-neutrino scattering off a nucleon (not a heavy nculei). If we use the analysis of Sec. I, we can write $F_2^{\bar{\nu}p} - F_2^{\nu p} = F_2^{\bar{\nu}p} - F_2^{\bar{\nu}n} = F_2^{\nu n} - F_2^{\nu p}$, and therefore instead measure the difference between neutrino-proton and neutrino neutron scattering. To date there has been no test of this sum rule.

2) *Space-Space ETCR's*

To test the space-space ETCR's we must resort again to the BJL limit. As before we compare (4.21) with (4.20) using the ETCR's given by (4.37). We have 2 sum rules:

$$\int_0^1 d\omega[F_1^{\bar{\nu}}(\omega)-F_1^{\nu}(\omega)] = +\frac{M}{2E} < p|4[V_0^3(0)]|p >_{AVE}$$

$$(4.40) \qquad = 2 < I_3 > \text{ [Bjorken Backward Sum Rule]}$$

and

$$i\int_0^1 d\omega[F_3^{\nu}(\omega)+F_3^{\bar{\nu}}(\omega)]\,\epsilon_{ijk}\frac{p_k}{M} = -4i\epsilon_{ijk} < p|[(\frac{2}{3})^{1/2}(V_k^{\,0}-A_k^{\,0}) + \frac{1}{(3)^{1/2}}(V_k^{\,8}-A_k^{\,8})]|p >_{AVE}$$

or, expressing the baryon-number current as $B_\mu = (\frac{2}{3})^{1/2} V_\mu^{\,0}$ and the hyper-charge current as $Y_\mu = \frac{2}{(3)^{1/2}} V_\mu^{\,8}$, we have

$$\int_0^1 d\omega[F_3^{\nu}(\omega)+F_3^{\bar{\nu}}(\omega)] = -4[< B + \frac{1}{2}Y >] = \left\{-6 \text{ for nucleons}\right\}$$

(4.41) [Gross-Llewellyn Smith Sum Rule] .

The sum rule (4.40) is not actually independent of the Adler sum rule. This is because in the quark model, as we shall see below, $F_2 = 2\omega F_1$. It can be rotated in iso-space into an inequality for e–N scattering

$$(4.42) \qquad \int_0^1 \frac{d\omega}{\omega}[F_2^{\gamma p}+F_2^{\gamma n}] \geq \frac{1}{2} \quad .$$

This inequality is probably trivial, since F_2 seems to approach a constant for small ω and the left hand side therefore diverges logarithmically.

The sum rule (4.41) is useful since it involves isoscalar combination of the F's:

$$F_3^{\nu p}+F_3^{\bar{\nu} p} = F_3^{\bar{\nu} n}+F_3^{\nu n} = F_3^{\nu p}+F_3^{\nu n} \quad ,$$

and can therefore be tested by scattering neutrinos off heavy nuclei. The right hand side of this sum rule is extremely large compared to, for example

$$\left(\int_0^1 d\omega \ F_2^{\gamma p}(\omega)\right)_{\text{exp}} \cong 0.36.$$

It depends crucially, in contract to the other current-current ECTR's, upon the specific representation of the currents; since it involves the representation dependent d_{abc}'s.

In the field algebra the right hand side of (4.40) and (4.41) should be replaced by zero. However, in this case, both F_1 and F_3 vanish identically.

3) $[J_i,J_k]$ *Sum Rules*

As in the case of electron-nucleon scattering, we derive sum rules for W_i in terms of $[\frac{\partial}{\partial t} J_i,J_k]$. These sum rules are derived exactly as before. The result is that:

$F_1 = F_3 = 0$ or

$$\text{(4.43)} \qquad \frac{\sigma_R(q^2,\omega)+\sigma_L(q^2,\omega)}{\sigma_S(q^2,\omega)} \xrightarrow[q^2 \to -\infty]{} \text{Algebra of Fields}$$

$F_2 - 2\omega F_1 = 0$ or

$$\frac{\sigma_S(q^2,\omega)}{\sigma_R(q^2,\omega)+\sigma_L(q^2,\omega)} \xrightarrow[q^2 \to -\infty]{} \text{Quark Model.} \tag{4.44}$$

4) Since the isovector part of the electromagnetic current and the $\Delta S = 0$ part of the weak current presumably are related by Chiral $SU(2) \times SU(2)$ one might hope to relate the magnitude of the structure functions for electron and neutrino scattering. In the limit of exact symmetry of course, the A–A contributions of F_i^{ν} are identical with the V–V contributions. However, even in the presence of symmetry breaking one can relate the two. If one assumes (a la Bjorken and Paschos) that the symmetry breaking term transforms like $(\underset{\approx}{\overline{3}},\underset{\approx}{3}) + (\underset{\approx}{3},\underset{\approx}{\overline{3}})$, i.e., as a quark mass term in the gluon model, then one can easily show that

$$< p|[\frac{\partial}{\partial t}(V_i - A_i),\ V_i + A_i]|p>_{AVE} = 0\ . \tag{4.45}$$

It then follows that the vector and axial vector contributions to the moments of F_2 or $F_2 - 2\omega F_1$ which are given by the ETCR of the time derivitive of the current with itself are equal. This allows us to derive an inequality:

$$\int_0^1 d\omega \left(\frac{F_2^{\nu p}(\omega)+F_2^{\nu k}(\omega)}{2}\right)^{\Delta S=0} = 2\int_0^1 d\omega \left[\frac{F_2^{\nu p}(\omega)+F_2^{\nu n}}{2}\right]^{\text{VECTOR}}_{\Delta S=0}$$

and by isotopic spin rotation

$$= 4\int_0^1 d\omega \left(\frac{F_2^{\gamma p}(\omega)+F_2^{\gamma n}}{2}\right)^{\text{ISOVECTOR}} \leq 4\int_0^1 d\omega \left(\frac{F_2^{\gamma p}(\omega)+F_2^{\gamma n}(\omega)}{2}\right). \tag{4.46}$$

The left hand side has not been determined experimentally, but from knowledge of the total neutrino cross section $\sigma_{tot}^{\nu A} = \frac{G^2 ME}{\pi}(0.6 \pm 0.15)$, (see below) it must lie between 0.6 and 1.8. Assuming $F_2^{\gamma n} < F_2^{\gamma p}$ the right hand side is less than $4 \times 0.18 = 0.72$, so that the inequality is consistent with the data.

In the gluon model itself an additional relation can be derived, relating the commutator for $[J_i, J_k]$ appearing in electron and neutrino scattering. This relationship [Llewellyn-Smith] yields the sum rule:

$$(4.47) \quad 12 \int d\omega\, \omega[F_1^{\gamma p}(\omega) - F_1^{\gamma n}(\omega)] = \int d\omega\, \omega[F_3^{\nu p}(\omega) - F_3^{\nu n}(\omega)] \,.$$

Finally we can express the total neutrino-nucleon cross-section in terms of moments of the structure functions. We recall the expression for $\frac{d\sigma\overset{(-)}{\nu}}{dQ^2 d\nu}$ in terms of the structure functions, (1.11), and rewrite this in terms of the dimensionless variables: $\omega = \frac{Q^2}{2\nu}$ and $X = \frac{\nu}{ME}$. We have

$$\frac{d\sigma\overset{(-)}{\nu}}{dX d\omega} = \frac{G^2 ME}{\pi}\left[\left(1 - X - \frac{M X\omega}{2E}\right) F_2^{\overset{(-)}{\nu}} + X^2 \omega F_1^{\overset{(-)}{\nu}}\right.$$

$$(4.48) \qquad \left. \mp X\left(1 - \frac{1}{2}X\right)\omega F_3^{\overset{(-)}{\nu}}\right] \,.$$

In the region where the incident neutrino energy E is much greater than M, the physical region of scattering is given by $0 \lesssim x \lesssim 1$, $0 \lesssim \omega \lesssim 1$. In this limit the total cross section is:

$$(4.49) \quad \sigma_{tot}^{\overset{(-)}{\nu}}(E) = \frac{G^2 ME}{\pi} \int_0^1 d\omega \left[\frac{1}{2} F_2^{\overset{(-)}{\nu}}(\omega) + \frac{1}{2} F_1^{\overset{(-)}{\nu}}(\omega)\cdot\omega \mp \frac{1}{3} F_3^{\overset{(-)}{\nu}}(\omega)\omega\right]$$

or

$$\sigma^{\binom{\nu}{\bar{\nu}}}(E) = \frac{G^2ME}{\pi}\int_0^1 d\omega F_2(\omega)A^{\binom{\nu}{\bar{\nu}}}(\omega)$$

where

$$(4.50) \qquad \frac{1}{3} \leq A^{\binom{\nu}{\bar{\nu}}} = \frac{1}{2}\left(1 + \frac{\sigma\binom{L}{R} - \frac{1}{3}\sigma\binom{R}{L}}{\sigma_R + \sigma_L + 2\sigma_S}\right) \leq 1 \quad .$$

The fact the $\sigma_{tot}(E)$ is a linear function of the neutrino energy is a direct consequence of scaling, and conversely provides the best experimental evidence, at the present time, for scaling in neutrino scattering. The coefficient of E is a moment of the F_i's, and can be expressed in terms of current commutators. We note that the different $\sigma^{\nu N} - \sigma^{\bar{\nu}N}$ is given in terms of F_3, and in general is not expected to vanish. Finally if there exists a W-meson, then at large energies the linear rise of σ_{tot} will be cut off and asymptotically

$$(4.51) \qquad \sigma_{tot}^{\nu} \approx \frac{G^2 M_W^2}{2\pi} \ell n \left(\frac{2EM}{M_W^2}\right) F_2^{\nu}(0) \quad .$$

Thus the linearity of σ_{tot}^{ν} as a function of E is evidence against the existence of a W.

g) Sum Rules for Annihilation

The annihilation of electrons and positrons into hadrons provides a sum rule for the Schwinger term in the ETCR of the electromagnetic current. Consider the vacuum expectation value of the commutator:

$$(4.52) \qquad \frac{1}{2\pi}\int d^4x\, e^{iq\cdot x} < 0|[J_\mu(x), J_0(0)]|0 > = \epsilon(q_0)(g_{\mu\nu}q^2 - q_\mu q_\nu)\rho(q^2) \quad .$$

If we insert intermediate hadron states into (4.52), we see that $\rho(q^2)$ is proportional to the total cross section for $e^+ + e^- \to \gamma \to$ hadrons:

$$\sigma_{tot}^{\gamma \to \text{had.}}(q^2) = \frac{32\pi^3\alpha^2\rho(q^2)}{q^2} \quad . \tag{4.53}$$

The integral of the space-time component of (4.52) gives:

$$-q_i \int_0^\infty dq_0{}^2 \rho(q_0{}^2 - \underset{\sim}{q}^2) = \int d^3x\, e^{-i\underset{\sim}{q}\cdot\underset{\sim}{x}} < 0|[J_0(\underset{\sim}{x},0)J_i(0)]|0 >$$

$$= -iq_i < 0|S(0)|0 > \tag{4.54}$$

where S(0) is the (spin zero) Schwinger term:

$$[J_0(\underset{\sim}{x},0),J_i(0)] = iS(0)\partial_i\delta^{(3)}(\underset{\sim}{x}) \quad .$$

We therefore have a sum rule for the annihilation cross section:

$$\int_0^\infty dq^2 \frac{q^2\sigma_{tot}(q^2)}{32\pi^3 \ \alpha^2} = < 0|S(0)|0 > \quad . \tag{4.55}$$

What can we say about the vacuum expectation value of the Schwinger term? In the algebra of fields the Schwinger term is a finite c-number. In this case we must have $\sigma_{tot}^{\gamma \to \text{had.}} \underset{q^2 \to \infty}{\lesssim} \frac{M_\nu{}^2}{q^4}$. This is a very rapid fall-off compared to the cross section for annihilation into leptons which behave as $\frac{1}{q^2}$. In the quark model the Schwinger term cannot be calculated canonically and thus one must resort to perturbation theory. In perturbation theory the cross section has a power behavior of $\frac{1}{q^2}$ and correspondingly the Schwinger term diverges quadratically. This in fact is what we might expect from the point of view of "scale invariance", since in a theory without dimensional parameters the total cross section must behave like $\frac{1}{q^2}$.

V. *SCALING, ANOMALIES AND LIGHT CONE BEHAVIOR*

a) Scaling and the BJL Limit

The original suggestion that the structure functions should scale was made by Bjorken on the basis of an analysis of the current algebra sum rules for electro-production. Consider the sum rule for W_2 derived in the previous section (4.22):

$$\lim_{q^2 \to -\infty} \int_0^1 d\omega F_2(\omega,a^2) = \lim_{|\vec{p}| \to \infty} \left\{ \frac{Mi}{4|\vec{p}|^2} \quad d^3x < \vec{p}|[\dot{J}_{||}{}^{EM}(\underline{x},0), J_{||}{}^{EM}(0)] - [\dot{J}_T{}^{EM}(\underline{x},0), J_T{}^{EM}(0)]|\vec{p} > \right\} \tag{5.1}$$

where $J_{||}(J_T)$ is the spatial component of J_i parallel (transverse) to $\vec{p}$. Now let us assume that the ETCR appearing in the sum rule has a finite matrix element, and that therefore the left hand side of (5.1) has a finite limit. One can then show that all the higher order sum rules, i.e., those involving the highest spin component of $[(\frac{\partial}{\partial t})^{2n+1} J^{EM}(\underline{x},0), J^{EM}(0)]$, are valid. These sum rules are of the form:

$$\lim_{q^2 \to -\infty} \int_0^1 d\omega F_2(\omega,q^2)\omega^{2n} \sim \lim_{|\vec{p}| \to \infty} \frac{1}{|\vec{p}|^{2+n}} < \vec{p}|[(\frac{\partial}{\partial t})^{2n+1} J^{EM}(\underline{x},0), J^{EM}(0)]|\vec{p} > \tag{5.2}$$

Since $F_2(\omega,q^2)$ is positive and ω is never greater than one:

$$\int_0^1 d\omega\ \omega^{2n}\ F_2(\omega,q^2) \leq \int_0^1 d\omega\ F_2(\omega,q^2)\ ;$$

and therefore the existence of the limit in (5.1) guarantees that the limit in (5.2) must be finite (or at worst oscillatory).

Bjorken then argued that the existence of (5.1) implies scaling. This is not strictly correct, for it assumes sufficient uniformity so that one interchange the limit and the integration. However, it is extremely plausible that if (5.1) has a finite, non vanishing, limit then so does $F_2(\omega,q^2)$. This argument can be reversed, namely if the Bjorken scaling limit exists then it follows that the BJL limit is valid in the $\vec{P} = \infty$ frame. To see this consider the limit of the dispersion relation for $T_2(q^2,\omega)$, as $|q_0| \to \infty$, and keep only the leading terms in $P_0 = -\dfrac{q_0}{2\omega}$:

$$(5.3) \qquad T_2(q^2,\omega) \xrightarrow[|q_0|\to\infty]{} -\frac{4M}{{q_0}^2}\sum_{n=0}^{\infty}\left(\frac{2P_0}{q_0}\right)^{2n}\int_0^1 F_2(\omega)\ \omega^{2n}d\omega\ .$$

Scaling, together with the assumption of an unsubtracted dispersion relation, leads to the existence of an expansion of $T_2(q^2,q\cdot p)$ in inverse powers of q_0 in the $\vec{P} = \infty$ frame. The coefficients of this expansion, which by BJL are the highest spin parts of the ETCR of the time derivitives of J_μ with itself, are all finite. Even if we were forced to make a finite number of subtractions, say if $F_2(\omega) \underset{\omega\approx 0}{\approx} \omega^{-N}$, then the first few sum rules would not exist, however for $2n > N-1$, all the sum rules $\int_0^1 d\omega\ \omega^{2n}F_1(\omega)$ exist. This is quite different, as we shall presently see, from the expectations one might have from perturbation theory.

b) Anomalies in Perturbation Theory

In the previous section we ignored all possible difficulties in deriving current algebra sum rules. We assumed the existence of an asymptotic expansion of $T_{\mu\nu}(q,p)$ in inverse powers of q_0 (the BJL limit), and assumed that the coefficients of this expansion were the ETCR's derivable from the equations of motion and the canonical ETCR's of the fields. Are these assumptions correct? The only place we can test these assumptions is perturbation theory. The original investigations of the BJL limit in perturbation theory were carried out by Johnson and Low, who in fact proposed the limit as a way to define the ETCR. They discovered that the ETCR defined in this manner did not, in general, coincide with that naively calculated by the use of canonical ETCR's. Soon after the derivation of current algebra sum rules for lepton hadron scattering and the proposal of scaling this question was reexamined. Since these anomalies in perturbation theory are thoroughly discussed in the accompanying lectures I will merely review the results.

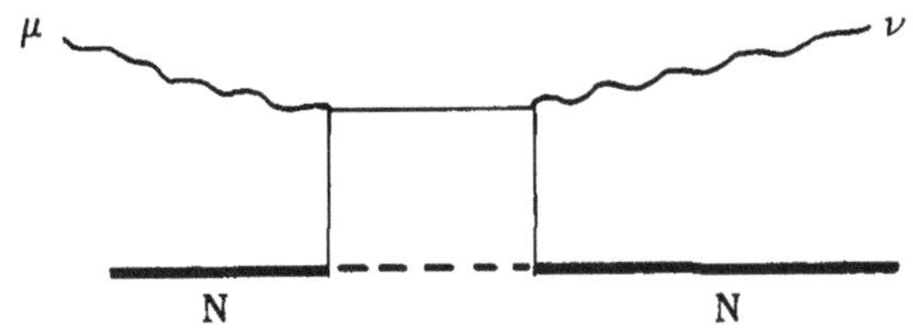

The anomalies of perturbation theory are already present when one calculates the structure functions to second order in perturbation theory. If one evaluates the imaginary part of the box diagram (the dotted line is a neutral vector meson or a pion) one finds that νW_2 does not scale as $q^2 \to -\infty$ for fixed w. The violation of scaling is logarithmic; i.e., $\ell n(\frac{q^2}{M^2})$. This means that the sum rule (5.1) is invalid to second order in perturbation theory, and that the time ordered product does not fall off as q_0^{-2} for large q_0. Furthermore it is not correct, to this order in perturbation theory, that the naive evaluation of the ETCR of the currents is corrent. Thus, for example, the tensorial structure of the singular product $\bar{\psi}\,\gamma_i\,\overleftrightarrow{\partial}_j\psi$ is different from that one would naively expect. Therefore it is not correct that $\frac{\sigma_L}{\sigma_T}$ vanishes, in the interacting quark model, as $\frac{1}{q^2}$. Another example of such a disease is the calculation of space-space ETCR's which differ from their canonical values even though they are finite.

Super renormalizable theories are slightly better than the above renormalizable ones. The structure functions in a $\lambda\phi^3$ theory do scale, and do preserve the naive tensorial structure of the canonical ETCR's. However even in a super renormalizable model all is not well. The sum rule, (5.1), is still divergent, not because $F(\omega,q^2)$ does not scale for large q^2 but due to the appearance of a pole at $\omega = 1$ for infinite q^2. This is a reflection of the fact that the matrix elements of $[\dot{J}_i, J_K] \sim \partial_i\phi\partial_K\phi$ are logarithmically divergent in perturbation theory.

The real world, on the other hand, appears to be much less singular than this perturbation theory would indicate. Of course the possibility always exists that experiments at higher energies will begin to deviate from scaling; however straight-forward extrapolation of the present experimental data suggest that the BJL limit is indeed correct, and that the current algebra sum rules are finite.

Nature appears to be behaving in some sense like a free field theory and shows no evidence of the singular nature of interacting field theories. It is possible to construct field theoretic models that do exhibit scaling and in which these anomalies are absent. One merely inserts, by hand, an ultraviolet cutoff into the theory. This is in essence what is done in the parton model, which is based on the idea that in the scaling region the nucleon can be considered as a collection of free particles.

In order to reconcile this idea with local field theory it is necessary to introduce a fixed momentum cutoff. If this is done one then recovers all the naive predictions of current algebra.

The important lesson to be drawn from the above discussion of perturbation theory is that the validity of current algebra sum rules for inelastic lepton-hadron scattering depends crucially upon the dynamics. Since we have no reason to suspect that perturbation theory provides an adequate dynamical approximation in the highly inelastic region, and since the experiments seem to be consistent with the assumption of a BJL limit, one can perhaps justify the existence of such sum rules. However, when one tests such a sum rule one is merely testing a hypothesis made about the structure of an ETCR. This structure is perhaps motivated by the ETCR of a free field theory (i.e., free quark model) but it is a non-trivial dynamical question whether one is testing the validity of an assumed structure for the currents within an interacting field theory. Thus if $\sigma_L/\sigma_T = 0$ this might mean that the electromagnetic current is constituted from spin-½ fields, but only if the dynamics is sufficiently non singular.

c) The Light Cone and the Operator Product Expansion

Let us consider the meaning of the scaling limit is position space, namely what region of space-time do we probe in the scaling region? To answer this question we apply the scaling limit to the current commutator $\int d^4x\, e^{iq\cdot x} < p|[J_\mu(x),J_\nu(0)]|P >$ in the rest frame of the target; $p = (M,0,0,0)$ $q = (q_0,0,0,q_3)$. In this frame, in the scaling limit,

$$q_0 = \frac{\nu}{M}\,,\quad q_3 = \left(\frac{\nu^2}{M^2} - q^2\right)^{1/2} = \frac{\nu}{M} + Mw \tag{5.4}$$

and thus the commutator can be written as

$$\int d^4x\, e^{i\left[(\frac{2\nu}{M}+Mw)\,(t-z)+Mw(t+z)\right]}\, C_{\mu\nu}(x^2,x_0M) \tag{5.5}$$

Therefore, for large ν, this Fourier transform will probe the region where:

$$t-z \lesssim \frac{M}{2\nu} \qquad \text{and} \qquad t+z \lesssim \frac{1}{2Mw}$$

so that

$$x^2 \le t^2-z^2 \lesssim \frac{1}{4\nu w} = -\frac{1}{2q^2} \; . \tag{5.6}$$

The commutator vanishes for the space like ($x^2 < 0$) separation of the currents, and therefore in the scaling limit one is probing the light cone behavior of the current commutator. This is in contrast to the region that is probed at high energies for fixed external masses (the Regge limit), where

$$x^2 \le t^2-z^2 \lesssim -\frac{1}{2q^2} = \text{fixed} \quad , \tag{5.7}$$

so regions where x is a finite distance from the light cone contribute to the structure functions.

An alternate way of exhibiting the approach to the light cone is Bjorken's representation of the structure functions in terms of "almost equal-time commutators". Consider the structure function F_1, which in the frame where $p = (p_0,0,0,p_3)$ and $q = (q_0,0,0,0)$ is given by:

$$F_1(\omega,q^2) = \frac{1}{M}\int \frac{d^4x}{(2\pi)}\, e^{iqx} < p|[J_x(\underset{\sim}{x},t),J_x(0)]|p> \tag{5.8}$$

The scaling limit can be reached by letting $q_0 \to \infty$, $p_0 \to \infty$, and keeping $\omega = -\dfrac{q^2}{2q\cdot p} = -\dfrac{q_0}{2p_0}$ fixed. In this limit we can write F_1 as

$$F_1(\omega) = \lim_{p\to\infty} \frac{1}{M}\int \frac{d\tau d^3x}{(2\pi)}\, e^{-i2\omega\tau}\, \frac{1}{p_0} < p|[J_x(\underset{\sim}{x}, \frac{\tau}{p_0})J_x(0)]|p> , \tag{5.9}$$

where $\tau = \frac{t}{p_0}$.

From this expression one can reproduce all the current algebra sum rules by inserting the Fourier transform and expanding both sides in powers of τ. Furthermore we see the approach to the light cone since effectively $(\frac{\tau}{p_0})^2 - \underset{\sim}{x}^2$ (which must be positive) vanishes for large p_0.

K. Wilson has proposed a generalization of ETCR's that is of particular relevance to the light cone behavior of current commutators. The proposal generalization is that the product of two local operators for small values of the space-time separation has an expansion in terms of a set of independent local operators of the form

$$A(\tfrac{x}{2})B(-\tfrac{x}{2}) \underset{x_\mu \approx 0}{\cong} \sum_n \mathcal{C}_n(x) 0_n(0) \quad . \tag{5.10}$$

The functions $\mathcal{C}_n(x)$ depend on the four vector x and have singularities for $x \approx 0$ of the form $[x^2 - i\epsilon x_0]^{-1}$ and perhaps logarithms of $x^2(\text{MASS})^2$. An essential ingredient in Wilson's scheme is the concept of scale invariance. In an operational sense one assumes that one can associate with each local operator 0_n a "dimension" d_n. These dimensions fix the dependence of $\mathcal{C}_n$ upon x, up to logarithms, as if the theory were exactly scale invariant. If scale invariance was a good symmetry there would exist a unitary operator U(s) such that the theory would be invariant when:

$$0_n(x) \to U^+(s) 0_n(x) U(s) = s^{d_n} 0_n(sx) \tag{5.11}$$

Applied to the expansion (5.10) this would imply that

$$\mathcal{C}_n(sx) = s^{d_n - d_A - d_B} \mathcal{C}_n(x) \tag{5.12}$$

namely that $\mathcal{C}_n(x)$ is homogeneous of order $d_n - d_A - d_B$ in x.

In a free field theory the above operator produce expansion is certainly correct and the operators are just Wick ordered products of the fields and their derivitives. The dimension of an operators, in this case, equals its physical dimension in mass units [scalar field ~ m^1, spinor field ~ $m^{3/2}$, current ~ m^3, etc.]. The dimension, however, may be changed by interactions. In fact Wilson argues that the only operators whose dimension will be canonical (equal to their free field theory values) are those which satisfy non linear ETCR's that force that result – namely vector currents and the energy momentum tensor.

The operator product expansion is a very useful extension of ETCR's. It, of course, determines all the ETCR's of the two operators; however, even if these are infinite the expansion exists for small but finite x_μ (the divergences appear only in the c-number functions $\mathcal{C}_n(x)$). Examples of terms in the expansion of the commutator of two currents are:

$$\left[J_\mu{}^a(\tfrac{x}{2}),\, J_\nu{}^b(-\tfrac{x}{2})\right]_{x_\mu \approx 0} = \delta_{ab} x_\mu x_\nu \left[\frac{1}{(x^2 - i\epsilon x_0)^8} - \frac{1}{(x^2 + i\epsilon x_0)^8}\right]$$

$$+ f_{abc}\left[\frac{1}{(x^2 - i\epsilon x_0)^4} - \frac{1}{(x^2 + i\epsilon x_0)^4}\right](x^\mu J_\nu{}^c(0) + x^\nu J_\mu{}^c(0)) \quad , \tag{5.13}$$

where the first term contributes a quadratically divergent c-number Schwinger term to $[\dot{J}_0{}^a(\underset{\sim}{x},0),\, J_i{}^a(0)]$, and the second contributes $f_{abc}J_\nu{}^c(0)$ to $[J_0{}^a(\underset{\sim}{x},0), J_\nu{}^b(0)]$ [using $\partial_0\delta(x^2) \underset{x_0 \approx 0}{\propto} \delta^{(3)}(\vec{x})$].

It is useful to exhibit the Lorentz character of the various operators 0_n appearing in the expansion. We can take 0^n to be a symmetric traceless n^{th} rank tensor so that:

$$A(\tfrac{x}{2})B(-\tfrac{x}{2}) \underset{x \approx 0}{\approx} \sum_n \mathcal{C}_n(x^2)x^{\mu_1}\ldots x^{\mu_n} 0^n_{\mu_1\ldots\mu_n} \quad . \tag{5.14}$$

For short distances a given tensor $0^n_{\mu_1 \ldots \mu_n}$, of dimension D_n, will contribute a term to the product which behaves as $(x)^{D_n - d_A - d_B}$. Therefore only operators with dimension D_n less than $d_A + d_B + 1$ survive.

Let us now extend the operator product expansion to the light cone. In this case x_μ does not necessarily vanish so that the $0^n_{\mu_1 \ldots \mu_n}$ contributes a term which behaves as $(x^2)^{D_n - n - d_A - d_B}$ for small x^2. Thus one can get contributions to the light cone behavior from an infinite number of operators whose dimension minus spin is less than $d_A + d_B + 1$. In a free field theory there is a definite relation between the lowest dimension of an operator and its spin. In a free quark model, for example, the operators of lowest dimension (excluding derivatives of local operators) are: $\bar{\psi}\gamma_\mu\psi$, $\bar{\psi}[\gamma_\mu \overleftrightarrow{\partial}_\nu + \gamma_\nu \overleftrightarrow{\partial}_\mu[\psi, \ldots \sum_{\text{PERM}} \bar{\psi}\gamma_{\mu_1}\overleftrightarrow{\partial}_{\mu_2} \ldots \overleftrightarrow{\partial}_{\mu_n}\psi$; and these satisfy the relation:

$$\text{dimension} = \text{spin} + 2 \ . \tag{5.15}$$

Therefore they would contribute equally to the light cone expansion.

Let us now apply the light cone expansion to the diagonal matrix element of the current commutator, isolating the coefficient of $P_\mu P_\nu$:

$$< p|[J_\mu(\tfrac{x}{2}), J_\nu(-\tfrac{x}{2})]|p > \underset{x^2 \approx 0}{\approx}$$

$$\simeq \sum_n C_n(x^2) x^{\mu_1} x^{\mu_2} \ldots x^{\mu_n} < P|0^{n+2}_{\mu,\nu,\mu_1 \ldots \mu_n}|p > \ . \tag{5.16}$$

If the operators in the expansion obeyed the canonical relation (5.15) then $C_n(x^2)$ would have dimension $3+3+n-(n+4) = 2$. This would imply that

$$C_n(x^2) = C_n \left[\frac{1}{x^2 - i\epsilon x_0} - \frac{1}{x^2 + i\epsilon x_0} \right] = C_n \delta(x^2) \epsilon(x_0) \ .$$

Therefore canonical dimensions imply that the singularity in the light cone is a delta-function. When the above expression is Fourier transformed, such a light cone singularity leads to scaling structure functions. Conversely the existence of a non-trivial scaling limit demands the existence of an infinite set of operators of increasing spin n and dimension = spin + 2.

What are these operators? If we write the matrix element of 0^{n+2} as

$$< p|0^{n+2}_{\mu,\nu,\mu_1 \ldots \mu_n}|p> = 0^n(0)\, p_\mu p_\nu p_{\mu_1} \cdots p_{\mu_n} + \ldots ,$$

then

$$< p|[J_\mu(\tfrac{x}{2}), J_\nu(-\tfrac{x}{2})]|p>$$

$$= p_\mu p_\nu \delta(x^2)\epsilon(x_0) \sum C_n (x \cdot p)^n 0^{n+2}(0) + \ldots . \tag{5.17}$$

When this expression is Fourier transformed we obtain

$$0^{n+2}(0) = \text{const.} \int_0^1 d\omega\, \omega^n\, F_2(\omega) \tag{5.18}$$

The right hand side of (5.18) we recognize as the sum rule for the highest spin components of the ETCR $\left[(\frac{\partial}{\partial x_0})^{n+1} J_\mu(\underset{\sim}{x},0), J_\nu(0)\right]$. It is possible to make this identification directly by taking the equal time limit of successive time derivitives of Eq. (5.17).

Within the framework of the operator product expansions one can make assumptions which lead directly to the sum rules derived previously. Thus if we assume that the tensorial structure of the product near the light-cone is that given by a free quark model we recover $\sigma_L/\sigma_T = 0$. This expansion appears to provide a natural framework for discussion of light cone behavior. Much work remains to be done on developing additional applications of such an expansion and on understanding why the behavior on the light cone is that of free fields.

VI. PARTON MODELS

a) Feynman's Parton Model

The parton model is motivated by the point like behavior of the structure functions as observed at SLAC. As discussed previously, any model in which the nucleon is considered to be a composite state of point like (structureless, i.e., bare form factors) constituents leads to scaling if not to the observed behavior of νW_2. The main ideas behind the parton model were conceived by R. P. Feynman and elaborated on by Bjorken and Paschos and by Drell, Levy and Yan. The most important feature of Feynman's model for relativistic nucleon lepton scattering is the emphasis placed on the infinite momentum frame of reference. It is in this frame that the nucleon has a simple structure and in which the impulse approximation is valid.

Feynman's argument is that when viewed at infinite momentum the motion of the constituents (hereafter called partons) of the nucleon is sufficiently slowed down by the time dilatation effect that the incident lepton scatters instantaneously and incoherently (since there is no time for rearrangement) off the partons which make up the nucleon.

Consider the infinite momentum frame, $\vec{P} \to \infty$,

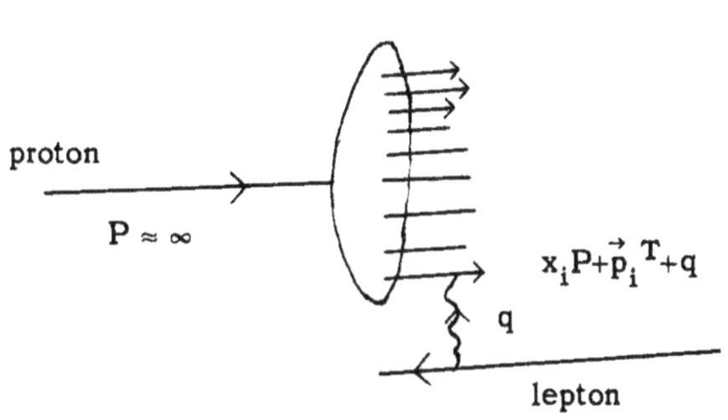

which is equivalent to the hadron-lepton centre of mass frame at asymptotic energies. Assume that the proton is composed of constituents, partons, which are structureless (i.e., free particles). A particular virtual state of the proton will consist of N partons where i^{th} parton has longitudinal momentum $x_i \vec{P} \left(\sum_{i=1}^{N} x_i = 1 \right)$ and transverse momentum $\vec{p}_i^{\,T} \left(\sum_{i=1}^{N} \vec{p}_i^{\,T} = 0 \right)$. If the interaction time is very small compared to the lifetime of this virtual state, then the partons can be considered to be bare particles. By the Uncertainty Principle, the lifetime of the virtual states is:

$$T_V^{-1} \approx E_f - E_p$$

$$= \sum_{i=1}^{N} ((x_i \vec{P} + \vec{P}_i^{\,T})^2 + \mu_i^2)^{1/2} - (P^2 + m_P^2)^{1/2}$$

$$= P \left[-1 + \sum_{i=1}^{N} x_i \right] + \sum_{i=1}^{N} \frac{\mu_i^2 + (p_i^T)^2}{2|x_i P|} - \frac{M_P^2}{2|P|}$$

where μ_i is the "mass" of the ith parton. If we make the following assumptions:

1) The transverse momentum distribution is bounded,
2) The masses of the partons are bounded, then:

$$T_v \approx \frac{2P}{(\text{Mass})^2} \tag{6.1}$$

On the other hand, the time of interaction with the lepton is estimated to be in the COM frame:

(6.2) $$T_{INT} \approx \frac{1}{q_0} \approx \frac{4P}{2\nu+q^2} = \frac{4P}{2\nu(1-\omega)} .$$

Thus:

(6.3) $$\frac{T_{INT}}{T_V} = \frac{(\text{characteristic mass})^2}{2\nu(1-\omega)}$$

So that if $2\nu(1-\omega) >> (\text{mass})^2$ then the partons can be considered to be bare particles. Note that this is not reliable near threshold $(\omega \approx 1)$.

Therefore one can hopefully assume that during the time of interaction the partons can be considered to be free particles. If $-q^2 >> m^2$ then the exited parton is kicked out with a large momentum transfer $q_T \approx (-q^2)^{1/2}$ and we would expect the scattering to be incoherent.

To calculate the structure functions in the parton model we assume:

1) At $P = \infty$ the nucleon is in a state of N free partons with probability P_N, such that $\sum_n P_N = 1$.

2) The longitudinal momentum of the ith parton is $P_i = x_i P$ with a probability distribution $F_i{}^N(x_i)$.

3) The "mass" of the partons, before and after the collision, is small.

4) The transverse momentum before collision is much less than $(-q^2)^{1/2}$, the transverse momentum imparted by the collision. Then we add the cross sections corresponding to the lepton scattering off each parton in each possible configuration.

In a given configuration, each parton with momentum xP contributes to $W_{\mu\nu}$ a term

$$\int < xP|J_\mu(y)J_\nu(0)|xP >_{AV.} \frac{e^{iq\cdot y}}{(2\pi)} dy = (2\pi)^3 \{ <xP|J_\mu(0)|xP+q >$$

(6.2a) $$< xP+q|J_\nu(0)|xP >_{AV} \} \delta[(xP+q)^2-\mu^2] .$$

Since we have assumed the parton to behave like a bare particle only one parton intermediate states will contribute. Then

(6.3a) $$W^{i}_{\mu\nu} = \frac{2P_\mu P_\nu \, x_i^{\,2} Q_i^{\,2} M}{x_i} \, \delta(q^2 + 2x_i\nu) + \ldots\ldots$$

where the deleted terms contribute to W_1 and depend on the spin of the parton. Note the factor x_i which appears in the denominator. This factor must be put in since we have normalized states so that $\langle xP|xP\rangle = x\langle P|P\rangle$. A check of the above formula is provided by showing that the differential cross section $\frac{d\sigma}{dq^2}$, at infinite incident lepton energy, approaches the Mott cross section $\frac{4\pi\alpha^2}{q^4} Q_i^{\,2}$. We have then, for large q^2 and ν,

(6.4) $$W_2(\omega,\nu) = \sum_N P(N) \int_0^1 dx_i \sum_{i=1}^{N} \frac{M x_i Q_i^{\,2}}{\nu} \, \delta(x_i - \omega) f_N^{\,i}(x_i)$$

or

(6.5) $$F_2(\omega) = \omega \sum_N P(N) \sum_{i=1}^{N} Q_i^{\,2} f_N^{\,i}(\omega) \; .$$

An analogous calculation of W_1 will convince the reader that, as might be expected, spin zero partons do not contribute to W_1 (i.e., $\sigma_T = 0$) and spin one half partons lead to $W_1 = \frac{1}{2\omega} W_2$ (i.e., $\sigma_L = 0$). The experimental value of R seem to be small indicating spin one half partons (quarks?).

The fact that we have derived a scaling limit in this model should come as no surprise. We have in effect built scaling into the model by assuming bounded transverse momenta and masses, and by assuming bare pointlike form factors. One of the nice features of the parton model is the physical interpretation it gives to the scale variable $\omega = \frac{Q^2}{2\nu}$ as the fraction of the longitudinal momentum carried by the parton which interacts with the lepton in the infinite momentum frame. Thus $\frac{F_2(x)}{x}$ measures the longitudinal momentum distribution of the charged partons in this frame.

To proceed further, one must make assumptions about the actual distribution of $f_N{}^i(x)$ within a given configuration of N partons and about the distribution of the nucleon into a given configuration. To gain insight let us consider the general features of (6.5)

1) *High Energy Limit*

If one had only a finite number of configurations then it is obvious from (6.5) that as $\omega \to 0$ (i.e., $\nu >> Q^2$) $F_2(\omega)$ vanishes like ω, since $F_N{}^i(x)$ is normalizable $\int_0^1 f_N{}^i(x)dx = 1$. This behavior (Fig. 23) of the

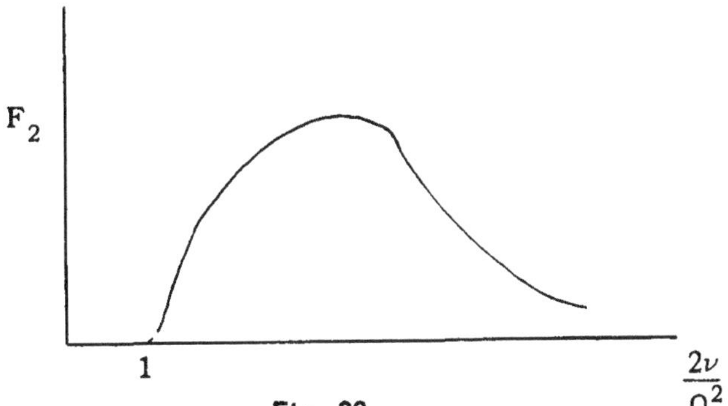

Fig. 23

structure function exhibits the expected quasielastic peak and is contradicted by the experiments which are consistent with $F_2(\omega) \xrightarrow[\omega \to 0]{} \text{const.}$ The only way of circumventing the vanishing of $F_2(0)$ is to allow the number of partons to be infinite. If $F_2(\omega) \xrightarrow[\omega \to 0]{} \text{const}$ then, in fact:

$$\int_{\epsilon}^{1} \frac{F_2(\omega)}{\omega} d\omega \approx \ell n \epsilon \simeq \sum_{N=1}^{\infty} P(N) \sum_{i=1}^{N} Q_i^2 \quad .$$

Now $\sum_{i=1}^{N} Q_i^2 \approx N$ so that $\sum_{N=1}^{\infty} P(N) \cdot N$ must diverge logarithmically. Therefore one requires configurations with as arbitrary number of partons and:

$$P(N) \approx \frac{1}{N^2} \quad N \to \infty \quad . \tag{5.6}$$

2) *Average Parton Charge*

We can derive a sum rule for the structure functions without any detailed information as to the nature of the momentum distribution. If we assume that each parton has an identical momentum distribution, then $f_N^{\,1}(x_1) = f_N(x_1) = \int dx_2 \ldots dx_N \, f_N(x_1, x_2, \ldots x_N)\, \delta(1-\Sigma x_i)$, where $f_N(x_1 \ldots x_N)$ is the probability of finding partons with longitudinal momenta $x_1 P, \ldots, x_N P$. It follows that $f_N(x_1 \ldots x_N)$ is a symmetric function of $x_1, \ldots x_N$ and thus

$$\int_0^1 x_1 f_N(x_1) dx_1 = \frac{1}{N} \int dx_1 \ldots dx_N \, \Sigma x_i \, f_N(x_1 \ldots x_N) \delta(1-\Sigma x_i)$$

$$= \frac{1}{N} \quad . \tag{6.7}$$

Finally we have:

$$\int_0^1 F_2(\omega) d\omega = \sum_N P(N) \sum_{i=1}^{N} \frac{Q_i^2}{N} = \langle Q^2 \rangle = \text{mean-square charge per parton} \tag{6.8}$$

The experimental value of the left hand sice of (6.8) is roughly 0.16. This appears rather low if we were to think of the partons as quarks. For example, if the proton consisted of three quarks with the usual charges ($\frac{2}{3}, \frac{1}{3}, \frac{1}{3}$ for p,n,λ quarks), then

$$< Q^2 >_{proton} = \frac{1}{3}[(\frac{2}{3})^2 + (\frac{2}{3})^2 + (\frac{1}{3})^2] = \frac{1}{3}$$

(6.9) $$< Q^2 >_{neutron} = \frac{1}{3}[(\frac{1}{3})^2 + (\frac{1}{3})^2 + (\frac{2}{3})^2] = \frac{2}{9} .$$

However, we have already seen that configurations with many partons are necessary to reproduce the correct energy dependence. If we assume, therefore, that a configuration of N+3 partons consists of 3 quarks, ϵN charged quarks [$q\bar{q}$ pairs] and $(1-\epsilon)N$ neutral partons and that the mean-square charge of the quark-anti-quark cloud is statistical

$$\frac{(\Sigma Q_i^2)}{N\epsilon} = \frac{1}{3}[(\frac{2}{3})^2 + (\frac{1}{3})^2 + (\frac{1}{3})^2] = \frac{2}{9} \quad \text{then}$$

$(\Sigma Q_i^2)_N = 1 + \frac{2}{9}(N\epsilon - 3)$. The mean-square charge per parton is then

(6.10) $$< Q^2 > = \sum_N P(N)[\frac{1}{3N} + \frac{2\epsilon}{9}] = \frac{2\epsilon}{q} + \frac{1}{3} < \frac{1}{N} > = 0.16$$

Therefore $\epsilon \lesssim 0.71$ and $< \frac{1}{N} > \lesssim 0.48$, so that one must have configurations with many neutral partons.

At this point it is of value to take a fresh look at the parton model. We have seen that experiment forces us to consider configurations with an infinite number of charged and neutral partons. The resulting picture is quite different from the description of inelastic lepton scattering off nuclei as quasielastic scattering off the consistuent nucleons. There, of course, one can "see" the partons (nucleons) by simply observing the particles present in the final state of the collisions. The decay products in inelastic lepton-hadron collisions are, of course, hadrons; which behave quite differently from partons. What then are the partons? Perhaps the best way to think of them is as bare hadronic states. The nucleon can be considered to be a superposition of an infinite number of configurations of bare particles, which after the instantaneous, incoherent collision with the lepton reassemble to form the physical particles in the final state. This is, in fact, the way that the parton model is realized by Drell, Levy and Yan. This picture, however, makes it difficult to develop a feeling for the momentum distribution functions, except on the basis of specific field theoretic models.

There are consequences of the general parton model which follow merely from assumptions as to the quantum numbers of the partons. Thus if one assumes that the partons consist of quarks and perhaps an SU(3) singlet (a gluon), then one has, for the structure function relevant to a SU(3) current, J:

$$F_2 = 2\omega F_1 = \omega \sum_N P(N) \sum_q f_q^N(\omega) n_q^N Q_q^2 \quad , \tag{6.11}$$

where one has in the Nth configuration n_q^N partons of type $q(q = p,n,\lambda)$ and $Q_q = \langle q|J|q\rangle$ is the expectation value of the current for the q-type bare quark. Since there are six independent measurable F_2's for lepton scattering off nucleons, (i.e., γp, γn, γP, γn, $\bar{\nu} p$, $\bar{\nu} n$) and only five unknown functions in a given configuration (i.e., $f_p^N n_p^N$, $f_n^N n_n^n$, $f_{\bar{p}}^N n_{\bar{p}}^N$, $f_{\bar{n}}^N n_{\bar{n}}^N$ and $(f_\lambda^N + f_{\bar{\lambda}}^N) n_\lambda^N$ since $n_\lambda^N = n_{\bar{\lambda}}^N$ due to the zero strangeness of the nucleon) there exists one relation between the structure functions:

$$6[F_2^{\gamma p} - F_2^{\gamma n}] = \omega[F_3^{\nu p} - F_3^{\nu n}] \tag{6.12}$$

The first moment of this relation is derivable in the gluon model (4.47).

Furthermore if one imposes the constraint that the quantum numbers within each configuration be those of the nucleon (baryon number and charge) and uses the normalization of $f_q^N(\omega)$, one can derive (as expected) the Adler sum rule (4.3a) and the Gross-Llewellyn Smith sum rule (4.41). It is instructive to consider W_3 in the parton model. We have seen that an integral of W_3 (4.41) is related to the baryon number of the target. This emerges in a nice way within the parton quark model.

Consider the scattering of a neutrino or an anti-neutrino off a parton in the C.O.M. If we note that neutrinos and partons are left-handed (due to the V-A interaction), we see from Fig. 24 that angular momentum cannot be conserved for anti-neutrino parton scattering in the backward direction. (The arrows in Fig. 24 indicate the helicities of the parton and the neutrino). Therefore anti-neutrino parton scattering, and similarly neutrino-anti parton scattering, must vanish in the backward direction. If we transform to the lab. frame this implies that the parton contributions to anti-neutrino nucleon scattering vanishes when $\theta_{LAB} = \pi$ and $E' = 0$, $\nu = E \to \infty$. If we recall (1.11) and (1.16) we see that partons contribute only to σ_L and anti-partons only to σ_R. The structure function W_3, which is proportional to $\sigma_R - \sigma_L$, is therefore a measure of baryon number.

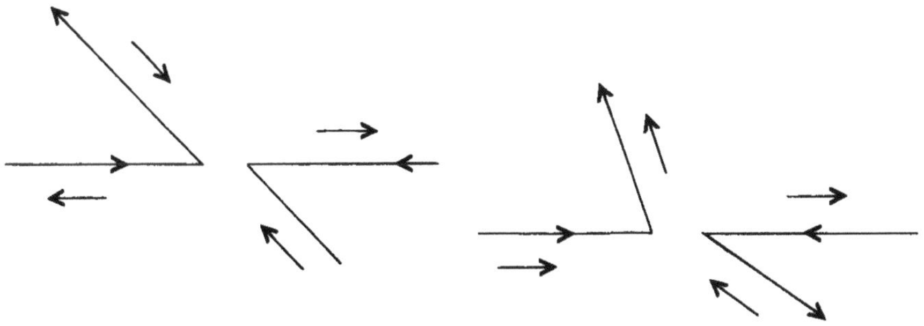

Neutrino-Parton Scattering Anti-Neutrino Parton Scattering

Fig. 24

If only partons (baryons) were to contribute to the structure functions one would have:

$$F_3^{\binom{\nu}{\bar{\nu}}} = \mp 2F_1^{\binom{\nu}{\bar{\nu}}} = \mp \frac{1}{\omega} F_2^{\binom{\nu}{\bar{\nu}}} \tag{6.13}$$

This would have rather interesting consequences. It implies that if $W_2 \sim \nu^{\alpha-2}$ then $W_3 \sim \nu^{\alpha-1}$. However the vacuum trajectory ($\alpha = 1$) can contribute to W_3 but not to W_3, so that such a relation is inconsistent with Regge behavior. Furthermore, it follows from the above and (4.50) that

$$\sigma^{\nu}(E) = 3\sigma^{\bar{\nu}}(E) \quad . \tag{6.14}$$

However, we might expect anti-partons (anti-baryons) to change these conclusions. In particular the "sea" of $q\bar{q}$ pairs, does not contribute to W_3 (since q and $\bar{q}$ contribute equal and opposite amounts) whereas it does contribute to W_1 and W_2. When this is taken into account, W_3 is suppressed, and (6.14) is no longer correct, although it is always the case that $\sigma^{\nu} > \sigma^{\bar{\nu}}$.

There have been some other applications of the parton model to weak processes. A particularly interesting one is a suggestion of Bjorken and Paschos that one might see pointlike structure in inclusive Compton scattering: $\gamma + p \to \gamma +$ anything. Roughly their prediction is that

$$\frac{\left(\dfrac{d^2\sigma}{d\Omega dE'}\right)_{\gamma p}}{\left(\dfrac{d^2\sigma}{d\Omega dE'}\right)_{ep}} = (\text{kinematical factor}) \times \frac{\langle \Sigma Q_i^{\,4} \rangle}{\langle \Sigma Q_i^{\,2} \rangle}$$

b) Field Theoretic Realization of the Parton Model

In the previous section we have explored the main features of the parton model. In order to make the model consistent with experiment it was necessary to consider the nucleon to be in configurations with an indefinite number of pointlike partons. This could be understood if the partons are the bare states of a relativistic quantum field theory. In this section we will describe such a realization of the parton model, as derived by Drell, Levy and Yan (DLY).

DLY consider a specific Lagrangian field theory of nucleons and pions whose interaction is described by the Lagrangian $\mathcal{L}_I = g\bar{\psi}\gamma_5 \underset{\sim}{\tau} \psi \underset{\sim}{\pi}$. They then discuss the amplitude for $\gamma(q^2)+P \to \gamma(q^2)+P$ in the Bjorken scaling limit. As we indicated, the basic idea of Feynman's parton model is that the nucleon can be considered to be composed of free bare constituents in the $P = \infty$ frame. In the DLY model the partons are identified with the bare nucleon and pion. To exhibit these states one takes the Heisenberg current $J_\mu{}^H(x)$ and "undresses" it by transforming to the interaction picture.

$$(6.15) \qquad J_\mu{}^H(x) = U^{-1}(t) j_\mu(x) U(t)$$

where j_μ is the interaction picture current,

$$(6.16) \qquad j_\mu = \bar{\psi}_p \gamma_\mu \psi_p + i\pi^+ \overleftrightarrow{\partial}_\mu \pi^-$$

written in terms of ψ_p and π the free ("in") fields. The transformation matrix U(t) is, of course, $U(t) = T\{e^{-i\int_{-\infty}^{t} H_I(t)dt}\}$. We then have for $W_{\mu\nu}$

$$(6.17) \qquad W_{\mu\nu} = \sum_n < UP|j_\mu(0)U(0)|n > < n|U^{-1}(0)j_\nu|UP > \delta\,(P+q-P_n)$$

where $|UP> = U(0)|P>$.

The above expression for $W_{\mu\nu}$ is superficially frame dependent. We utilize this by attempting to simplify (6.17) in the $P = \infty$ frame. However we must first introduce a transverse momentum cutoff which we recall was necessary so that the partons could be regarded as frozen during the time of interaction with the lepton. Specifically DLY *postulate* the existence of a region where $-q^2$ is much larger than the transverse momentum of all virtual constituents of the proton, which here are the states in $U|P>$. This is technically achieved by introducing a transverse momentum cutoff, $k_{\perp max}$, at each strong vertex, so that $U|P> \cong |P> + g|P\pi> + g^2|P\pi\pi> + \ldots$ consists of particles of bounded momentum transverse to $\vec{P}$ ($\simeq \infty$). They then argue that in the scaling limit the states $U|P>$ and $U|n>$ are eigenstates of the Hamiltonian with eigenvalue E_p and E_n. Consider a particular contribution to $< UP|j_\mu(0)U(0)|n>$ pictured in Fig. 25.

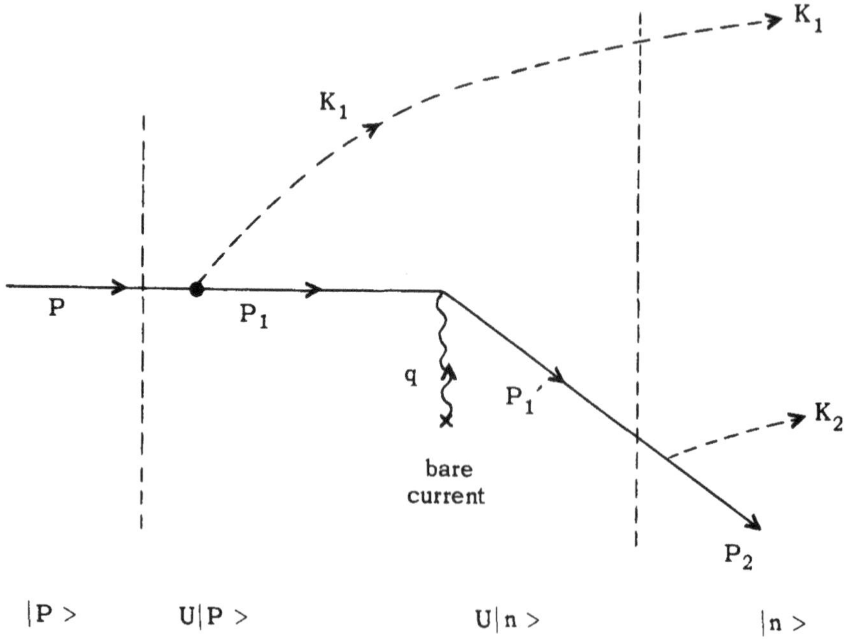

Fig. 25

The incoming nucleon with momentum P undresses into a state, $U|P>$, containing a nucleon with momentum $x_1 P_\perp^1$ and a pion with momentum $(1-x_1)P+K_\perp^1$. The lepton then interacts with the nucleon, via the bare current, transferring to it a momentum $q[q_{||} \approx \frac{1}{P}, q_\perp \sim (Q^2)^{1/2}]$ and taking it to the state $U|n>$. Finally, the nucleon and the pion go into the state $|n>$ consisting of a nucleon with momentum $x_2 P_1^{\ 1}+P_\perp^2$ and a pion with momentum $(1-x_2)P_1^{\ 1}+K_\perp^2$ and a pion with momentum K_1. If we evaluate

$$E_P - E_{U|P>} = E_P - E_{P_1} - E_{K_1} = (\underset{\sim}{P} + \frac{M^2}{2P}) - |x_1|P - (1-x_1)P - \frac{(P_\perp^{\ 1})^2 + N^2 + (K_\perp^{\ 1})^2 + \mu^2}{2P} \quad ,$$

we see that as $P \to \infty$, and if $P_\perp^{\ 1}$ and $K_T^{\ 1}$ are indeed bounded

$$E_P - E_{U|P>} \approx P(1-|x_1|-(1-x_1)) + 0(\frac{1}{P}) \quad . \tag{6.18}$$

Since this energy difference appears in an energy denominator in the above amplitude, due to the time integrals in the expansion of U(t), this particular state will give a negligible contribution unless $1 \geq x_1 \geq 0$, in which case $E_P - E_{U|P>} \underset{P\to\infty}{=} 0(\frac{1}{P})$. The same holds for $E_n - E_{U|n>} = P[1-|x_2|-|1-x_2|] \simeq 0(\frac{1}{P})$. Thus both $U|n>$ and $U|>$ are eigenstates of H. We can then write $W_{\mu\nu}$ in the $P=\infty$ frame as:

$$W_{\mu\nu} = \int d^4x e^{iq\cdot x} \sum_n < UP| e^{iP\cdot x} j_\mu(0) e^{-i\ P(x)} U(0)|n> < n|U^{-1}(0) j_\nu(0)|UP>$$

$$= \int d^4x\ e^{iq\cdot x} < UP| j_\mu(x) j_\nu(0)| UP > \quad . \tag{6.19}$$

[This only works for the so-called "good" components of J_μ, i.e., the time component or the component parallel to $\underset{\sim}{P}$. However, these are sufficient to recover $W_{\mu\nu}$].

Equation (6.19) is the field theoretic realization of the parton model. The nucleon before the interaction emits and absorbs pions and nucleon-anti-nucleon pairs in a cluster with momentum centered about the momentum of the nucleon. The bare current scatters off one of these bare constituents transferring to it a very large transverse momentum. The scattered particle then emits and reabsorbs pions and nucleon-anti-nucleon pairs, but does not interfere with the other constituents.

More detailed information can be derived for large values of ν/q^2. In this region DLY show that:

1) To leading order in $g^2 \ell n\, \nu/q^2$ the ladder graphs (Fig. 26) dominate the amplitude.

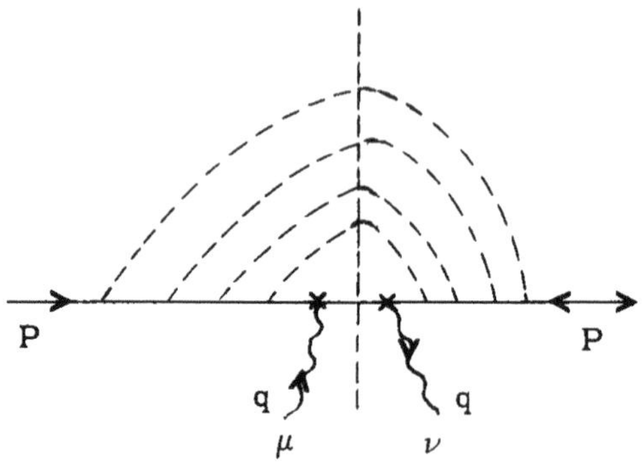

Fig. 26

In other words, vertex and self energy corrections contribute to lower powers of $\ell n\, \nu/q^2$ to a given order in g.

2) Furthermore, in the same approximation, the current only interacts with the spin-one half parton. Therefore

$$\sigma_S = 0, \text{ i.e., } F_1(\omega) = \frac{1}{2\omega} F_2(\omega) .$$

3) These graphs can be summed to yield a power behavior for νW_2

$$\nu W_2 = \text{const.}\,\omega^{1-\lambda}$$

$$\lambda = \frac{3}{4\pi}\frac{g^2}{4\pi}\ell n\left[1+\frac{K_{\perp\,MAX}^{\;2}}{M^2}\right] \tag{6.20}$$

4) In the case of neutrino-nucleon scattering identical arguments lead to $\sigma_S = 0$ (the current interacts only with the spin-one half nucleon) and $\sigma_R = 0$ (the current interacts only with the baryons). Thus (6.13) holds in this model for large ν/q^2 where the baryon-anti baryon sea is suppressed by factors of $\ell n\, \nu/q^2$.

The most serious difficulty with the DLY model is the ad hoc introduction of a transverse momentum cutoff. If one removes this cutoff then scaling is immediately lost, along with the other features of the parton model. Even though it is unlikely that this cutoff could emerge from the field theory itself, the model can still be a useful heuristic tool. In particular, it allows one to make detailed applications of the parton idea to other reactions. DLY have, in fact, applied their model to annihilation processes, to the relation between elastic electromagnetic form factors and the threshold dependence of νW_2, and to the Lederman experiment.

VII. REGGE POLES AND DUALITY IN THE SCALING REGION

a) Regge Poles

As we have seen previously it is not obvious that Regge poles determine the behavior of photo absorption cross sections in the kinematic region where the virtual mass of the photon is very large. Even though the scattering angle in the t-channel, for forward Compton scattering, grows like $\cos\theta \approx \frac{\nu}{p_t q_t} \sim \frac{\nu}{M(Q^2)^{1/2}}$, when $\nu \to \infty$, $Q^2 \to \infty$ and $\frac{\nu}{Q^2}$ = fixed; one cannot deduce anything about the behavior of

$$(7.1) \qquad F_2\left(\frac{\nu}{Q^2}, Q^2\right) = \frac{\nu}{M} W_2 \simeq \sum_i \nu\beta_i(Q^2)\left(\frac{\nu}{M(Q^2)^{1/2}}\right)^{a_i-2} + $$

$$+ \text{Background} \ldots\ldots,$$

unless one has information regarding the Q^2 dependence of the Regge residues and of the background. Note that it is entirely possible to have both scaling, i.e., $\lim_{Q^2 \to \infty} F_2\left(\frac{\nu}{Q^2}, Q^2\right) = F_2\left(\frac{\nu}{Q^2}\right)$ and Regge behavior,

i.e., $\lim_{\nu \to \infty} F_2\left(\frac{\nu}{Q^2}, Q^2\right) = \gamma(Q^2)\nu^{a-1}$, without having the position of the leading Regge trajectory (a) determine the behavior of $F_2\left(\frac{\nu}{Q^2}\right)$ for large $\frac{\nu}{Q^2}$. An example of such a function is

$$F_2\left(\frac{\nu}{Q^2}, Q^2\right) = \frac{(Q^2)\left(\frac{\nu}{Q^2}\right)^{a-1}}{1+Q^2\left(\frac{Q^2}{\nu}\right)^{\beta}} ;$$

which approaches $(Q^2)^{2-a}\,\nu^{a-1}$ for large ν, fixed Q^2; and approaches $\left(\frac{\nu}{Q^2}\right)^{a+\beta-1}$ for large (Q^2), fixed $\frac{\nu}{Q^2}$.

The experimental evidence presented in the previous sections appears to be consistent with Regge asymptotic behavior in the variable $\frac{\nu}{Q^2}$. The structure functions νW_2 and W_1 are seen to behave as $\left(\frac{\nu}{Q^2}\right)^0$ and $\left(\frac{\nu}{Q^2}\right)^1$ respectively, as if they were dominated by a scaling vacuum (Pomeron) Regge pole. It is, therefore, not unreasonable to assume that for large values of Q^2 Regge poles scale and that for large $\frac{\nu}{Q^2}$ they constitute the dominent contribution to the structure functions. With these assumptions we then can make predictions as to the asymptotic behavior of the structure functions, for fixed, but large, values of $\frac{\nu}{Q^2}$. Thus the difference between the proton and neutron structure functions should be dominated by A_2 Regge trajectory [I=1, G=−]:

$$(7.2) \qquad \nu W_2^{\gamma p} - \nu W_2^{\gamma n} \underset{\frac{\nu}{Q^2} \gg 1}{\approx} \left(\frac{\nu}{Q^2}\right)^{\alpha_{A_2}(0)-1} \approx \left(\frac{\nu}{Q^2}\right)^{-1/2}$$

and the structure function W_3 for neutrino scattering off nuclei should be dominated by the ω-trajectory [I=0, G=−] :

$$(7.3) \qquad \nu W_3^{\nu(p+n)} \underset{\frac{\nu}{Q^2} \gg 1}{\approx} \left(\frac{\nu}{Q^2}\right)^{\alpha_\omega(0)} \approx \left(\frac{\nu}{Q^2}\right)^{1/2} .$$

The preliminary SLAC data on electron-deuteron scattering seems to confirm (7.2), at least insofar as it indicates that $\nu W_2^{\gamma p} - \nu W_2^{\nu n}$ scales for fixed $\frac{\nu}{Q^2}$ and vanishes when $\frac{\nu}{Q^2} \to \infty$ (Fig. 21).

The assumption of Regge pole dominance in the scaling region leads to a seemingly strange dependence of Regge residues upon the mass of the virtual photons. The residue $\beta_i(Q^2)$, as defined in (7.1), must behave as

$$(7.4) \qquad \beta_i(Q^2) \underset{Q^2 \to \infty}{\approx} \left(\frac{1}{(Q^2)^{1/2}}\right)^{\alpha(0)}$$

Is there any reason to suppose that the external mass dependence of Regge residues will be correlated with the value of the trajectory in this fashion? This question was posed by Abarbanel, Goldberger and Treiman (AGT). It clearly can only be answered if one possesses a detailed model of Regge behavior. Accordingly AGT analysed the mass dependence for a sum of ladder diagrams in theory of scalar mesons with a $g\phi^3$ interaction. It is well known that the sum of these ladders Reggeizes for fixed external mass

(7.5) $\sum_{\text{ladders}}$

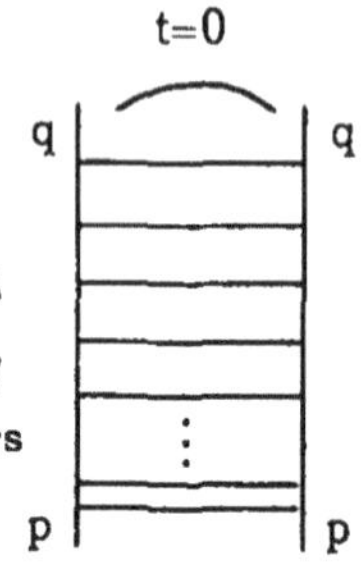

$$\approx \beta(q^2)\,(p\cdot q)^{\alpha(0)}$$

$$p\cdot q \to \infty\,;\quad p^2, q^2 \text{ fixed}$$

AGT then considered the imaginary part of the sum over ladders, thus replacing the photon by a scalar meson, in the limit $q^2 \to -\infty$, $\left(\frac{\nu}{-q^2}\right) \gg 1$ and found that:

$\mathrm{Im} \sum_{\text{ladders}}$

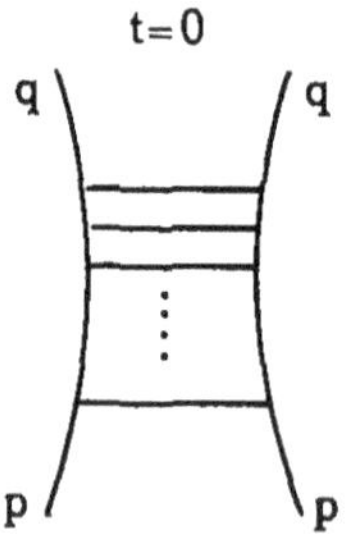

$$\approx \frac{1}{(-q^2)}\left(\frac{p\cdot q}{-q^2}\right)^{\alpha(0)}$$

$$p^2 = m^2,\ p\cdot q \gg m^2,\ -q^2 \gg m^2,$$

$$\frac{p\cdot q}{-q^2} \gg 1\,.$$

(For $m = 0$ the sum can be explicitly performed). This is not exactly scaling due to the additional $\frac{1}{q^2}$ dependence, however they conjectured that this is merely a reflection of the fact that the photon was replaced by a *scalar* meson. This conjecture has been verified by Altarelli and

Rubinstein who calculated the imaginary part of photon-scalar meson scattering, in the ladder approximation and found:

(7.6)

$$\mathrm{Im}\left\{\sum_{\text{ladders}}\right.$$

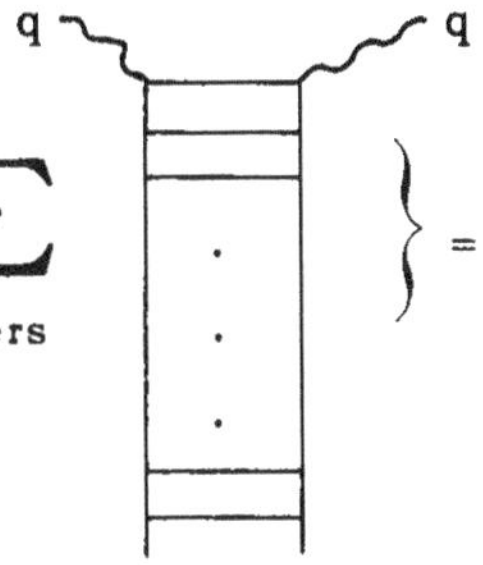

$$\left.\right\} = \begin{cases} \nu W_2(\nu,q^2) \to \left(\frac{\nu}{-q^2}\right)^{\alpha-1} \\ W_1(\nu,q^2) \to \frac{1}{(-q^2)}\left(\frac{\nu}{-q^2}\right)^{\alpha} \ell n\left(\frac{-q^2}{M^2}\right) \end{cases}$$

For fixed $\left(\frac{\nu}{-q^2}\right)$ we then have scaling and as expected for a current coupled to spinless particles $\frac{\sigma_T}{\sigma_L} \xrightarrow[-q^2 \to \infty]{} 0$ (as $\frac{\ell n q^2}{q^2}$) .

This is reassuring if we wish to apply Regge theory to the scaling region; however the following objections can be raised to the above arguments. First, while it is known that for fixed q^2 the ladder graph is dominant over all other graphs of the same order in g, it has not been shown that this is the case in the scaling limit. An additional problem is that the set of graphs considered is not manifestly gauge invariant. One can, however, show that in the scaling region each ladder graph is gauge invariant up to terms of order $[\ell n(\frac{\nu}{q^2})]^{-1}$. Perhaps the most serious objection is that these results are restricted to the super renormalizable $\lambda\phi^3$ theory. Since the current interacts in this model with scalar mesons along, one necessarily has $\frac{\sigma_T}{\sigma_S} = 0$. A more realistic model would be a renormalizable theory of nucleons and mesons. However in such a model, unless some sort of cutoff is introduced, scaling is violated in perturbation theory.

There have been other attempts to derive the asymptotic behavior of the scaling structure functions, $F_i(\frac{\nu}{Q^2})$, for large $\frac{\nu}{Q^2}$, using causal

[DGS or Jost-Lehmann-Dyson] representations for the current commutator. These investigations, however, only show that these representations of $F_i(\frac{\nu}{Q^2})$ are consistent with Regge asymptotic behavior. The assumptions required to prove that Regge poles dominate F_i are essentially equivalent to assuming that Regge behavior is uniform in Q^2, an assumption equivalent to the desired result.

b) Duality

In the realm of strong interactions the combination of Regge asymptotic behavior and narrow resonance saturation of dispersion relations has led to the very useful concept of duality. Let us recall the derivation of the finite energy sum rules (FESR). If a scattering amplitude $A(\nu,t)$ is dominated by Regge poles, then if these are subtracted from it, one has an amplitude that vanishes as $\nu \to \infty$, namely $A(\nu,t) - \sum_i \nu^{\alpha_i(t)} \beta_i(t)$. This combination therefore satisfies a superconvergence relation (assume we have removed enough Regge poles so that the above combination vanishes faster than $\frac{1}{\nu}$):

$$\lim_{N \to \infty} \int_0^N \mathrm{Im} \left[A(\nu,t) - \sum_i \nu^{\alpha_i(t)} \beta_i(t) \right] d\nu = 0 \tag{7.7}$$

and therefore

$$\int_0^N \mathrm{Im}\, A_B(\nu,t)d\nu + \int_0^N \mathrm{Im}\, A_R(\nu,t)d\nu \underset{N \to \infty}{=} \text{Pomeron} + \sum_i \frac{N^{\alpha_i+1}}{(\alpha_i+1)} \beta_i(t) \tag{7.8}$$

In the above FESR we have separated the imaginary part of the amplitude into a piece dominated by narrow resonances, A_R, and a piece containing the non-resonating continuum. A corresponding separation is made in the crossed channel between the Pomeron and all other Regge poles. Duality is essentially the hypothesis that the FESR is approximately correct for moderate values of N, that the contribution of the Pomeron is equal to the integral of the imaginary part of the continuum, whereas all other Regge poles are given by an average of the contributions of narrow resonances. These assumptions have proved quite useful in correlating high energy behavior with the nature of the allowed resonances in the direct channel and have lead to many interesting and successful predictions.

Let us now examine the FESR for the electroproduction structure functions, assuming that these are asymptotically dominated by Regge poles. If we consider only the non-Pomeron, non-continuum contributions we have:

$$\int_{\nu_0}^{\nu_M} [\nu W_2(\nu, q^2)]_{\text{Res.}} d\nu \cong \sum_i \frac{\nu_M^{\,a_i} \beta_i(q^2)}{a_i} \tag{7.9}$$

Now the q^2 dependence of each resonance contribution to the left hand side of (7.9) is given by the square of a transition electromagnetic form factor. These are rapidly decreasing functions of q^2. One therefore might conclude that the residues of all Regge poles, except the Pomeron, are rapidly decreasing functions of q^2. This, in fact, is the hypothesis advanced by Harari who further assumes that the q^2 dependence of the Pomeron is different from that of the other trajectories, and could in fact be consistent with scaling. Of course, this hypothesis does not explain scaling, it merely adds another mysterious feature to the behavior of the Pomeron. It has rather drastic consequences, however, for the non-diffractive parts of the structure functions – namely they should vanish rapidly for fixed $\omega = -q^2/2\nu$ and large q^2.

This is in contradiction to almost all current algebra sum rules. For example, the Adler sum rule (4.39)

$$\int_0^1 \frac{d\omega}{\omega} [F_2^{\bar{\nu}}(\omega,q^2) - F_2^{\nu}(\omega,q^2)] = 4 < I_3 > ,$$

would be invalid, since the integrand receives contributions from $I = 1$ trajectories. Similarly other sum rule such as (4.27), (4.41) are probably violated since the appropriate commutator has both $I = 0$ and $I = 1$ components. More significant, however, is the fact that experiment seems to indicate that the difference between the proton and the neutron structure functions scales. Let us therefore re-examine the finite energy sum rule.

First we note if we cut off the finite energy sum rule at a given value of $S = M^2+2\nu+q^2$, say above the resonance region, $(S_M)^{1/2} \approx 3$ BeV, then $\nu_M = \dfrac{S_M - M^2 - q^2}{2}$. If we then rewrite (7.9) in terms of the scaling $x = \dfrac{1}{\omega} = \dfrac{2\omega}{Q^2}$ we have:

$$\frac{Q^2}{2} \int_1^{1+\frac{S_M - M^2}{Q^2}} F_2(x,Q^2)dx = \sum_i \frac{\nu_M{}^{a_i} \beta_i(Q^2)}{a_i} \tag{7.10}$$

Thus if we keep the cutoff S_M fixed as Q^2 increases the left hand side can be saturated by a finite number of resonances; however these contribute for large Q^2, only to the region where $x \approx 1$. Since the structure function vanishes at threshold, $x = 1$, there is no contradiction between the rapid falloff in Q^2 of the resonance form factors, and scaling in the region of $x \approx 1$. If we wish to investigate the region where $x > 1$, we must let the cutoff in (7.10) be fixed, which implies that $S_M(Q^2) \simeq Q^2$. In this case as Q^2 is increased more and more resonances contribute to the FESR, and the net contribution of all the resonances, each of which might be a rapidly decreasing function of Q^2, could very well scale.

Bloom and Gilman, in a recent paper, argue that the resonances in fact do scale near threshold. They show that if one plots the experimental data for νW_2 vs. $x' = \frac{2\nu + M^2}{Q^2} = x + \frac{M^2}{Q^2}$ (which a priori is as good a scaling variable as x) then the resonances move, as Q^2 is increased, toward $x' = 1$, each resonance following in magnitude the smooth scaling limit curve. They conclude that the resonances are an intrinsic part of the scaling behavior of νW_2, a fact consistent with duality and the existence of non vacuum components of νW_2. They then assume further that local averages over resonances build up the scaling limit case for νW_2. The contribution of a given resonance is

$$\nu W_2 = 2\nu[G(q^2)]^2 \delta(s - M_R{}^2)$$

$$\underset{Q^2 \to \infty}{\approx} Q^2[G(Q^2)]^2\, \delta(s - M_R{}^2) \quad , \tag{7.11}$$

where the transition form factor falls off as $G(Q^2) \sim (\frac{1}{Q^2})^{\frac{n}{2}}$. This can be consistent with the threshold behavior of

$$\nu W_2 \underset{x' \approx 1}{\approx} (x' - 1)^P \text{ if}$$

$$n = P + 1 \ . \tag{7.12}$$

Therefore if the resonances are to scale all transition form factors must fall with some power, and this power must be related to power with which νW_2 vanishes at threshold. The experimental evidence is consistent with $n = 4$, $p = 3$. Bloom and Gilman then apply this assumption to the contribution of the nucleon elastic peak itself. Since the magnetic form factor dominates over the electric form factor for large Q^2 one deduces that $R = \frac{\sigma_S}{\sigma_T}$ vanishes at threshold and that the ratios of neutron to proton cross sections is equal to the ratio of the square of the magnetic moments at threshold:

$$\frac{\nu W_{2n}}{\nu W_{2p}} \underset{x' \approx 1}{\approx} \left(\frac{\mu_n}{\mu_p}\right)^2 \approx \frac{1}{2} \quad . \tag{7.13}$$

Both of these predictions are consistent with experiment.

In conclusion it appears that the dynamical concepts, developed for hadron scattering, of Regge asymptotic behavior, finite energy sum rules and duality are indeed applicable to inelastic lepton scattering in the scaling region.

Princeton University

BIBLIOGRAPHY

I. Experimental Papers

[1] E. D. Bloom, et al, *Phys. Rev. Letters* 23, 930 (1969) (Electron scattering Data at 6° and 10°).

[2] E. D. Bloom, et al, *Phys. Rev.* 23, 935 (1969) (Discussion of Data).

[3] I. Bugadov, et al, *Phys. Letters* 30B, 364 (1969) (Neutrino-Nucleon Scattering Data).

[4] E. D. Bloom, et al.

[5] *Review* (a) *R. E. Taylor,* Invited paper at International Conference for Particle Reactions at the New Accelerators (SLAC-PUB-746).
(b) Proceedings of 4th International Conference on Electron and Photon Interactions at High Energies, Liverpool (1969).

II. Applications of Current Algebra

[1] S. Adler, *Phys. Rev.* 143, 1144 (1969) (Derivation of sum rule for W_2 for neutrino-nucleon scattering).

[2] J. D. Bjorken, *Phys. Rev.* 148, 1467 (1969) (BJL limit, application to lepton-hadron scattering and to radiative corrections).

[3] J. D. Bjorken, *Phys. Rev.* 163, 1767 (1969) (Derivation of backward sum rule).

[4] J. D. Bjorken, *Phys. Rev.* 179, 1547 (1969) (Scaling in the Bjorken limit, almost equal-time commutator).

[5] K. Johnson and F. E. Low, *Progr. Theor. Physics* Kyoto Suppl. No. 37, 38 (1969) (BJL limit, anomalies from triangle graphs).

[6] D. Gross and R. Jackiw, *Nucl. Phys.* B14, 269 (1969) (Necessary and sufficient conditions for construction of T* products and the cancellation of seagulls and Schwinger terms).

[7] J. Cornwall and R. Norton, *Phys. Rev.* 177, 2587 (1969) (BJL limit and sum rules).

[8] C. G. Callan and D. Gross, *Phys. Rev. Letters* 21, 311 (1968) (Derivation of sum rules using $[\dot{J},J]$ given by Sugawara model).

[9] C. G. Callan and D. Gross, *Phys. Rev. Letters* 22, 156 (1969) (Derivation of σ_T/σ_L sum rules in quark model and field algebra).

[10] D. Gross and C. H. Llewellyn-Smith, *Nucl. Phys.* B14, 337 (1969) W_3 sum rule, partons and current algebra in neutrino scattering).

[11] R. Jackiw and G. Preparata, *Phys. Rev. Letters* 22, 975 (1969); *Phys. Rev.* 185, 1748 (1969); S. Adler and W. Tung, *Phys. Rev. Letters* 22, 978 (1969) (Breakdown of BJL limit in perturbation theory).

[12] C. H. Llewellyn-Smith, *Nucl. Phys.* (to be published) (Derivation of sum rules in parton model and current algebra).

[13] J. D. Bjorken and E. A. Paschos, *Phys. Rev.* D1, 3151 (1970) (Kinematic analysis of neutrino scattering and review of sum rules).

[14] J. Cornwall (UCLA preprint (1970) (Model for higher order equal-time commutators).

III. Operator Products at Small Distances and Light Cone Singularities

[1] K. Wilson, *Phys. Rev.* 179, 1499 (1969) (Proposal of "operator product expansions" for products of operators at short distances).

[2] S. Ciccariello, R. Gatto, G. Sartori and M. Tonin, *Phys. Letters* 30B, 549 (1969); G. Mack, University of Miami preprint (Sum rules based on specific assumptions about the operators appearing in the operator expansion).

[3] R. Brandt, *Phys. Rev. Letters* 23, 1260 (1969) (Model for light cone behavior of current commutators).

[4] H. Leutwyler and J. Stern, CERN TH-1138 (1970); R. Jackiw, J. West and Van Royen, MIT-CTP 118 (1970); J. Cornwall, D. Corrigan and R. Norton, *Phys. Rev. Letters* 24, 1141 (1970) (Light cone singularity from the scaling, sum rule for F_{long}).

IV. Parton Model

[1] J. D. Bjorken and E. Paschos, *Phys. Rev.* 185, 1975 (1969) (Parton model applied to electroproduction).

[2] S. Drell, D. Levy and T. M. Yan, *Phys. Rev. Letters* 22, 744 (1969); *Phys. Rev.* 187, 2159 (1969) (Outline of a field theoretic realization of the parton model); *Phys. Rev.* D1, 1035 (1970); D1, 1617 (1970); D1, 2402 (1970) (details of model); *Phys. Rev. Letters* 24, 855 (1970) (Nature of final state in electroproduction).

[3] H. Cheng and T. T. Wu, *Phys. Rev. Letters* 22, 1409 (1969) (Structure functions in Q.E.D., Lack of scaling).

[4] S. T. Chang and P. Fishbane, *Phys. Rev. Letters* 24, 847 (1969) (Criticism of DLY model, lack of scale invariance when cutoff is removed).

V. Regge Poles and Duality

[1] H. Abarbanel, M. Goldberger and S. Treiman, *Phys. Rev. Letters* 22, 500 (1969) (Propose that Regge poles dominate scaling limit, and argue that Regge residues might scale).

[2] H. Harari, *Phys. Rev. Letters* 22, 1078 (1969); *Phys. Rev. Letters* 24, 286 (1970) (Application of duality to scaling region).

[3] G. Altarelli and H. R. Rubinstein, *Phys. Rev.* 187, 2111 (1969) (Investigation of W_1 and W_2 in ϕ^3 ladders).

VI. Vector Dominance

[1] J. J. Sakurai, *Phys. Rev. Letters* 22, 981 (1969); and review talk at Liverpool (Vector dominance predictions of electroproduction structure functions).

VII. Miscellaneous

[1] H. Abarbanel, Review paper on J-plane phenomena in Weak Interactions Conference on Regge poles, Irvine, California (1969).

[2] Review talks by Bjorken and Pais at Conference on "Expectations for Particle Reactions at the New Accelerators", University of Wisconsin, Madison, Wisconsin (1970).

Lightning Source UK Ltd.
Milton Keynes UK
UKHW022138230922
409344UK00004B/223

9 780691 619828